A NEW
BIOLOGY

NEW SCHOOL SERIES

General Editor

R. STONE, M.A., A.Inst.P.
Second Master, Manchester Grammar School

A NEW SCHOOL GEOGRAPHY
Part 1: The Elements of Geography
Part 2: A Physical and Human Geography of the British Isles
Part 3: Map Reading and Local Studies
Part 4: North West Europe
Part 5: Canada and the United States
F. R. DOBSON, B.A., and H. E. VIRGO, M.A.

A NEW GEOLOGY
M. J. BRADSHAW, M.A.

READING GEOLOGICAL MAPS
and
GEOLOGICAL MAP EXERCISES
M. J. BRADSHAW, M.A., and E. A. JARMAN, B.Sc.

GENERAL SCHOOL CHEMISTRY
S. CLYNES, M.A., A.Inst.P., and D. J. W. WILLIAMS, M.A.

GENERAL SCHOOL PHYSICS
Volume 1: Heat and Light
Volume 2: Mechanics, Hydrostatics and Sound
Volume 3: Magnetism and Electricity
R. STONE, M.A., A.Inst.P., and N. BRONNER, M.A.

STARTING POINTS
B. A. PHYTHIAN, M.A., B.Litt., and G. P. FOX, M.A.

LISTEN
Dramatic Monologues and Dialogues
A. THOMPSON, M.A.

CREEDS AND CONTROVERSIES
P. F. MILLER, M.A., and K. S. POUND, M.A.

A NEW
BIOLOGY

K. G. BROCKLEHURST
M.A., M.I.Biol.
Senior Biology Master, Warwick School
Formerly Examiner to the Oxford and Cambridge Schools Examinations Board

HELEN WARD
B.Sc., F.I.Biol.
Formerly Head of the Science Department
North Manchester Grammar School for Girls

THE ENGLISH UNIVERSITIES PRESS LTD

ISBN 0 340 05243 0

First printed 1968. Reprinted 1968, 1969,
1970 (corrected and in S.I. units), 1971

The English Universities Press Ltd
St Paul's House, Warwick Lane, London EC4P 4AH

Printed and bound in Great Britain
by Butler & Tanner Ltd, Frome and London

Editor's Introduction

We live in a rapidly changing world: some might say in a too-rapidly changing one. Today ideas can alter as much in a decade as in the whole of the preceding century. As the kaleidoscopic patterns of syllabus and method pass before us, it is not always easy to distinguish the important and permanent from the ephemeral. It is our hope that, in the New School Series, we shall be able to produce a group of books which will help colleagues in the classroom, and their pupils, to meet the challenge of the next ten years.

Our main purpose is to provide a wide range of books, covering both the Arts and the Sciences at a number of levels, which will allow the teacher increased latitude in his approach, and offer him full scope to develop the creative aspects of his work. It must not be forgotten that the suitability of a particular text for a given form depends largely on the way in which the teacher uses it: the speed and depth of the work will be dictated by the abilities and interests of the pupils rather than by the wishes of the teacher or the nature of the text. Many of the new books will be suitable for all but the lower streams of the comprehensive schools: others will express the newer conceptual, as opposed to factual, approach to teaching but may be contained within somewhat narrower academic boundaries. We shall be greatly in debt to the many teachers, professional bodies, and others whose untiring efforts have done so much to change the pattern of teaching in recent years. Nor must we forget those pupils who, whether consciously or not, have been the guinea pigs in the experiments which were necessary to prove the new ideas.

We hope to arrange for books to be written by teams of two or more experienced teachers who have tried out the new methods and syllabuses in the classroom, and who will be able to engender in their readers the same enthusiasm which they have instilled into their pupils. It is this dual interest of those who teach and those who are taught which is the key to all successful learning, a process in which we hope to play our part.

R. STONE
Manchester Grammar School

Acknowledgements

The authors wish to express their thanks to the following for providing photographic material for some of the chapter headings:

Chapter 4 (part)   E. Hosking, F.R.P.S.
Chapter 8 (part)   Flatters & Garnett Ltd.
Chapter 17   Dr. T. Mann, C.B.E., F.R.S.
Chapters 5 and 21   Ransomes Sims & Jefferies Ltd.

Thanks are also due to others who have provided material to illustrate the text. Acknowledgement of these photographs is made in the captions to each one.

Permission to publish questions from recent examination papers is gratefully acknowledged to the following Examining Boards:

Associated Examining Board. [A]

University of Cambridge Local Examinations Syndicate. [C]

University of London, University Entrance and School Examinations Council. [L]

Joint Matriculation Board. [N]

Oxford Local Examinations. [O]

Oxford and Cambridge Schools Examination Board. [O & C]

Southern Universities Joint Board for School Examinations. [S]

Welsh Joint Education Committee. [W]

Associated Lancashire Schools Examining Board. [A]

Southern Regional Examinations Board. [S]

West Midlands Examinations Board. [M]

Authors' Preface

This book has been written for secondary school students in their last two or three years of a Biology course. It can be used either as an up-to date conventional text-book for the Ordinary level syllabuses of the main examining bodies for General Certificate of Education and Certificate of Secondary Education, or as a companion reference for the more experimental types of courses.

The introductory chapters discuss the characteristics of living organisms, the general features of the flowering plant, the mammal, and some simple organisms; the main part of the book covers the syllabus matter through functional morphology, and the later chapters deal with topics of more general interest such as disease, heredity, evolution, and food production. As no two teachers of Biology take topics in quite the same order, we have tried to make the book versatile in use by dividing chapters into numbered sections, by including adequate cross-references, a glossary, and a full index.

Organic evolution is known to be a difficult topic for students at this level and we have attempted a fairly full treatment of this subject, approaching it from a genetic basis. We recommend, however, that this part of the course should be left until the final year.

The illustrations are an integral part of the book and are intended to be read carefully rather than glanced at quickly. We wish to emphasise that the reading of this textbook is in no way a substitute for a proper course of practical work. We have stressed the need for personal experiment and observation but have made full use of photographic material (most of which has been specially prepared for this book) to stimulate and guide the student, to act as a corrective to the simplification inherent in all diagrams, and to show detailed structures which cannot always be readily demonstrated. We hope that the arrangement of a photomicrograph alongside a drawing made from it will encourage students to make similar sketches from their own material.

We have to acknowledge our general debt to the many authors whose standard works of reference we have consulted, and our particular debt to the many teachers who have made useful suggestions for improving the clarity of both illustrations and text, and to our own colleagues, students, and technicians, many of whom have helped more than they realised. We were indeed fortunate to enlist the photographic skill of Mr M. I. Walker, who very kindly prepared photographs of certain difficult subjects for the book. We are greatly indebted to Mrs. Brocklehurst and Miss I. A. Webster, B.A., for their continuous help and encouragement throughout the preparation of the manuscript, and especially for their critical reading of the typescript and illustrations. Finally, we wish to record our thanks to the General Editor, who has given us much valuable advice and criticism, and to the officials of The English Universities Press, who have given us every possible assistance.

K. G. B. & H. W.

Contents

1 The Living Organism

Biology (*Gk. bios, life; logos, discourse*) is the science of living things or organisms. This book is intended as a companion for you during a serious practical course of study in this subject. You may rightly ask why you should follow this course, where it is leading and how you will benefit from it. It will create for you an awareness of the vast array of forms of life which normally go unseen, of the similarities which exist even in organisms as different as apple trees and earthworms or moulds and man. It will give you an insight into the work which individual organisms have to do to maintain themselves in an unfriendly environment, into the way in which species perpetuate themselves by reproduction, perhaps the most impressive of all life's characteristics, and into the way in which populations maintain themselves by adaptation.

If you are an enquiring student you will surely ask questions to which no answer has as yet been found, and so you will come to realise that biology is a relatively young experimental science and one which still offers unlimited scope for original work. It is the opinion of many experienced scientists that the next hundred years will see such new facts about living organisms brought to light that man's society will experience yet another revolution. Perhaps in this course you will also acquire an insight into scientific work and become infected with the enthusiasm and excitement which is characteristic of today's biology. Such feelings stem from the recent discoveries with electronmicroscopes of the finest structures of cells, of their chemical nature, and the roles they play in life's activities. Your most valuable acquisition from this course may well be a wider knowledge of the living world and its problems, which will make you a better and more understanding member of human society.

It is well known that animals such as cats, robins, herrings, snails, spiders, wasps, and earthworms, and plants such as oak trees, daisies, tulips, ferns, mosses, mushrooms, and seaweeds, are living; that is, they all possess a quality or nature which we call life. Further, no one doubts that rocks, motor-vehicles, and crystals of salt are non-living. Life itself cannot be seen or even defined, and is far from being understood, so how is this distinction made? It is by noticing whether an object moves, and feeds, and grows, that we are able to distinguish the living from the non-living. In addition to these three characteristic activities of living organisms, there are several others which may not be so familiar to you. They are all described in this chapter.

CHARACTERISTIC ACTIVITIES

1.1 Making Proteins, Nucleic Acids and Membranes

The living substance of organisms is organised into units, as can be seen in Figs. 1.1, 1.8, 1.9. These units are called **cells** and each contains a single body known as a **nucleus**. The cell contents

ANIMAL CELLS

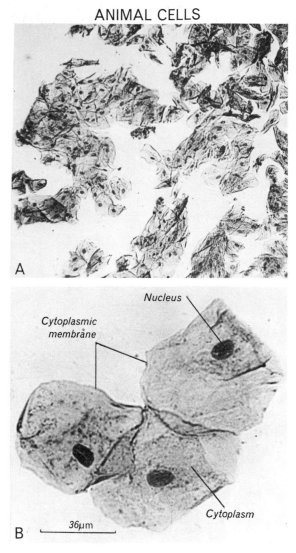

A

Nucleus

Cytoplasmic membrane

Cytoplasm

B |— 36µm —|

FIG. 1.1 Photographs of cells from the lining of the human mouth as seen through a microscope—a photomicrograph. The cells in B are more highly magnified than those in A. 1 µm or micrometre = $\frac{1}{1000}$ mm or 10^{-6} m.

used to be called protoplasm and were at one time thought to be made of a uniform solution of **proteins** and many other substances in water. Proteins are a group of complex compounds containing carbon, hydrogen, oxygen, nitrogen, and usually sulphur; they are known only from living organisms which make them. Some examples such as albumen (in the 'white' of a hen's egg), gelatin (in jellies) and casein (in milk) may be known to you already. The nucleus has for many years been known to differ from the rest of the protoplasm in that it contains long thread-like bodies not present elsewhere. These threads have now been

shown to contain a special type of **nucleic acid** not found in the rest of the cell. Nucleic acids are another group of complex organic compounds containing carbon, hydrogen, oxygen, nitrogen, and phosphorus. Furthermore, the part of the protoplasm surrounding the nucleus has been found through electronmicroscope studies to contain a vast array of structural features such as **membranes,** made of fat and protein, not known to occur within the nucleus. Such membranes are not visible with an optical microscope (compare Figs. 1.1 and 1.2). It is, therefore, better to avoid the word protoplasm and to refer to either the nucleus or the **cytoplasm** in which it is embedded. The outer membrane of the cell is known as the **cytoplasmic membrane,** and there is also present a **nuclear membrane** which marks the boundary between the nucleus and the cytoplasm. There seems no doubt that the ability to make these membranes and the two very complex groups of compounds, proteins and nucleic acids, is a fundamental chara - teristic of living organisms.

1.2 Feeding

Every organism must take in food from its surroundings if it is to remain alive for long. Once consumed, the food undergoes a series of chemical changes until finally part of it is incorporated in the material of the body; for example, as cytoplasm in the muscles of an animal or in the new leaves of a plant. The intake of food together with these changes is known as nutrition, but the intake alone is feeding.

Types of food are as varied as the organisms themselves. Animals, for instance, generally take most of their food in the form of complex organic compounds and their nutrition is said to be **holozoic** (*Gk. holos, whole; zoon, animal*). Such compounds can be obtained only from other organisms either living or dead. In contrast, green plants, whose nutrition is said to be **holophytic** *(Gk. phyton, plant)*, utilise gaseous and liquid foods such as carbon dioxide from the air, and water and dissolved mineral salts (e.g. potassium nitrate and ammonium sulphate) from the soil. These are inorganic compounds.

1.3 Respiration

Energy required for carrying out vital processes, such as movement and growth, is obtained by all organisms from the breakdown of complex organic substances like fat or sugar. This energy production is called respiration; it never stops during life, even in dry seeds and hibernating mammals, though in these cases we believe it proceeds so

STRUCTURES IN CYTOPLASM

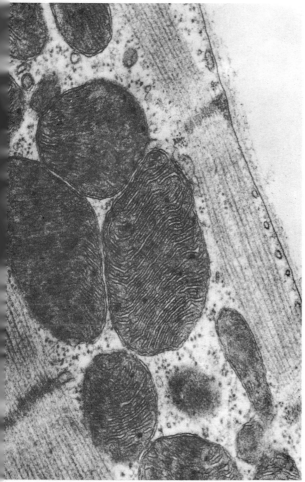

FIG. 1.2 An electronmicroscope picture (electronmicrograph) of part of the cytoplasm of a cell from rat heart muscle ($\times 40\,000$). The cytoplasmic membrane is the sharp black line; the dark central bodies are composed of double membranes and are known as mitochondria. (*Electronmicrograph by Prof. A. R. Muir.*)

slowly that it is difficult to detect. It may be summarised thus:

$$C_6H_{12}O_6 + 6O_2 \longrightarrow 6CO_2 + 6H_2O + \text{Energy}$$

Sugar   Oxygen   Carbon Water Energy
                       dioxide

Gaseous oxygen is usually necessary, and is obtained from the air or from solution in the water round the organism; the waste carbon dioxide is produced and removed continuously. In respiration there is, therefore, an exchange of gases between the organism and its surroundings, which is known to most as breathing, since it is given this name in man. It is, however, misleading to say that plants and earthworms breathe, for in them there is no active movement of part of the body causing the exchange of gases to take place. Instead of breathing, the term respiration (in a restricted sense) is frequently used—man's respiration rate is fifteen times per minute—but it is important to realise that respiration (in its wide sense) includes all the processes involved in energy production.

CARBON DIOXIDE FROM SEEDS

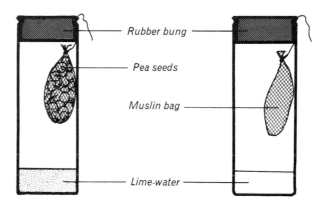

FIG. 1.3.

A simple apparatus (see Fig. 1.3) can be used to demonstrate carbon dioxide production in small organisms. Pea seeds, which have been soaked in water for a day, when enclosed in a tube with lime-water as shown, cause it to turn chalky in a few hours, thus indicating that they are giving off carbon dioxide. It is important to carry out an identical experiment at the same time, but without the pea seeds. If the lime-water in this experiment remains clear, then it is certain that it was the seeds, and not any carbon dioxide from the air in the tube at the start of the experiment, which made the lime-water chalky in the first case. The second experiment (without the seeds) acts as a **control** for the first and illustrates a principle we shall meet many times. If the tubes are left for longer than a few hours, and the gases remaining are tested for oxygen with a burning splint, it will be found that the splint burns for some time in the control tube, but is rapidly extinguished in the other. These results show that the seeds have used up oxygen.

Obviously other small organisms could have been substituted for the seeds in the above experiment, and even the micro-organisms in a bag of soil can be shown in this way to produce carbon dioxide.

Using the apparatus in Fig. 1.4 it can also be

CARBON DIOXIDE FROM MAN

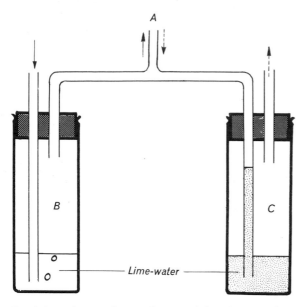

Fig. 1.4 Apply mouth to tube A and breathe in and out gently.

shown that human beings produce carbon dioxide. The demonstrator applies his mouth to the tube A and breathes in and out *gently*. As he breathes in, a stream of air bubbles is drawn through the lime-water in tube B and enters his lungs. As he breathes out, a stream of expired air is forced through the lime-water in tube C. This lime-water rapidly turns chalky, but the lime-water in tube B shows only a very slight cloudiness, which is due to the small amount of carbon dioxide in normal air.

Respiration, then, in its wide sense, is a universal process in living organisms, involving the intake of oxygen, the breakdown of complex substances with the liberation of energy, and the production of simpler substances such as carbon dioxide and water.

1.4 Excretion

In the cells of any organism many hundreds of chemical changes are taking place. These are collectively known as **metabolism**. Some of the substances produced are of no use to the organism; they may even be harmful if they are allowed to accumulate. The discharging of these waste products of metabolism from the body is excretion.

Mammals excrete when they discharge urine, but much of the material passed out at the anus is not excretory, since it is undigested food which has never formed part of the cells of the body.

The excretion of plants is less obvious than that of animals and is described in chapter 10.

A somewhat similar activity of living organisms is **secretion**. This is the formation of a substance useful to an organism, and its discharge from the cells where it was produced. The saliva of man and the nectar of flowers are well-known secretions.

1.5 Growth

No one who has seen babies change gradually into adults, foals into mares, and the sticky buds of horse-chestnut twigs transform themselves into new leafy stems often with a large terminal mass of flowers (see Figs. 1.5 and 1.6), can fail to appreciate two important features of growth—increase in size and increase in complexity. You may have noticed this activity only in domesticated animals and plants which you see frequently, but it does take place in all organisms.

The growth of crystals, which involves the addition of new material on the outside, does not strictly resemble that of living organisms where the new materials are made within the body.

1.6 Movement

Animals, which nearly all search for their food, are able to move the whole body about; plants are surrounded by their food, are almost always fixed in one place, and their obvious movements are restricted to organs such as stem tips and leaves. One example of this type of movement is responsible for the attachment of the leaf tendrils of pea

ANIMAL GROWTH

Fig. 1.5 New Forest mare and foal.

GROWTH - OPENING OF BUDS

FIG. 1.6 Photographs of horse-chestnut buds opening. The bud on the right shows a mass of flowers as well as leaves.

plants to supports, as shown in Fig. 1.7. The tendrils, which are stroked with the rough end of a used match, gradually curl toward the side which has been stroked.

You may find it interesting to look for movement of cytoplasm inside cells of the leaves of Canadian pondweed (see Fig. 1.8) or in hairs from the stamens of Tradescantia, a commonly cultivated plant.

1.7 Sensitivity (Irritability)

The ability of organisms to detect or perceive changes in their surroundings and to respond to them is known as sensitivity. An example will make this clearer. When a bright light is shone into a person's eye (this is the change or **stimulus**), the diameter of the pupil in the centre of the coloured iris immediately becomes smaller (this is the **response**), thus admitting less light and protecting the delicate parts inside from damage.

An example from plants is the slow bending of stems toward a lighted window in a dark room.

Here, the stimulus is the light coming from one side and the response is the bending toward the light. This bending is too slow to be detected by eye. Consider again the experiment illustrated in Fig. 1.7. What is the stimulus, and what response does it produce? Is it advantageous for an organism to respond to stimuli?

1.8 Reproduction

Perhaps the most remarkable activity of living organisms is their ability to create new ones like themselves, i.e. to reproduce their kind. At first the offspring are often quite unlike the parent or parents, e.g. the acorns of an oak tree, or the spawn of a frog, but, gradually, as they grow, they come to resemble the parents, and in their turn reproduce. It is also characteristic of organisms that, in the end, if they are not killed in the struggle for existence, they die. It seems that life can remain in one particular organism only for a limited period of time, and we do not really understand why.

MOVEMENT IN TENDRILS

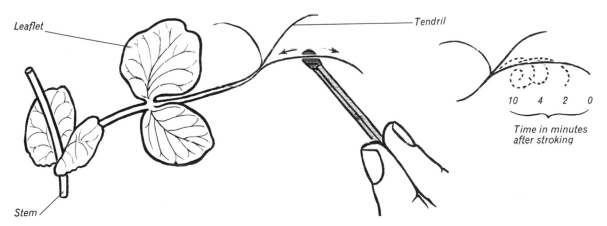

FIG. 1.7 Stroke gently to and fro under the tendril for about one minute, then watch for curvature.

The difference between living and non-living things can now be stated more fully. Living things have all the activities listed below, but non-living things exhibit none or only a few of these. It is a useful exercise to check the following list to see which of the activities are carried out by a motor-car, a television receiver, a copper sulphate crystal and perhaps a computer.

Making membranes, proteins, nucleic acids

Feeding

Respiration

Excretion

Growth

Movement

Sensitivity (response to stimulus)

Reproduction

In addition to these characteristics every organism has a built-in system of checks and balances: such a **self-regulating mechanism** helps it to survive in a changing and often unfriendly environment.

NAMING AND CLASSIFYING ORGANISMS

1.9

More than a million different species of organisms are known today. Each **species** or type is given a double biological name, but only the more familiar ones have acquired common names as well. Some examples are given below:

Quercus robur—oak tree.
Bellis perennis—daisy.
Ranunculus acris—field buttercup.
Sciurus carolinensis—grey squirrel.
Talpa europaea—mole.
Passer domesticus—house sparrow.

Within both the plant and the animal kingdoms one can arrange groups of organisms in order, starting with those whose structure is relatively simple and ending with those whose structure is highly complex. For example, the blue-green algae,

PLANT CELLS

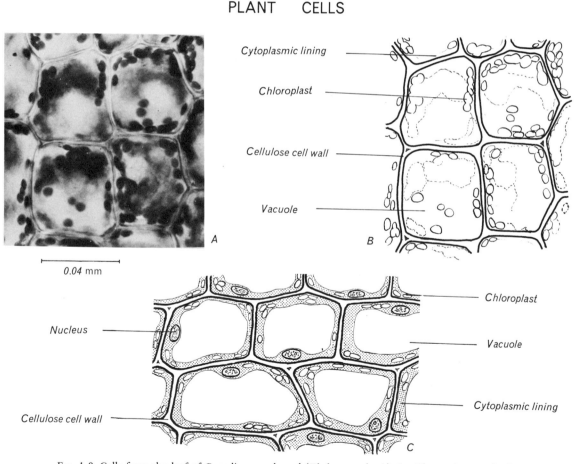

Fig. 1.8 Cells from the leaf of Canadian pondweed (*Elodea canadensis*). A—Photomicrograph of living cells. B—Key to photomicrograph. C—Diagram of cells as seen in section under the microscope. Not all the structures drawn are readily visible in untreated living cells.

a group of microscopic organisms, some of which form a gelatinous film on wet stones, are among the simplest plants known, and the flowering plants are among the most complex. Biologists have good reason for believing that such an arrangement is only possible because organisms have been undergoing a process of very slow change or **evolution** for more than 2,000 million years. They believe that the first organisms on this planet were the simplest, and that some of these gave rise to more complicated ones, while others retained almost the same structure up to the present day. Some of these more complicated ones changed further, and so on until the most complex ones were formed.

A simplified classification of organisms will be found at the end of this book. You should not attempt to memorise this in one or two efforts, but refer to it whenever you meet a new organism.

PLANT–ANIMAL DIFFERENCES

1.10

Although you may be able to place organisms without difficulty in either the plant or the animal kingdom, it is essential that you should know the basic nutritional differences between these two groups.

Plants require sunlight energy in order to build up their complex organic compounds (starch, for example) from the simpler inorganic substances they absorb. This process, **photosynthesis**, is not only dependent on **sunlight**, but requires the presence of the green pigment **chlorophyll**, and a supply of **water** and **carbon dioxide** within the plant, before it can take place. Animals do not carry out such a synthesis, they take in ready-made organic compounds by feeding on other organisms either living or dead. Thus plants are the only organisms capable of making organic compounds, and animals are dependent on them, either directly or indirectly, for their supply of these substances.

Animals also differ from plants in the structure of their cells. Those of animals are bounded by a very thin layer of special cytoplasm (a **cytoplasmic membrane**); those of plants have a much thicker layer of non-living material outside the cytoplasmic membrane (see Figs. 1.1 and 1.8). This thick layer, known as the **cell wall**, is usually made of an organic substance called **cellulose**. Furthermore, plant cells, unlike those of animals, generally have spaces or **vacuoles** full of cell sap within their cytoplasm. Chlorophyll is not found in animal cells, but is present in small disc-shaped bodies or

PLANT CELLS

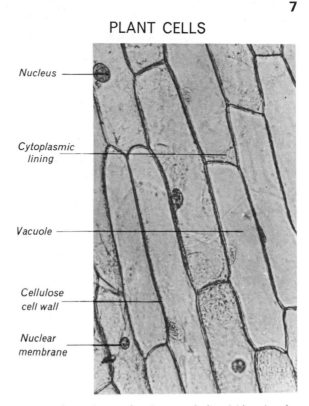

Nucleus

Cytoplasmic lining

Vacuole

Cellulose cell wall

Nuclear membrane

Fig. 1.9 Photomicrograph of part of the epidermis of a scale leaf of an onion bulb.

chloroplasts in the cytoplasm of many plant cells. The nucleus is a more or less spherical body present in both cell types.

Plants	*Animals*
Feed on inorganic substances	Feed on complex organic compounds
Photosynthesise	Do not photosynthesise
Possess chlorophyll	Have no chlorophyll
Are fixed, but move parts of the body	Move the whole body
Branching body	Compact body
Are less sensitive and respond slowly	Are highly sensitive and respond quickly
Cellulose cell walls usually present	Have no cellulose cell walls

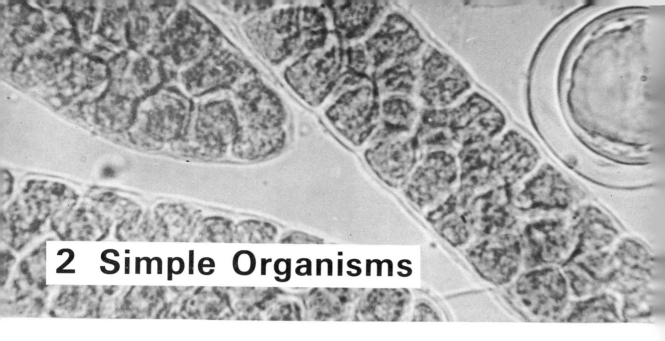

2 Simple Organisms

Many small and microscopic organisms are found in water. Tiny specks of grey matter moving over the surface of most pond mud may be living animals such as **Amoebae**, and the green colour of shallow pools of water in ditches and farmyards may be caused by minute green organisms such as **Euglena**, which swim in the water. The bright green mass of tangled slimy threads found in ditches and ponds may be the green alga **Spirogyra**. If in the summer you leave some pondweed in a jar overnight you may find, attached to the jar, several pale brown animals with long tentacles spread out in the water like a net. This is the usual way of collecting the fresh-water polyp, **Hydra**, our commonest two-layered animal.

At some time most people must have seen a

AMOEBA

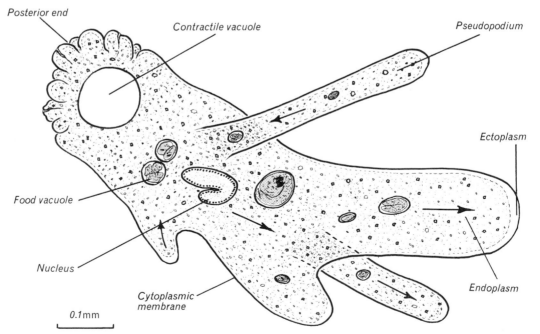

Fig. 2.1 *Amoeba proteus.* A young adult as seen under the microscope. Arrows indicate the direction of flow of endoplasm.

8

dense felt of white threads covering parts of compost heaps, horse dung, tomatoes, or pieces of old bread which have not been allowed to dry out. Such a growth is part of the body of one of the pin-moulds, possibly **Mucor**, the one we are going to study.

In this chapter you can read about *Amoeba, Euglena, Spirogyra, Hydra,* and *Mucor,* their structure and how they live. These five organisms, because of their simplicity, form a suitable introduction to the highly complex flowering plants and mammals.

AMOEBA—A NON-CELLULAR ANIMAL

2.1 Structure

The largest Amoebae can just be seen with the naked eye, but they must be examined with the aid of a microscope before their structure can be made out. Their shapes change as they move and they usually possess one or more blunt processes called **pseudopodia** (*Gk. pseudos, false; podos, foot*) into which, or from which, the cytoplasm is flowing. Their surfaces are bounded by a cytoplasmic membrane and are perfectly smooth except at one place, the posterior end, where they are wrinkled during active movement. Inside the membrane is a clear layer of firm cytoplasm, the **ectoplasm**, which is distinct from the inner fluid cytoplasm or **endoplasm**, since this contains granules of reserve food. Within the endoplasm as it flows, the single nucleus, a transparent disc-shaped body, can be seen being carried along. Also in the endoplasm, but usually at the posterior end, is the **contractile vacuole**, which rapidly fills with water until it is a large sphere, and then disappears by expelling its contents only to reappear as it fills once more. (You could use the second hand on your watch to time this cycle of events. Does it vary in different specimens?) This vacuole has not been proved to get rid of excretory products and probably only serves the purpose of regulating the water content of the body. After an Amoeba has been well fed there are, in the endoplasm, several green or brown **food vacuoles** containing organisms which have been eaten. You may already have realised that this animal is not made up of cells—it is non-cellular, and in its whole body there is usually only one nucleus.

2.2 Movement and Feeding

Amoeba moves in contact with a solid surface by putting out new pseudopodia and withdrawing others (see Fig. 2.2). New pseudopodia are formed

FIG. 2.2 Photomicrographs of Amoeba taken at 30-second intervals. Note the vertical line, which is a reference point.

AMOEBA FEEDS ON A CILIATE

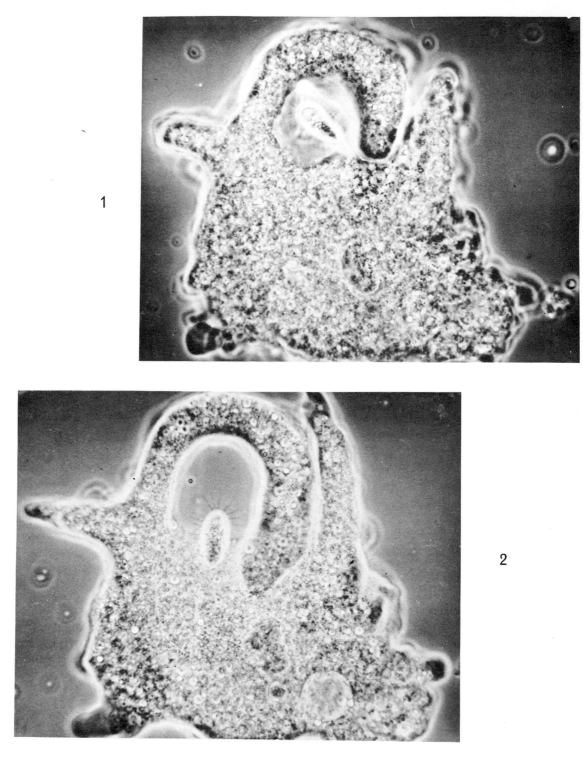

FIG. 2.3 *Amoeba discoides* capturing a ciliate. (*Courtesy P. Harris Ltd. Photomicrograph by M. I. Walker.*)

MOVEMENT IN AMOEBA

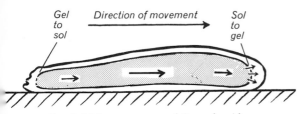

FIG. 2.4 *Movement viewed from the side.*

from a central stream of endoplasm which flows out from the rest of the animal. Such endoplasm is in the sol state as opposed to the firmer ecto-plasm in the gel state. A layer of clear ectoplasm is always present on an advancing pseudopodium, so it is reasonable to assume that there must be a sol to gel change at its rear (see Fig. 2.4). You should realise that this is far from being a complete ex-planation of Amoeba's movement; it does not, for instance, explain what force drives the endoplasm forward. It is interesting to note that movement ceases in an Amoeba from which the nucleus has been removed, but it can be restarted if the nucleus is replaced by one from another Amoeba.

Amoebae do put out pseudopodia into the water, but these cannot be effective in locomotion like those which adhere to solid surfaces. If stirred up in water they put out many fine pseudopodia in all directions (this can be regarded as their effort to contact something solid), and when one of them touches any solid object it adheres and the rest of the body flows into it. Anyone who has tried to dislodge Amoebae from the bottom of a dish will know that they can adhere to it, but how they do it has never been satisfactorily explained.

Feeding is best observed in a young Amoeba; it involves the engulfing of a tiny pond organism such as a bacterium or ciliate (see glossary) by pseudopodia which form round it (see Fig. 2.3). Within the food vacuole so formed the organism is digested, any undigested remains being extruded from the body at the posterior end as the animal moves (see Fig. 2.5).

2.3 Life-history and a Culture Method

Amoeba starts life as a tiny spherical **spore** not more than 5 µm (or 5×10^{-6} m) in diameter. At this stage there is a protective wall of dead material round it, which allows it to resist the drying up of the pond or stream. This ability to withstand desiccation provides the animal with a means of dispersal to other places, for the spores are easily blown about by the wind.

PARAMECIUM EMPTYING A FOOD VACUOLE

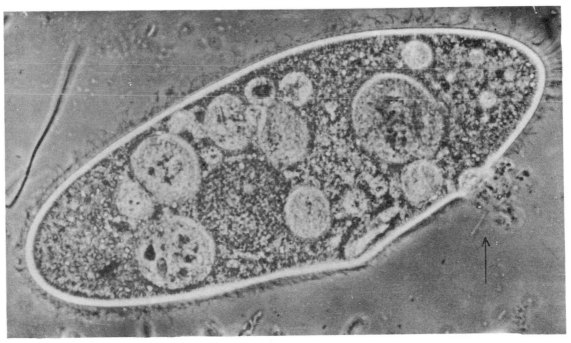

FIG. 2.5 *Paramecium caudatum* emptying a food vacuole (*Courtesy P. Harris Ltd. Photomicro-graph by M. I. Walker.*)

LIFE-HISTORY OF AMOEBA

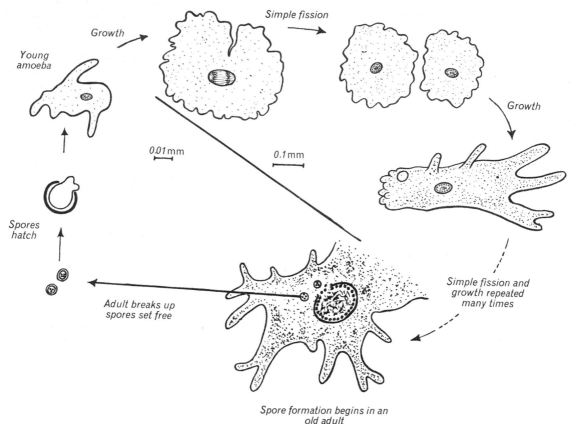

FIG. 2.6 (*Modified from M. Taylor.*)

On the return of the spores to water, tiny Amoebae hatch out of the protective cases and start to move actively, at the same time feeding on minute organisms, especially bacteria. At first the young Amoebae are quite transparent and without granules in the endoplasm, but these appear in increasing numbers and get larger as the animals get older. Every few days the young Amoebae stop feeding, become more or less spherical, and divide by **simple fission** to form two individuals (see Fig. 2.6). They also have periods when they stop 'creeping' and remain inactive. After many divisions the Amoebae reach the adult stage with their cytoplasm very full of granules and no longer able to undergo simple fission. At this time spores start to form in the body of the adults, each spore containing a portion of the old nucleus and some cytoplasm. During this period of **spore formation** the store of food granules is used up, and finally the old adult breaks up setting free the spores which may either hatch directly or remain dormant during a period of drought. The whole cycle takes from four to six months when food is plentiful. Amoeba can also form a **cyst** around itself when adult (see Fig. 2.7), which allows it to survive in adverse conditions.

To cultivate Amoeba, first boil some pond water to kill the organisms in it and put it in a clean shallow dish (0.15 m diameter ×0.04 m high is a convenient size). Add boiled wheat grains (22 per litre) and cover with a glass plate to keep out dust and reduce evaporation. Obtain a sample of a well-established culture, let it stand until the Amoebae are at the bottom, and add the top third of it to the dish. Leave for a few days to allow the food organisms to increase their numbers, when the Amoebae can be safely added. It will probably be necessary to feed such a culture with more wheat grains every three months, and to divide it into two after a year, adding more boiled pond water to dilute the waste products. Do not throw away the cultures if they contain no large Amoebae; these will reappear from the spores if conditions are right. It is essential that all glassware, especially pipettes for taking out samples, should be sterilised in strong hydrochloric acid and well washed in water before use, since the accidental introduction of water-fleas and similar animals or their eggs would lead to the death of the Amoebae.

Amoeba, being small, non-cellular, and possessing only one nucleus, is certainly simple in its structure. It is probably not so simple in the way it carries out life's activities, for they must all go on in the one mass of cytoplasm.

EUGLENA—A NON-CELLULAR PLANT/ANIMAL

2.4 Structure

Many species of Euglena are common in every sort of fresh water. They have a spindle-shaped body which, in the largest species, reaches a length of 0·5 mm. Externally the body is covered by a flexible **pellicle** within which is the cytoplasm. The cytoplasm of *Euglena deses*, the species illustrated (Fig. 2.8), contains about 30 to 40 green disc-shaped **chloroplasts** each with a central body known as a **pyrenoid**. The chloroplasts are concerned in photosynthesis, but the precise function of the pyrenoid is unknown. There is a single large nucleus in the cytoplasm (not visible in the photomicrograph) and also a large number of granules which contain a food store known as paramylum. The anterior end is easily distinguished by its tubular 'gullet' which opens into a spherical cavity, the **reservoir**. Also marking the anterior end is an obvious collection of red pigment, the **stigma**. Two cytoplasmic threads or **flagella** (sing. flagellum) have their origin in the reservoir. One is short and does not emerge from the reservoir, the other is long and whip-like.

2.5 Movement, Nutrition, Reproduction

Euglena, by lashing its flagellum, swims through the water rotating on its long axis as it goes. It exhibits changes of shape as shown in Fig. 2.8 but only assumes the near-spherical form when at rest. It is very sensitive to changes in light intensity, and if a pool or dish containing it is partly shaded it will be found to collect in the unshaded part, provided the light is not too strong. The eyespot plays some part in this response.

AMOEBA ENCYSTED

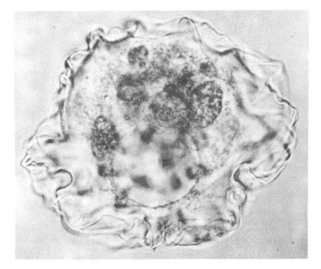

Fig. 2.7 Photomicrograph of an encysted Amoeba.

The nutrition of *Euglena deses* is similar to that of the higher plants, since it carries out photosynthesis, but most species are probably unable to live without a supply of organic compounds of nitrogen, whereas higher plants utilise simple nitrogenous compounds such as nitrates. When there is a plentiful supply of organic compounds dissolved in the water, some species of Euglena can live without light, in which case they lose their green colour. Thus Euglena is both holophytic and saprophytic (see sec. 2.14), but it has never been seen to take in solid food through the gullet and is, therefore, not holozoic.

Reproduction takes place by simple fission (lengthways) and it may occur in an active individual or in one which has made a thin-walled cyst round itself.

To cultivate Euglena put a layer about one inch deep of good garden loam in a wide-mouthed bottle. Damp well and put a foil cap loosely over the bottle. Heat in a pressure cooker at 1 atm or 103 kN/m² above atmospheric pressure for 15 minutes. When required, add distilled water so that the bottle is about two-thirds full and inoculate with Euglena. Keep at 21 °C in a strong light.

Euglena obviously exhibits a confusing mixture of animal and plant features (see sec. 1.10), but a full discussion of whether it is plant or animal is not possible in a book of this type. Try to make lists of the plant and animal features of this organism.

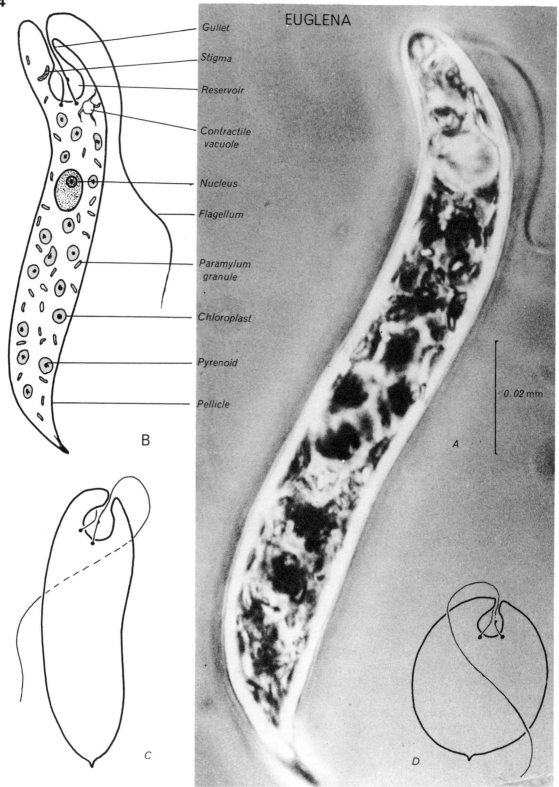

EUGLENA

Gullet

Stigma

Reservoir

Contractile
vacuole

Nucleus

Flagellum

Paramylum
granule

Chloroplast

Pyrenoid

Pellicle

0.02 mm

B

A

C

D

F<small>IG</small>. 2.8 A—Photomicrograph of living *Euglena deses*. (*Courtesy P. Harris Ltd. Kodachrome taken at 0.5 ms by M. I. Walker.*) B—Diagram of structural features of *Euglena deses*. Not all the structures shown are identifiable in a single photomicrograph. C and D—Show changes in shape of Euglena.

SPIROGYRA

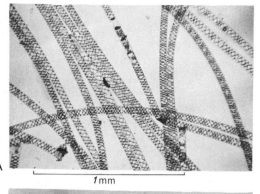

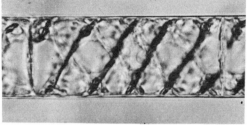

A

|____1mm____|

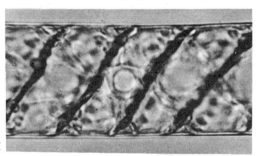

B

|____0.1mm____|

C

|_._._.|
0.01mm

SPIROGYRA—A THREAD-LIKE GREEN ALGA

2.6 Structure

Spirogyra is a green plant and, unlike Amoeba, it is made up of many cells (i.e. it is **multicellular**). A single individual is a cylindrical thread having a length of more than 100 mm and a diameter of about 0.1 mm. The thread or filament is made up of a few hundred similar cylindrical cells joined at their ends (see Fig. 2.9). Each cell has a wall of cellulose covered on the outside by **mucilage**, a slimy material, and on the inside by an extremely thin layer of cytoplasm which is difficult to observe. Running spirally in this layer of cytoplasm are long chloroplasts (the species shown in Fig. 2.9 has four) each containing several small spherical **pyrenoids**. No other alga has such chloroplasts, so they form a ready means of identification. The nucleus of each cell is suspended near its centre by fine bridles of cytoplasm which run from the layer inside the cell wall. The spaces between the bridles are vacuoles containing cell sap (see Fig. 2.10).

2.7 Nutrition

Spirogyra feeds by absorbing, from the water in which it floats, simple inorganic substances such as carbon dioxide, mineral salts, and water itself. It should be noted that the cellulose wall is no barrier to the entry of such substances. From them, using the energy of sunlight, the plant is able to build up complex organic compounds.

◄Fig. 2.9 A—Photomicrograph of filaments of Spirogyra. B—Single cells showing chloroplast and pyrenoids. C—Part of single cell showing nucleus and cytoplasmic bridles.

SPIROGYRA

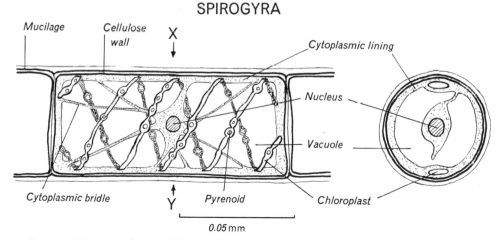

Fig. 2.10 Diagram of one cell from a filament of Spirogyra as seen under a microscope, and a section across it at XY.

CONJUGATION IN SPIROGYRA

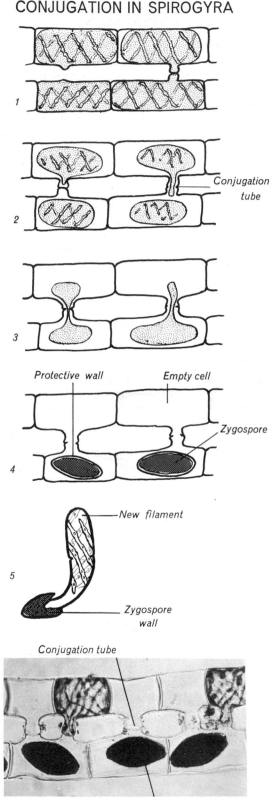

1

2 — Conjugation tube

3

Protective wall   Empty cell

Zygospore

4

New filament

5

Zygospore wall

Conjugation tube

Zygospore

Starch, which collects round the pyrenoids, is the most obvious product of this photosynthetic process. For further detail of photosynthesis see section 6.9.

2.8 Reproduction

The filaments increase their length by division and growth of their cells; their number is increased by **fragmentation**, a process involving the breakage of a filament between two cells.

Spirogyra also reproduces by **conjugation** (see Fig. 2.11). Here cells of two filaments lying next to each other, for at least part of their length, form tubular swellings which push the filaments apart as they grow. This results in a ladder-like arrangement of tubes connecting opposite cells, but at first there is a cross wall separating the cytoplasm of each cell from that of its opposite number. The cells reduce their volume by expelling water, the chloroplasts disintegrate, and when the cross walls are dissolved the contents of one cell pass through the tube into the other cell where the two nuclei fuse together. This fusion is known as **fertilisation**, a process whose significance will be considered in chapter 16. The products of fertilisation are green ellipsoidal **zygospores** (*Gk. zeugos, pair; sporos, fruit*) containing fat droplets and protected by a thick case which they have made (see Fig. 2.11). At first they lie within the old cell walls of one filament (the cells of the other filament are empty), but these soon decay, setting the zygospores free. They may then be carried far afield on the feet of animals or, if the ditch should dry up, they may be blown about by the wind. On their return to normal surroundings the zygospores are able to produce a new filament by cell division. Conjugation provides an explanation of how Spirogyra may become abundant in one place quite suddenly and then disappear equally quickly. Actually it will leave behind the minute zygospores.

In contrast to Amoeba, Spirogyra is multicellular and its cells, although structurally quite complex, are all similar; they can all feed, reproduce, and respire, and none is specialised to carry out one particular activity. In Hydra, our next simple organism, this is not the case, for most cells show some specialisation.

◄FIG. 2.11 1—Filaments together, swellings appear. 2—Conjugation tubes formed, cell contents contract. 3—Movement of cell contents and fertilisation. 4—Zygospores formed. 5—Zygospore germinates. The photomicrograph shows zygospores.

HYDRA

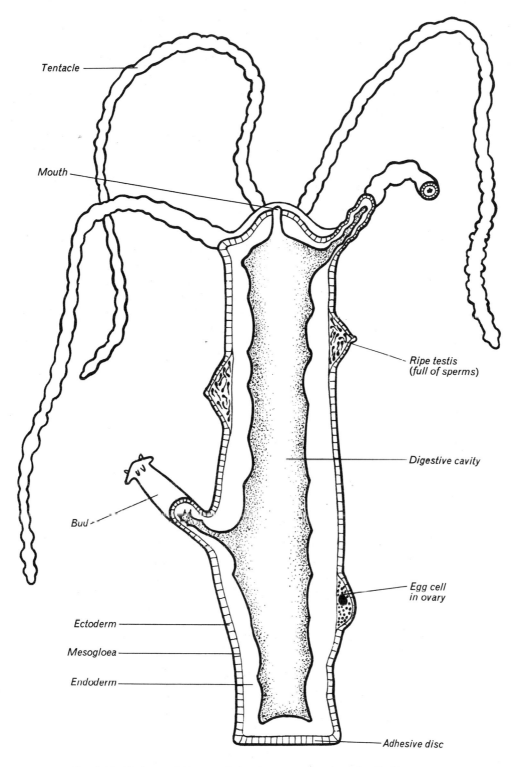

Tentacle

Mouth

Ripe testis
(*full of sperms*)

Digestive cavity

Bud

Egg cell
in ovary

Ectoderm

Mesogloea

Endoderm

Adhesive disc

FIG. 2.12 Hydra partially extended and cut away to show detail of its structure.

HYDRA—AN ANIMAL HAVING TWO LAYERS OF CELLS

2.9 Structure

Hydra is a common animal in our ponds throughout late spring and summer. Its body is shaped like a hollow cylinder and, when extended, may reach a length of 20 mm. When contracted, it becomes a small globular organism. At one end is a flattened **adhesive disc** with which it can attach itself to the surface film of water, to leaves or stones, or to any other solid object. From the other end projects a conical structure which bears the mouth at its apex (see Fig. 2.12). Round the base of this cone are six, seven, or eight hollow **tentacles** used for catching water-fleas and other small aquatic animals.

The body wall, surrounding the **digestive cavity** into which the mouth opens, is composed of two layers, each one cell thick. The outer layer or **ectoderm** (*Gk. ectos, outside; derma, skin*) is separated from the inner layer or **endoderm** (*Gk. endon, inside*) by a thin layer of jelly-like material, the **mesogloea** (*Gk. mesos, middle; gloia, jelly*).

2.10 Movement

The greater part of the ectoderm is made up of **muscle-tail cells** (see Fig. 2.13), so called because their inner ends are drawn out into long threads parallel to the long axis of either tentacle or body and lying close to the mesogloea. The cytoplasm of these threads is able to contract, so shortening

the tentacle or body, but it cannot cause the opposite effect when it relaxes. There are, however, some other cells in the endoderm (**nutritive cells**) which have muscle-tails arranged in a circular manner on the inner surface of the mesogloea: they appear as dots in Fig. 2.13. Their function is to reduce the diameter of the body and tentacles, so lengthening these structures when they contract. Thus it can be seen that the muscle-tails of the two layers work antagonistically.

Hydra, although it remains attached to one spot for long periods, moves its body and tentacles continuously. The body and tentacles are usually extended, the latter being spread like a net in the water, but when touched the whole animal often contracts into nearly spherical shape. It can move slowly from place to place by **looping** and somersaults on occasions (see Fig. 2.14).

2.11 Feeding

Hydra feeds on water-fleas and other small organisms which it catches with its tentacles. It is fascinating to watch a water-flea touch a tentacle and remain there 'stunned'. Other tentacles curl around the prey and draw it toward the mouth, which enlarges to allow it to enter the digestive cavity. How does Hydra manage to hold the water-fleas, and how does it immobilise them?

It has, on the tentacles, groups of cells which contain **thread-capsules** (see Fig. 2.15) and give the tentacles a knobbly appearance. The threads, before use, are 'turned outside in' and coiled spirally within the fluid-filled capsule, but when

CELLS OF HYDRA

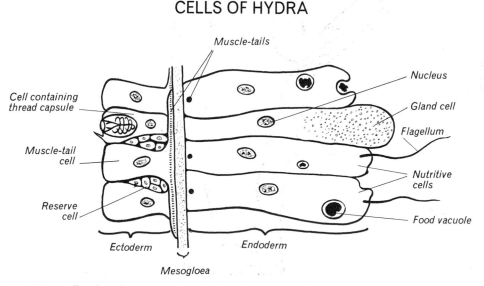

FIG. 2.13 Cells of Hydra as seen in a longitudinal section of the body. Note the direction in which the muscle-tails run.

MOVEMENT OF HYDRA

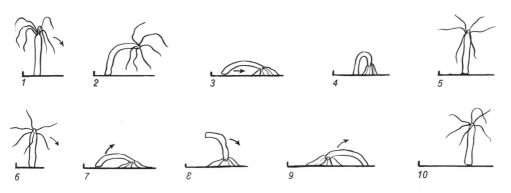

FIG. 2.14 Stages in the movement of Hydra. 1 to 5 looping, 6 to 10 somersaulting. Note the short vertical line which is a reference point.

a projecting hair (the trigger) is touched they can be shot out by being 'turned outside out'. If you cannot understand how these thread-capsules work, try blowing into a rubber glove when the fingers are tucked inside the hand. The largest of the capsules have barbs which pierce the prey as their threads are shot out and allow a paralysing fluid to be injected. Other smaller ones, when exploded, twist around the tiny bristles of the prey,

so helping to hold it. Each thread-capsule can explode only once and it is replaced from **reserve cells** which increase their number by division, and make a new capsule within themselves. The thread-capsule cells are clearly an example of specialisation for one function.

Once the prey is in the digestive cavity, **gland cells** of the endoderm pour out digestive juices which help to break it up into small fragments.

THREAD CAPSULES

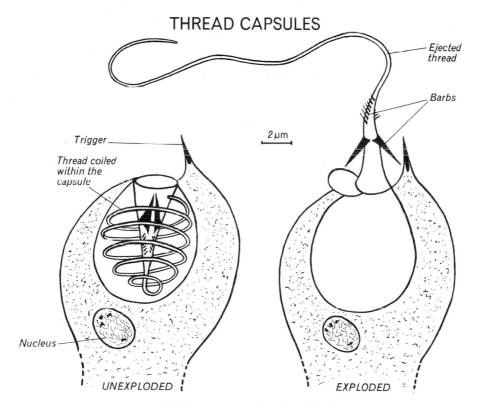

FIG. 2.15 The largest type of thread-capsule of Hydra.

Digestion is completed by nutritive cells which engulf fragments at their inner ends, forming food vacuoles in a similar way to Amoeba. These cells also possess flagella projecting into the fluid of the digestive cavity, which they are said to keep well mixed. Pieces of undigested matter are rejected through the mouth.

2.12 Life-history

When well fed, Hydra will readily produce new small individuals from bulges which develop on the body. These are known as **buds** (see Fig. 2.12). As these bulges grow they develop tentacles and finally a mouth at their free ends. The young Hydra can, at this stage, obtain food from the parent and also by catching prey with its own tentacles. Eventually the bud narrows at the base, the young Hydra clings to the pondweed to which the parent is fixed and breaks free from the parent to lead an independent existence. It is not unusual for a bud itself to commence budding before it is free from its parent, neither is it rare to see several buds attached to one individual. This reproductive method occurs throughout spring and summer when food supplies are good.

At certain times of the year, particularly at the onset of winter, another type of reproduction occurs. Hydra may be found with several **testes**, which are ectodermal swellings near the mouth. Within the testes develop many thousands of minute unicellular bodies, each with a whip-like cytoplasmic tail used in swimming. These sper-matozoa (sperms is an accepted abbreviation), as they are called, eventually burst out of the testes and swim freely in the water. Later, Hydra develops an ectodermal swelling, the **ovary**, nearer the adhesive disc than the testes, and one of its cells enlarges tremendously as more and more food is stored in it. Finally, this cell becomes spherical and large enough to be easily seen with the naked eye: it is a ripe **ovum** (*L. ovum, egg*) and can form a new individual only after a sperm has entered it and the two nuclei have fused together. The ovary ripens after the testes, thus making it impossible for one Hydra to fertilise its egg with one of its own sperms.

Once a sperm has swum to an ovary and fertilised its egg, the latter divides into two cells, these two into four, and so on until there is a large number of cells. The organism, then known as an embryo, makes a horny case round itself before it drops off the parent into the mud, where it may lie dormant for several months. In the spring a young Hydra emerges from the case, starts to feed and grow and reproduce by budding.

It is a simple matter to collect Hydra and keep them at home or at school in a shallow dish containing pondweed, but they must be kept cool. They can be fed on water-fleas collected by filtering pond water through a net of fine mesh.

The table below summarises the activities of the different cells of Hydra, and emphasises that this organism shows an advance over Spirogyra, where all the cells are similar and not specialised to fulfil one main function.

Layer	Cell Types	Specialisation	Special Activities
Ectoderm	Muscle-tail cells	Contractile tails	Movement—shortening body and tentacles
	Nerve cell	Fine branched processes which join processes from similar cells to form a network	Conduct stimuli
	Thread-capsule cells	Explosive capsule and trigger	Catching food and protection
	Reserve cells	Not specialised	Replacement of other cells
Endoderm	Gland cells	Ability to make digestive juice	First stage of digestion
	Nutritive cells	Contractile tails, ability to ingest food fragments	Movement—Lengthening body and tentacles, second stage of digestion

MUCOR (PIN-MOULD)—A SAPROPHYTIC FUNGUS

2.13 Structure

If you expose a piece of bread to the air for about a day and moisten it slightly before placing it on a plate covered by a glass dish, you may find after a few days that a white felt develops on it. If this felt, when examined closely, is found to consist of parallel white threads some of which end in spherical black heads, then you have produced a growth of pin-mould which may be Mucor. It is rare to obtain such a growth without seeing some

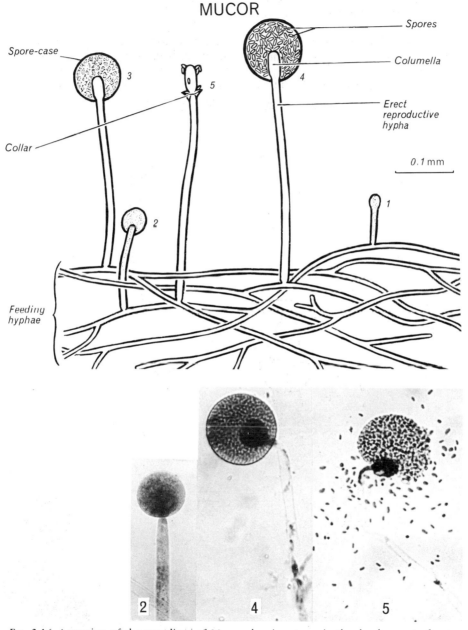

FIG. 2.16 A portion of the mycelium of Mucor showing stages in the development of spore-cases. 1—Erect hypha swells at the tip. 2—The swollen tip is cut off by a dome-shaped cross wall. 3—Cytoplasm in the spore-case starts to form spores. 4—Spore-case contains mature spores each with its own protective wall. 5—Spore case ruptured: a few spores still remain stuck to the columella and collar. in the photomicrographs of the spore-cases of Mucor, the numbers refer to the stages in the upper diagram.

blue-green and green patches of other fungi, such as Penicillium or Aspergillus, amongst the pin-mould. Pin-moulds also occur commonly on leather, jam, oranges, horse-dung, leaf litter, and a wide variety of cooked foods.

The structure of the body or **mycelium** (*Gk. mykes, fungus*) of Mucor is unlike anything we have previously met, for it consists of a chaotic tangle of threads branching amongst the material on which it feeds. A single thread is known as a **hypha**; it is bounded by a cylindrical wall of non-living material inside which is a lining of cytoplasm enclosing a large central vacuole. These hyphae have many small nuclei scattered in their cytoplasm and are not divided into cells by cross walls (compare Spirogyra). The aerial hyphae branch off from the feeding hyphae roughly at right angles to the surface of the food and are much thicker (see Fig. 2.16).

Mucor, like all other fungi, does not possess chlorophyll and so is unable to feed in the way typical of plants. Why then is it placed in the plant kingdom? Because it has walls of non-living material, because it has the diffuse branching form and continuous growth of plants, and because it cannot move except by growth.

CONJUGATION IN MUCOR

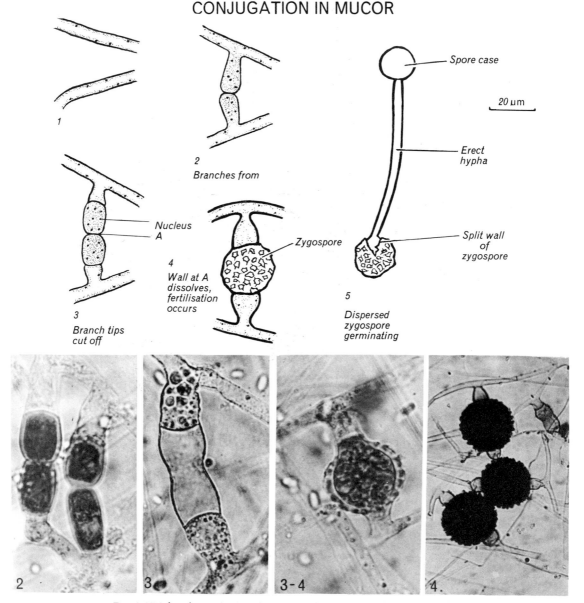

FIG. 2.17 The photomicrographs are of material stained with aniline blue.

2.14 Nutrition

The thin 'feeding' hyphae grow by increasing their length at the tip, so pushing themselves through the bread. The tips are helped in this by their producing a secretion capable of changing the solid starch of the bread into a sugary solution which is absorbed through the hypha walls into the cytoplasm. Also from the food, Mucor needs to absorb mineral salts, and nitrogen in the form of amino-acids, before it can grow. By feeding in this way Mucor causes the decay of dead organic matter and produces the musty odour characteristic of mouldy food. It is described as a **saprophyte** (*Gk. sapros, putrid; phyton, plant*) (see chapter 8, p. 82).

Through its very large surface Mucor both absorbs food in solution and, when conditions are dry, loses water by evaporation. Death is normally brought about by lack of water or food, but before this occurs reproductive bodies are usually produced.

All fungi are not saprophytes—in fact many live as **parasites** and obtain food from living organisms. They cause many crop diseases such as 'rusts' in wheat, etc., and 'mildews' on the leaves of fruit trees, potatoes, etc. The disastrous Irish Potato Famine of 1846–1847 was caused by a fungus— *Phytophthora infestans*—which spread rapidly in the warm damp weather of those summers. Relatively few fungi are parasitic on animals, though ringworm and athlete's foot are both diseases of fungal origin.

2.15 Reproduction

The black spherical structures previously mentioned are **spore-cases**, each of which contains several hundred elliptical multinucleate spores. They develop at the tips of erect reproductive hyphae which contain dense cytoplasm without the large vacuoles of the slender 'feeding hyphae'. At first the spore-cases are white, but as they become mature, they change colour through yellow and brown to black. Details of their growth are given in Fig. 2.16.

When the elliptical spores are ripe the spore-cases absorb water and dissolve except for a small collar left at the base of the **columella**. This leaves the spores as a glistening black mass embedded in a slimy material which allows them to adhere to various objects. It is not known what method of spore dispersal is the most important, but insects have been shown to carry the spores on their bodies. If the spores reach a suitable food material they germinate by putting out a new hypha which grows and branches in all directions to form a new mycelium. These spores must be present almost everywhere, otherwise the method of obtaining Mucor already described would not be successful so often.

Mucor also reproduces by conjugation (see Fig. 2.17), which results in the formation of zygospores which can resist drought for long periods. They form from branches put out when two hyphae come in contact. The branches swell, their tips containing several nuclei are cut off by a new wall, and eventually the cytoplasm mixes when the intervening walls are dissolved. As this happens, the nuclei probably fuse in pairs, one nucleus of each pair coming from one branch and one from the other. The resulting zygospore is suspended between the two hyphae, and rapidly makes a thick rough wall round itself and becomes black and spherical. The zygospores contain fat droplets and will germinate only after a period of dormancy. They are dispersed in the dormant state. Germination involves the production of an erect hypha with a spore-case at its tip. The spores from this are shed and may give rise to new mycelia as described before. Thus, from one zygospore many mycelia can develop. It is often possible to obtain zygospores from the mycelia produced when spores are spread over the cut surface of an orange. In some species of Mucor zygospores are formed only when the hyphae from mycelia developed from different spores come in contact. The two types of mycelia are referred to as + and − strains.

Mucor differs from Spirogyra in not having any chlorophyll, in being non-cellular, and in having specialised parts of the body for reproduction.

QUESTIONS

1. Describe the structure and reproduction of Mucor. State briefly how the reproduction of Spirogyra differs from that of Mucor. [O]

2. (*a*) Draw a labelled diagram of a named unicellular animal. (*b*) Draw a labelled diagram of a named alga. (*c*) What difference do you observe between (*a*) and (*b*)? (*d*) Where would you look for specimens (*a*) and (*b*)? [M]

3. Make a large labelled diagram to show the structure of Euglena. How does it differ from Amoeba (*a*) in nutrition, (*b*) in locomotion? [O]

4. Make a fully labelled diagram to show the structure of Euglena. To which of the following do you consider Euglena most closely related: Amoeba, Mucor, Spirogyra? Give your reasons. [O]

5. Describe the structure and life-history of Mucor. How would you, unaided, proceed to obtain and grow a sample of Mucor? [O & C]

6. Make labelled diagrams to show the structure and life-history of Mucor. Explain briefly how its method of nutrition differs from that of a green plant. [O & C]

7. Draw a large labelled diagram to show the structure of Amoeba. Why is Amoeba said to be (*a*) a living organism and (*b*) an animal? [S]

8. In what ways do the structure and life-history of Spirogyra (or other named filamentous green alga) (*a*) resemble and (*b*) differ from the structure and life-history of Mucor (or other named mould)? [C]

9. Describe, with the help of diagrams, the cellular organisation of Spirogyra and of Hydra and its bearing on their respective modes of life. [O]

10. What are the differences between sexual and asexual reproduction? Show by means of diagrams, which must be labelled, the way in which asexual reproduction is brought about in (1) a named alga, (2) a named fungus, (3) a named coelenterate. [A]

3 Flowering Plants

A typical flowering plant (see Fig. 3.1) is divided into two main parts—the root system and the shoot system. The main functions of the root system are to anchor the plant in the ground and to absorb water and mineral salts; the main functions of the shoot system are to manufacture food and to produce flowers and seeds.

3.1 Root System

In order to examine the root system, pull up a wallflower (or shepherd's purse) plant carefully and wash the roots. You will then see that it has a **main root** which grows straight downwards into the soil, and branching from it numerous **lateral roots**. This whole mass of roots is termed the root system, and, though it may look like a tangled mass, a more careful examination will show that the lateral roots branch off from the main root in four rows.

In some plants the main root, or **tap root** as it is called, may serve to store food (see Fig. 3.2). In other plants such as the grasses and buttercup there is no clearly defined main root but a number of roots of the same size. Such a root system is described as **fibrous**.

Roots serve to anchor the plant in the soil;

they also carry out the absorption of water and mineral salts by means of **root hairs**. If you pulled up the wallflower plant very carefully you probably noticed that the ends of the youngest roots were covered with small particles of soil. This happened because these parts of the root system were covered with tiny root hairs, though most of the hairs are usually torn off when the soil is removed. On the older parts of the root system these hairs have withered. You can see root hairs more clearly if you put some cress or mustard seeds on damp cotton-wool and allow them to germinate in a moist atmosphere. The tips of the roots have no root hairs and are protected by a conical mass of cells termed the **root cap**. The slimy outer cells of the root cap help to lubricate the root's passage between the soil particles.

3.2 Shoot System

The shoot system of the wallflower consists of a branched stem bearing leaves, buds, and perhaps flowers and fruits.

The **stem** is erect, cylindrical, and strong. The lower part is woody while the upper part is more tender, green, slightly ridged, and hairy. For some distance above ground level there are no leaves but

FLOWERING PLANT

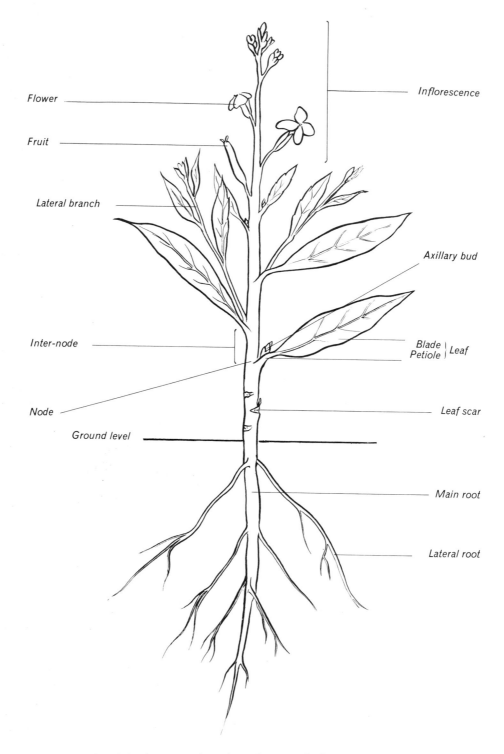

Flower

Fruit

Lateral branch

Inter-node

Node

Ground level

Inflorescence

Axillary bud

Blade | Leaf
Petiole |

Leaf scar

Main root

Lateral root

FIG. 3.1 Diagram to show the main parts of a flowering plant.

ROOT SYSTEMS

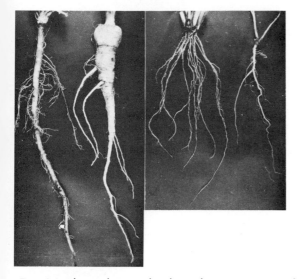

Fig. 3.2 These photographs show the root systems of dandelion and parsnip on the left (both modified for food storage), a fibrous root system of a grass and a typical root system of shepherd's purse on the right.

tiny **leaf scars** left when the leaves withered and fell off. The **leaves** are arranged spirally round the stem; the part of the stem to which the leaf is attached is called a **node** and the part between two leaves is an **internode**. The spiral arrangement of the leaves should be closely examined in the wall-flower plant. Start with the lowest leaf on the plant and follow the spiral round the stem. You will find that the fifth leaf you come to is above the one from which you started, and that in order to reach this leaf you have followed the spiral twice round the stem. The ridges on the stem make it easy to see which leaves come in the same

straight line. You will find a different arrangement of the leaves in other plants. Try to work out the arrangement in lime and privet.

At the tip of the stem, if there are no flowers, the leaves are small, very close together, and overlap one another to form a **terminal bud**. You will also find smaller buds in the angle or **axil** (*L. axilla, armpit*) between the stem and each leaf. These, because of their position, are called **axillary buds** and may either remain dormant or grow into side branches. These are shown in the photograph of a privet shoot which forms part of the heading to this chapter.

The main functions of the stem are to support the leaves and flowers and to carry materials from the roots to the leaves and vice versa. In some plants stems are also used to store food (see sec. 16.6).

Leaves are attached to the stem either by a narrow base as in the wallflower or by a short

COMPOUND LEAVES

Fig. 3.4 The two upper photographs are of the leaves of rose and creeping buttercup; the lower photograph is of a laburnum leaf.

SIMPLE LEAVES

Fig. 3.3 Photographs of the upper surface of a lilac leaf (left) and of the under surface of a lime leaf (right).

WALLFLOWER AND FRUIT

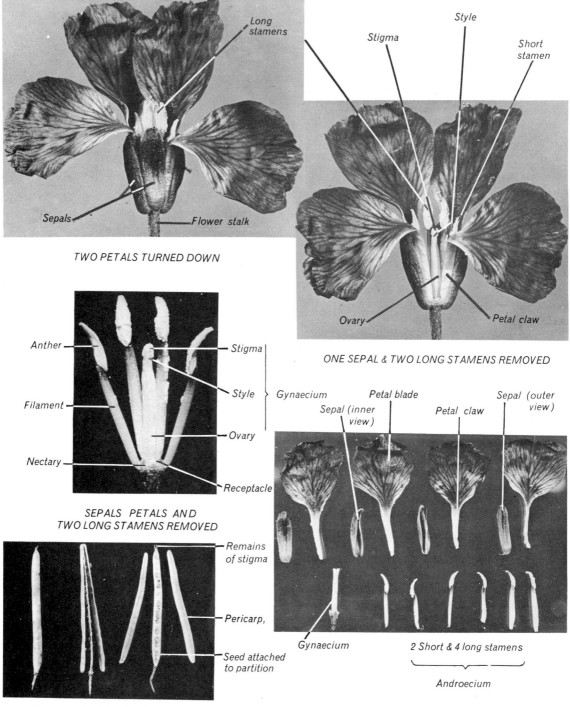

TWO PETALS TURNED DOWN

ONE SEPAL & TWO LONG STAMENS REMOVED

SEPALS PETALS AND
TWO LONG STAMENS REMOVED

FRUIT-WHOLE, SPLIT & DISSECTED

Fig. 3.5 Wallflower. The photographs show the whole flower and dissected parts; the fruit is also shown whole and dissected.

stalk or **petiole** (*L. petiolus, little foot*) as in lilac. The wallflower leaf blade has a main vein with smaller ones branching from it, but some leaves, like those of grass, have several parallel veins joined by short branches. Leaves like those of wall-flower (also lime, nasturtium) which have a single blade are said to be **simple leaves** (see Fig. 3.3); leaves such as those of lupin, rose, horse-chestnut, which have the blade divided into several parts or leaflets, are termed **compound leaves** (see Fig. 3.4). You should examine twigs of privet and ash carefully and make certain that you can distinguish between an entire simple leaf and a single leaflet of a compound leaf. A leaf always has a bud growing in its axil, but a leaflet never has one.

The main functions of the leaf are to manufacture food for the plant and to permit the exchange of gases between the plant and the air.

3.3 Flower and Fruit

The branches which bear flowers do not at first differ from ordinary branches, and like them arise from axillary buds. The production of leaves, however, soon ceases and then the branch bears flowers only. The group of flowers borne on the stem is called the **inflorescence**.

The flower is really a special part of the shoot in which the leaves are modified to help in the formation of seeds by which the plant reproduces itself. These floral leaves are joined to a very short stem termed the **receptacle** and they are arranged in four main rings or whorls.

Choose a flower which has only just opened from the wallflower plant, and look at the outside. You will see that the outer whorl or **calyx** (*Gk. kalyx, husk*) is made up of four narrow green or purple **sepals**. These are pointed at the tip and are arranged in opposite pairs; the two sepals of one pair bulge out at the base to form pouches. The sepals protect the flower when it is in bud.

Pull the sepals off your specimen and you will see the whole of the second whorl or **corolla** (*L. small crown*) made up of four brightly coloured **petals**. These are oval and are joined by a long stalk or claw to the receptacle. You will probably have noticed that the petals are arranged alternately with the sepals. The petals make the flower conspicuous and attract insects to it.

Inside the ring of petals you will find six **stamens**, and these form the third whorl or **androecium** (*Gk. andros, male; oikion, house*). Each stamen consists of a thin green stalk or **filament** and a yellow bi-lobed head or **anther** inside which form numerous tiny yellow pollen grains. The six stamens are arranged in two groups: there are two outer stamens, one opposite each pouched sepal and four slightly longer inner stamens arranged in pairs opposite the other two sepals. Use a hand lens to examine the base of one of the shorter stamens and you will see a green swelling. This is a **nectary** which secretes a sugar solution on which many insects feed. If you look inside one of the pouched sepals you may find a drop of this nectar.

In the middle of the flower is the **gynaecium** or pistil. This is a cylindrical green hollow structure with a forked outgrowth at the tip. The thicker part is the **ovary**; the forked outgrowth is the **stigma** and the tapering part which joins these is the **style**. The ovary is divided lengthways by a partition into two compartments containing small white ovules which may later grow into seeds. The floral leaves which form the gynaecium are called **carpels** and in the wallflower there are two which are joined together.

These four parts—the calyx, corolla, androecium, and gynaecium—which are all modified leaves, form the flower, and all play some part in producing the seeds which reproduce the plant.

The actual seeds are formed from the ovules, but they must first be **fertilised**, and for this pollen from the anthers is needed. This transference of pollen is called **pollination** and section 16.4 explains the part played by pollen in fertilising the ovule to form the seed. After pollination and fertilisation have taken place, the sepals, petals, and stamens wither and fall off the receptacle, leaving the gynaecium which by now has grown into the fruit. This, in the wallflower, is about two inches long and when fully grown the fruit wall or **pericarp** becomes dry and splits so that the two halves break away from the partition. The seeds remain joined to the partition and are eventually blown away by the wind.

If you want to discover the names of the flowers which you find, you should learn how to use a Flora. *A Flower Book for the Pocket* by MacGregor Skene and *An Excursion Flora of the British Isles* by Clapham, Tutin, and Warburg are both books which contain information about the method of identifying flowers by means of a key. *The Concise British Flora in Colour* by W. Keble Martin is another similar book.

4 Mammals

Man is a mammal; so also are his beasts of burden (horse, mule, yak, camel), and many of his domesticated animals (cat, dog, rabbit, sheep, goat, pig, and cattle). Everyone is familiar with at least some members of the class but not everyone knows why whales and bats are mammals. Many might wrongly classify them as fish and birds respectively. Wild mammals such as foxes, badgers, hedgehogs, field-mice, and weasels are rarely seen because they conceal themselves efficiently and are largely nocturnal. There can be no doubt that mammals form a highly complex group both structurally and functionally, and that their success in the struggle for existence has been largely due to their intelligent behaviour. Some of the features which distinguish them from other vertebrates (animals with backbones) are:

1. The young are nourished for some time after birth by milk sucked from the **mammary glands** of the mother.
2. They usually possess an almost complete covering of **hair**.
3. A muscular **diaphragm** divides their body cavity into two.
4. **Sweat glands** are usually present in their skin.
5. They have a **highly developed fore-brain**.

GENERAL STRUCTURE

Mammals contain several hundred different types of specialised cells which are aggregated together to form tissues. A **tissue**, muscle or bone for ex- ample, is a mass of cells which appears uniform in texture and chemical composition. Tissues frequently contain more than one type of cell.

Within a mammal's body there are many localised structures concerned with one main func- tion. Such structures are known as **organs** (e.g. kidneys, heart, lungs, brain) and may contain several different tissues. **Organ systems** (e.g. nervous system, blood system) differ from organs only in that they are not localised but are distributed over a large part of the body.

4.1 The Skeleton

Every mammal has an internal bony framework which gives support to the body, protects delicate organs, and provides anchorage for muscles so that when they contract parts of the body can be moved at the joints.

The central axis of the skeleton is the spin or **vertebral column** (see Fig. 4.1), which runs almost the whole length of the body from behind the skull to the tip of the tail. It is made up of a number of bones called **vertebrae** (sing. vertebra), each of which can be moved slightly on its neighbours, thus making the column, as a whole, flexible. In the thoracic (*Gk. thorax, chest*) region, jointed to each vertebra at the sides, is a pair of **ribs** which are curved and join the breastbone or **sternum** on the ventral (*L. venter, belly*) side to complete the bony cage of the thorax.

Jointed to the first vertebra is the **skull**, a very complicated mass of fused bones which encloses and protects the brain, the ear apparatus, the organs of smell, and to a lesser extent, the eyes.

RABBIT SKELETON

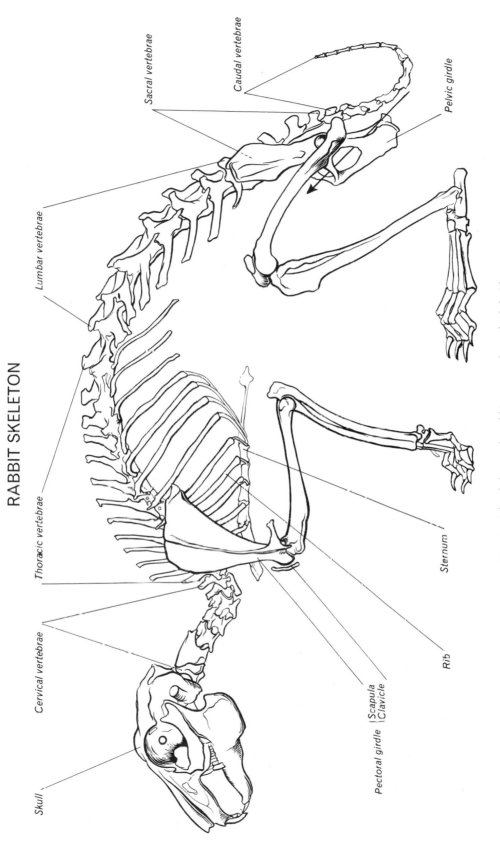

Skull

Cervical vertebrae

Thoracic vertebrae

Lumbar vertebrae

Sacral vertebrae

Caudal vertebrae

Pelvic girdle

Pectoral girdle | Scapula
                | Clavicle

Rib

Sternum

FIG. 4.1 The skeleton of a rabbit as seen from the left side.

The upper jaw is fixed immovably to the skull, but the lower jaw is jointed to it close to the ear region.

The limb bones are jointed to two incomplete girdles of bone round the vertebral column. The **pectoral girdle** is the anterior one, and, in the rabbit, is composed of two shoulder blades (scapulae) held in place by muscles at the sides of the thorax. There are also two tiny collar-bones (clavicles) which, although apparently useless in the rabbit, are well developed in man. At the ventral end of the shoulder blade is a socket which receives the upper bone of the fore-limb. Posteriorly, with the help of the vertebral column to which it is attached, the **pelvic girdle** (pelvis) forms a complete ring of bone in the sacral region. An arrow passes through this ring in Fig. 4.1. The fore-limb and hind-limb bones are similar in basic plan and are fully described in section 13.4.

There is a great deal of muscle in a mammal; it is mainly concentrated in the limbs, round the girdles, and along the vertebral column, which is only to be expected, since these are the parts used in locomotion.

4.2 The Body Cavities and their Contents

The soft parts of the body lie in two cavities, the **thoracic cavity** and the **abdominal cavity**, which are divided from one another by the diaphragm (see Fig. 4.2). A thin sheet of muscle covered externally by skin forms the body wall round these cavities.

In the thoracic cavity lie the **heart** and a pair of **lungs**, both of which are well protected by the ribs. The lungs contain a tremendous number of minute thin-walled sacs which are filled with air at every breath. The air enters through the mouth or nose and passes down the windpipe (or trachea), which leads through the neck into the thorax where it divides into two branches, one to each lung. The heart is a muscular four-chambered pump which forces the blood through tubes known as **arteries** to the narrow **capillaries** which branch and rejoin amongst every tissue of the body. From the capillaries the blood is collected in **veins** and returned to the heart.

In the abdominal cavity lies almost the whole **alimentary canal**. This is essentially a much-coiled tube down which food passes from the mouth to the anus, and during its passage digestion takes place. Only the buccal cavity and oesophagus, a narrow tube which conveys food from the mouth through the thorax to the stomach,

lie outside this cavity. The stomach is a large somewhat J-shaped sac lying close up against the diaphragm and surrounded by the lobes of the liver, which is the largest gland of the body. In the stomach, food is held for a short time (about one hour in man) before it passes on down the much-coiled intestine to undergo further digestion. Soluble food taken into the blood supplying the small intestine is circulated round the body. Undigested matter collects in the large intestine where it is partially dried by absorption of water into the blood and then passed out of the body at the anus.

Attached to the dorsal (*L. dorsum, back*) side of the body wall in the abdominal cavity is a pair of excretory organs, the **kidneys**, from each of which a narrow tube leads posteriorly to open into the bladder. The latter is a muscular sac in which the urine, made by the kidneys, collects before being expelled to the exterior through a single tube, the urethra.

Behind the kidneys in the female is a pair of small pink **ovaries** which produce egg cells. The corresponding organs in the male, the **testes**, move from this position early in life and come to lie in a scrotal sac outside and below the abdominal cavity. These reproductive organs pass their products to the exterior through a system of tubes, a part of which (the uterus) in the female is specialised to nourish and protect the embryos which develop from fertilised eggs. At birth the embryo is expelled from the uterus and then takes its first breath.

4.3 The Nervous System

The greater part of the nervous system is not in a body cavity. The central nervous system lies in the mid-dorsal line and is composed of the **brain** within the skull, and the **spinal cord** running posteriorly from the brain through a canal in the vertebrae. From the brain and spinal cord issue pairs of branching **nerves** which supply all parts of the body. Their function is to carry 'incoming' messages from sense organs to the central nervous system, or to convey 'outgoing' messages from the latter to organs such as muscles and glands which are able to respond. The 'messages' travel at high speeds (up to 120 metres per second in man) in the nervous system, which helps to make it efficient at co-ordinating a suitable series of responses to meet any situation.

Several other chapters in this book give further details of the structure and function of a mammal.

MAMMAL-INTERNAL ORGANS

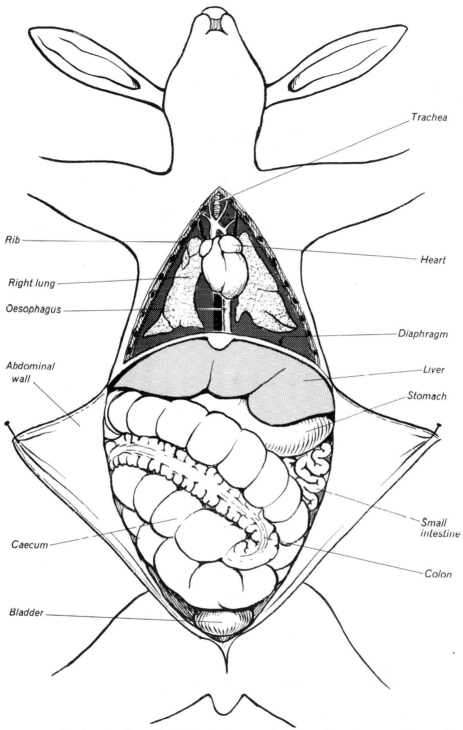

Fɪɢ. 4.2 A drawing of a dissected rabbit. The thorax and abdomen have been opened ventrally and the thymus gland has been removed to expose the heart and its blood vessels fully. Note that the lungs have collapsed and no longer fill the thorax. The abdominal organs are undisturbed.

RABBIT—NERVOUS SYSTEM

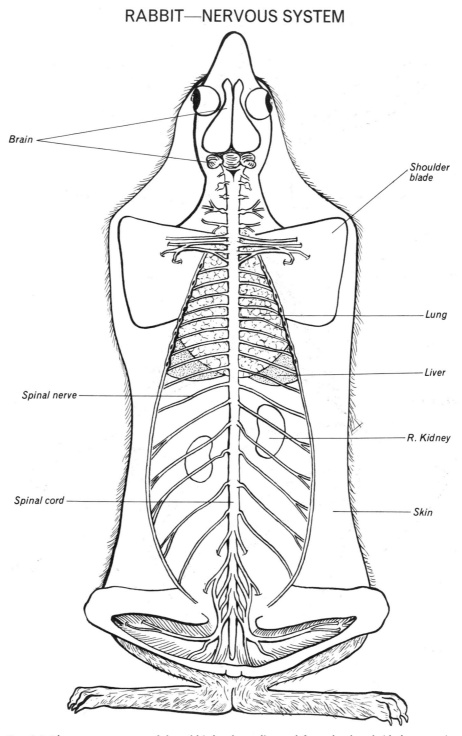

Brain

Shoulder blade

Lung

Liver

Spinal nerve

R. Kidney

Spinal cord

Skin

FIG. 4.3 The nervous system of the rabbit has been dissected from the dorsal side by removing the skull, vertebral column, and dorsal musculature. The shoulder blades have been pushed ventrally and away from the mid-line to display the nerves of the thoracic region. Much muscle has been cut away to display the nerves supplying the hind limbs. Lungs, liver, and kidneys are seen through the thin lining of the body cavities. No nerves from the brain are visible in this dissection.

A REVIEW OF THE MAMMALS OF GREAT BRITAIN

4.4

Living in the wild state in Great Britain or round its shores are over eighty species of mammal.

A group of small mammals represented in Britain by the mole, the hedgehog, and four species of shrew are the insectivores (*L. insectum, something cut in; voro, I devour*). Shrews are easily recognised as small mouse-like animals, but they have a long snout and all except the Scilly Shrew have orange enamel on their teeth. The group is by no means entirely insect-eating as its name suggests, for moles are known to lay up in their burrows stores of earthworms which they have immobilised by biting their anterior ends.

Bats, of which there are thirteen British species, are the only mammals to have developed the power of flight (see the photograph in the heading to this chapter). They are further remarkable for the way in which they avoid obstacles by using echoes of the ultra-high-pitched squeaks they emit during flight.

The rabbit, the brown hare, and the mountain hare are closely related to the large group of gnawing mammals known as rodents. This latter group includes the voles, the mice, the black and brown rats, the red squirrel, and the dormouse, which is our only native mammal to hibernate properly during the winter. Several rodents have been introduced by man, amongst them the grey squirrel, which is now a pest in some parts of the country, and the coypu, which was bred in captivity for its fur and now lives wild along the River Yare in Norfolk and in Hertfordshire.

Flesh-eating mammals (carnivores) are represented by the fox, badger, weasel, stoat, otter, polecat, wild cat, and pine-marten, of which the last two are now rare. Also in this group are the common seal and grey seal, which breed on our shores.

The red deer, which is now found mainly in Scotland, and the roe deer, which occurs more extensively, are the only naturally wild hoofed mammals, the group to which our cattle, sheep, goats, pigs, and horses belong.

Although six whalebone whales are listed below as British, they are rarely seen in our coastal waters but keep to the deeper waters of the Atlantic.

If you wish to know more of the natural history of British mammals you will find an excellent account in Dr. Harrison Matthews' book, *British Mammals*.

A List of Living British Mammals

(I) indicates those introduced intentionally by man.

Insectivores

Hedgehog	Common Shrew	Mole
	Pygmy Shrew	
	Water Shrew	
	Scilly Shrew	

Bats

Noctule	Daubenton's Bat	Greater Horseshoe
Leisler's Bat	Natterer's Bat	Bat
Pipistrelle	Whiskered Bat	Lesser Horseshoe
Serotine	Bechstein's Bat	Bat
Long-eared Bat	Mouse-eared Bat	
Barbastelle		

Lagomorphs

Rabbit	Brown Hare	Mountain Hare

Rodents

Bank Vole	Long-tailed Field	Dormouse
Orkney Vole	Mouse	Edible Dormouse (I)
Short-tailed Vole	Yellow-necked Field Mouse	Red Squirrel
Water Vole	Harvest Mouse	Grey Squirrel (I)
	House Mouse	
Musk Rat (I)	Black Rat	Coypu (I)
	Brown Rat	

Carnivores

Fox	Badger	Grey Seal
	Otter	Common Seal
Wild Cat	Pine Marten	
	Stoat	Four other seals and
	Weasel	the walrus are
	Polecat	occasional visitors.

Hoofed Mammals

Red Deer	Fallow Deer (I)	At least six other
Roe Deer	Sika (I)	escaped species of
	(Japanese Deer)	deer may be
		encountered.

Whales

Toothed Whales

Common Porpoise	False Killer	Sperm Whale
Common Dolphin	Black Fish or	Bottle-nosed
	Pilot Whale	Whale
White-sided Dolphin	Risso's Dolphin	Cuvier's Whale
White-beaked Dolphin	White Whale or	Sowerby's
	Beluga	Whale
Bottle-nosed Dolphin		
Killer	Narwhal	True's Whale

Whalebone Whales

Atlantic Right Whale	Blue Whale	Sei Whale
Lesser Rorqual or	Fin Whale	Humpback
Piked Whale		

5 Soil

Soil, which is present as a thin layer over a large area of the earth, provides the water and mineral salts which plants require as food material. It is through the roots, which are firmly embedded in the soil, that plants take up this food material. The soil is, therefore, highly important to man, who relies on it to produce so much of his food.

5.1 Formation

Soil is formed as a result of an interconnected series of slow physical, chemical and biological processes. The first stage consists of the physical breaking down or **weathering** of rocks, and is brought about mainly by the action of frost and water. When water freezes it expands, and when this happens in the crevice of a rock the crack may be enlarged or pieces of rock become separated and, where the rock face is sloping, fall to the ground (see Fig. 5.1). The alternate heating and cooling of the rocks causes expansion and contraction, which also help to break them up. At the foot of bare rock faces may be seen large stones which have been loosened in this way.

Rocks are not made of one material, and though much of the rock is insoluble, parts may be slowly dissolved by rain. This solvent action of the water is increased by the carbon dioxide which makes it acidic. Eventually the soil will contain organic acids in solution and so the chemical breakdown of the rocks will be speeded up. Besides breaking up the rocks this chemical action releases mineral salts from them.

As a result of these physical and chemical pro-cesses the **mineral skeleton** of the soil is formed. The fragments of rock may be further reduced in size by mechanical friction as the wind, rivers, and sea move them about. On these fragments small and simple plants such as mosses grow, since they are able to survive long periods of drought. Small flowering plants may also grow between the fragments. As these plants die they form dark brown **humus** (see sec. 5.2). This decomposition is a biological process brought about by the activity of fungi and bacteria (see sec. 5.5), and results in the addition of more and more humus to the soil, which becomes darker in colour. During this process of decomposition organic acids

WEATHERING

Soil —
Subsoil —
Parent rock —

Fig. 5.1 Photograph of the edge of a quarry.

may be produced which help in the chemical breakdown of the mineral fragments, and ammonium salts, nitrates, and phosphates are released. The growth of plant roots and the activities of burrowing animals play an important part in producing the friable texture characteristic of a good mature soil. The processes involved in the formation of such a soil are summarised in Fig. 5.2. It should be emphasised, however, that the soil once formed does not remain unchanged. The processes described are continuous, and the soil is always changing rather like a living organism.

If you look at a quarry or recent excavation you will see that below this layer of dark soil there is a light-coloured one; this is the **sub-soil**, its lighter colour indicating the absence of humus. Below the sub-soil is the actual rock.

Soils which have been built by the gradual weathering of the rocks are termed **sedentary soils**, for the soil lies on top of the rock from which it has been formed. Soils do not always have the same constituents as the rocks below

them, for the soil may be carried by rivers or glaciers and deposited some distance from the place where it was formed. Such soils are termed **sedimentary** or **alluvial soils**.

5.2 Composition

Soil is a complex mixture of various chemical substances together with a very large number of small plants and animals. It is easy to show the presence of the main constituents by a very simple experiment. A weighed sample of soil should be spread to dry on a tray, left near a radiator for about two days, and then re-weighed. You will find that there has been a loss of weight owing to the evaporation of water. If you put some of this partly dried soil into a crucible, weigh it, and then heat it strongly it will smoke and it may smoulder. When the crucible has cooled you will see that the soil is lighter in colour, and on re-weighing that there has been a further loss. This is due partly to the evaporation of more water and partly to

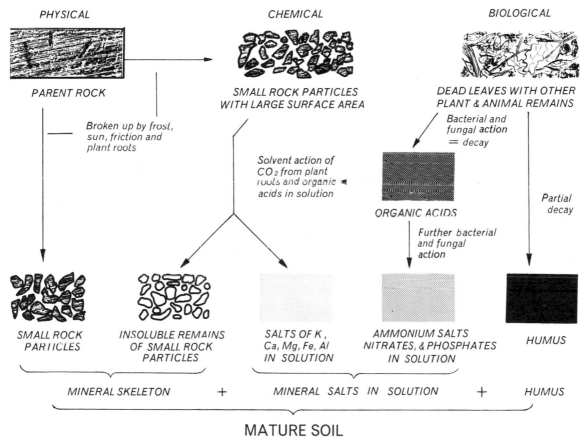

SOIL FORMATION

FIG. 5.2.

the burning off of the organic matter. The residue is mineral matter. Such an experiment shows that the soil contains water, organic matter or humus, and mineral matter.

It is the mineral matter, and particularly the size of the particles, which largely determines the character of the soil. If a sample of soil is well shaken with water in a measuring cylinder and then allowed to settle, some idea of the variety in size of these particles may be obtained. Small stones and gravel will settle first, then a layer of pale grains of sand, and over this a dark layer, so fine that the individual particles cannot be distinguished; this is the silt. All the soil does not settle and the water will remain cloudy because it is full of very fine particles which have not had time to settle; these particles may be clay or silt.

When the soil sample is shaken up with water in the measuring cylinder some dark-coloured material probably floats on the surface. This is the humus which has been formed from dead plant remains. When the gardener starts to make a compost heap it is very easy to separate out the individual leaves, but after a time the leaves become sodden, darker in colour, and fall to pieces when handled. When the rotting process is complete, the humus is a dark brown slime which coats the mineral particles. Humus is a mixture of decomposition products, dead soil organisms, fungal mycelia, and finely divided plant residues which have lost their structure.

These solid constituents of the soil—the mineral particles and the humus—are mixed together very thoroughly and in a special way. After the various particles have been allowed to separate in the measuring cylinder, if the water is poured off, the sediment emptied on to a tray, mixed up very thoroughly, and allowed to dry, it will be found that it becomes pasty and then cakes into a sticky mass quite unlike the original soil sample. Fresh soil has a crumbly texture and breaks up into small masses without becoming pasty. These small masses or **crumbs** cannot be made by mixing the separate solid particles with water. They consist of particles of sand, silt, clay, humus, etc., clinging together. These particles are held together by the surface tension of the water films around them. If you put two dry glass slides one on top of the other you will find that it is very easy to lift off the top one. If, however, you put a drop of water between the two slides and repeat the experiment both slides will probably be picked up, but if you put the slides under water you will find that they separate as easily as when they were dry. The slides were held together by the surface tension of

the thin film of water between them. (These same forces cause water to rise up a capillary tube and are sometimes referred to as 'capillary forces'.) It is this force which holds the particles together, but they can be separated by drying out or by wetting the soil.

This crumbly texture of the soil makes it porous, and the pores contain air. When a lump of soil is put into a beaker of water and then broken up bubbles of air escape from it. This air is necessary for the growth of roots and is also used by the micro-organisms living in the soil. These organisms can be shown to be present in the soil by the experiment described in section 1.3. They assist in converting the dead plants into humus and bring about many other changes in the soil. Some of their activities are described in section 5.5.

The soil water contains oxygen, dissolved from the air between the soil particles, and salts dissolved from the mineral particles and humus. The presence of these mineral salts can be shown by adding distilled water to a sample of soil, mixing very thoroughly, filtering, and evaporating the filtrate. The residue of mineral salts can be seen quite clearly, though it may be very small. These salts include the sulphates, phosphates, and nitrates of potassium, calcium, magnesium, and iron, many of which are essential for plant growth (see sec. 6.4).

A typical soil then contains the following six constituents.

1. Mineral particles. 4. Dissolved mineral salts.
2. Humus. 5. Air.
3. Water. 6. Micro-organisms.

5.3 Soil Types and their Properties

The nature of the soil is largely determined by the proportions of sand, silt, and clay that it contains, and for this reason soils are classified according to the size of their particles. The term sand is used to describe coarser particles (2.0–0.02 mm in diameter), clay to describe the finer particles (less than 0.002 mm in diameter), and silt to describe the intermediate particles. A **sandy soil** is one which the coarser particles predominate; a **clay soil** is one in which the finer particles predominate. A mixture of these two types is called a **loam**.

The size of the particles affects the amount of air in the soil, the rate and amount of drainage, and the rise of water in the soil. It is therefore responsible for the physical properties of the soil.

You can make a comparison of the degree of porosity (permeability to air) of sandy and clay soils with the apparatus shown in Fig. 5.3. The funnels must contain equal volumes of dry sandy soil and dry clay soil, and the tubes of water must be the same length and diameter. The funnels should be tapped gently to make sure that the soil has settled. When you open the clips at the bottom of the tubes, water will drain out of the tubes, dragging air through the sandy and clay soils. You will find that the water drains much more quickly from the tube joined to the funnel containing the sandy soil, showing that it is more porous than the clay soil. It is more convenient to carry out these experiments with sand and clay rather than with sandy and clay soils.

If the tubes which contained the water are removed you can use the same apparatus to compare the amount of water retained by the two types of soil. In this experiment you must pour the same volume of water into each funnel, and collect and measure the water which drains through the soil. You will find that nearly the whole of the water drains through the sandy soil, showing that little is retained, but the reverse is true of the clay soil.

The two funnels are now filled with thoroughly wetted soil and can be used to compare the rate

SOIL POROSITY

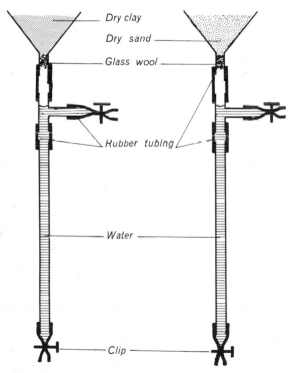

FIG. 5.3 Apparatus used to compare the permeability to air of dry sand and dry clay. The side tubes are useful in setting up the apparatus.

SOIL CAPILLARITY

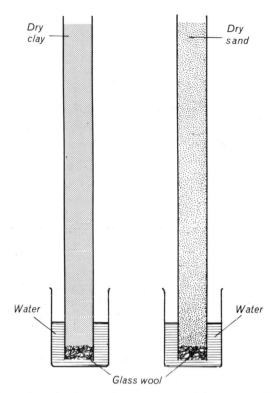

FIG. 5.4 Apparatus used to compare the rate at which water rises in sand and clay.

at which the water drains through the two types. Once again you must pour equal volumes of water into the two funnels, but this time record the time taken for the water to drain. You will find that the water drains fairly quickly through the sandy soil but it may be a few days before the last drop of water drains through the clay soil.

Soils vary considerably in their water content, since this depends not only on the type of soil but also on the amount of rainfall and the situation (e.g. altitude, slope of the land, etc.). Rain percolates through the soil and porous rocks until it reaches impervious rocks on which it accumulates and forms the **underground water table**. Even in dry weather plants are able to absorb water from the soil, for water from this underground water table can be drawn upwards. This rise is due to capillarity, and the height to which the water is raised depends on the size of the soil particles. You can demonstrate this by using the apparatus shown in Fig. 5.4. As you fill the tubes with dry sandy and dry clay soils you should tap them several times to make sure that the soil has settled properly. You must note the rate at which the

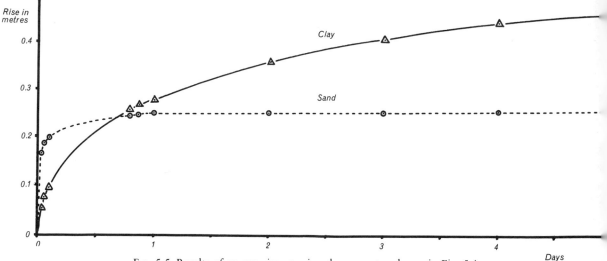

FIG. 5.5 Results of an experiment using the apparatus shown in Fig. 5.4.

water rises in both tubes by marking them at regular intervals with strips of gummed paper. You will find that the water rises very quickly in the sandy soil during the first few hours but stops after about two days, whereas in the clay soil it rises very slowly but continues to do so for several days and eventually reaches a much greater height. These facts will be shown more clearly if you plot the results as a graph (see Fig. 5.5).

The particles of soil are surrounded by thin films of water. This water film which connects

ROOT HAIR

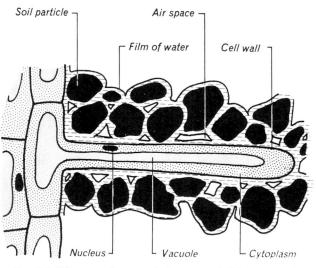

FIG. 5.6 Diagram of a root hair surrounded by water, air spaces, and soil particles.

the particles is unbroken, but between the particles there are minute spaces which are full of air. It is from these films that the root hairs of the plant absorb water. The films are elastic and the withdrawal of water from any area sets up a flow of water toward this area until the films are again of equal thickness. Thus evaporation of water from the surface of the soil, or the uptake of water by the root hairs, will start a flow of water through the soil from the water table.

In an acid soil the decomposition of the humus takes place very slowly if at all. (You can use a universal indicator to test the acidity of the soil.) Such a soil is usually waterlogged and consequently contains little air; it is also deficient in lime. The peaty soil found on moorlands contains a large amount of vegetable matter which has undergone very little change and is an excellent example of an acid soil.

The chemical properties of the soil are far less important than the physical ones. Excess salt (sodium chloride) in areas flooded by sea water has a marked effect on the flora as the plant is unable to absorb water from this concentrated solution and must obtain its water after heavy rain and store it within itself. Chalky soils derived from limestone also have a characteristic flora, as many plants cannot tolerate a large amount of lime in the soil.

The best type of soil is one which is light, warm, rich in mineral salts, and contains plenty of water without becoming waterlogged. No one soil type fulfils all these conditions, but a loam supplies most of them.

Sandy Soil	Clay Soil
Particles are large	Particles are small
Particles are easily separated	Particles cling together
Porous. Large air spaces	Not porous. Small air spaces
Drainage good	Drainage poor
Infertile soil—since the soluble salts are easily washed out	Fertile soil—since the soluble salts are not easily washed out
Retains little water	Retains much water
Dry and warm	Wet and cold
Light and easy to work	Heavy and difficult to work

5.4 Methods of Improvement

A sandy soil is well aerated but retains little water and is often poorly supplied with mineral salts. It can be improved by mixing it with clay, which helps it to retain more water, or by adding humus, which will have the same effect and make it more fertile. Once its water-retaining capacity has been increased it is worth while adding fertilisers.

A clay soil is badly drained and heavy to handle. Its drainage can be improved by mixing it with sand or, more easily, by the addition of slaked lime, which makes the small particles of clay join together. This **flocculation** can be shown by shaking some clay with water in a measuring cylinder and an equal volume of clay with lime water. You will find that the particles soon settle in the second cylinder but remain in suspension in the first for many days.

In a clay soil the poor aeration and drainage result in the accumulation of organic acids formed from the decaying plants, which make the soil 'sour'. The addition of lime neutralises this acid.

Addition of fertilisers

In the natural state soil remains fertile, but this is not true if the soil is cultivated. Plants take out of the soil various mineral salts which they use to make their food (see sec. 6.3) and if the soil is to be used to grow food crops it is necessary to see that a supply of these salts is maintained. For this reason the farmer or gardener adds fertilisers to the soil. Natural fertilisers such as stable manure, which is slow acting, should be added in the autumn so that the micro-organisms in the soil have time to release the mineral salts before the plants require them in the spring (see sec. 5.5). Compost and hop manure can be used as substitutes for stable manure. Such fertilisers also improve the water-holding capacity of sandy soils, as they contain much humus, and for the same reason they improve the drainage of clay soils.

Artificial fertilisers are used mainly to increase the supply of nitrogenous material; they are

soluble mineral salts such as sodium nitrate, potassium chloride, 'superphosphate' (a soluble compound of calcium phosphate and sulphuric acid), 'basic slag' (a compound of calcium phosphate and lime), and ammonium sulphate. These are quick acting and should be added in the spring just before they are required by the plants. If they are added earlier than this they are likely to be washed out of reach of the plant by the rain before they are required. This removal of soluble salts is called **leaching**, and will be more likely to occur in hill than in valley farms. Some of these salts, like

ROOT HAIRS

FIG. 5.7 These four-day-old cress seedlings were grown in a moist atmosphere. Note the root hairs forming a dense growth almost to the tip of the root.

ammonium sulphate, tend to make the soil acid and should not be used on 'sour' soil unless lime is added as well. Artificial fertilisers may upset the balance of the salts in the soil, so they must be used carefully; they do not provide any humus, so they do not affect the physical properties of the soil.

Cultivation

The condition of the soil can also be improved by good cultivation. Autumn **ploughing or digging** breaks up the soil and prevents it from becoming too compact or wet. The frost helps in this process too, for as the water freezes between the lumps of soil it expands and breaks them up (see sec. 5.1). In spring the soil should be broken down finely before the seed is sown. **Rolling** at this stage will press the soil particles together and make it easy for the water to rise by capillarity, so ensuring an adequate supply for the young seedlings. Once the seedlings are established the ground between them should be **hoed**—i.e. broken up. This prevents the growth of weeds which might compete with the food crop for supplies of water and mineral salts. Under certain conditions hoeing has been shown to reduce the amount of water lost by evaporation from the surface of the soil.

Rotation of Crops

Farmers may try to maintain the fertility of the soil by a rotation of crops. In the Norfolk plan there is a four-course rotation (see Fig. 5.8). The soil is manured before the sowing of the 'root' crop, and after the leguminous crop has been harvested the residue is ploughed in to increase the

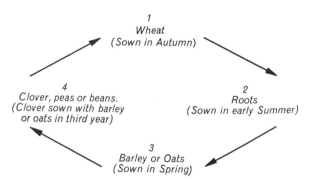

FIG. 5.8 The Norfolk plan for crop rotation.

amount of available nitrogenous compounds (see sec. 5.5). Such a rotation discourages the growth of weeds, as the 'root' crop can be easily hoed, and also reduces the spread of insect and fungal pests.

Modern farming demands a greater yield from the land than is given by this four-year rotation and its aim is to harvest more than one crop per year. Many of the crops, e.g. peas and beans, go straight from the farm to 'frozen food' and canning factories (see sec. 19.3), so that the time taken to harvest and dispose of the crop is short and the land is quickly made available for another crop. Even under these intensive methods of farming the principle of crop rotation is maintained.

5.5 Circulation of Nitrogen

One of the most important mineral substances which the green plant obtains from the soil is a supply of nitrates. The plant requires this to build up proteins (see sec. 6.9) from which cytoplasm is formed. Nitrogen is required for this, and although four-fifths of the air is nitrogen the plant is unable to use this in the synthesis of proteins but must have a supply of soluble nitrogenous compounds from the soil (see sec. 6.4).

Although green plants are continually removing nitrates from the soil the supply does not become exhausted, because in the normal course of events these nitrates are re-formed when the plant dies and decays. If, however, the plants are removed

from the soil (e.g. cultivated crops), then this general circulation of nitrogen is interrupted and special steps have to be taken to restore the nitrate if more crops are to be grown. Some of the methods by which this can be done were described in the previous section.

The green plant absorbs nitrates from the soil through its roots and uses them to synthesise proteins and finally cytoplasm. When these plants die they are attacked by **putrefying bacteria**. Since these contain no chlorophyll they are unable to manufacture their own food but feed saprophytically (see sec. 2.14) on the dead plants. Some of these bacteria break down the proteins and release the nitrogen in the form of ammonia which readily combines with other substances in the soil. Thus a supply of ammonium compounds is formed in the soil as a result of the decay of the green plant. These ammonium compounds will

ROOT NODULES

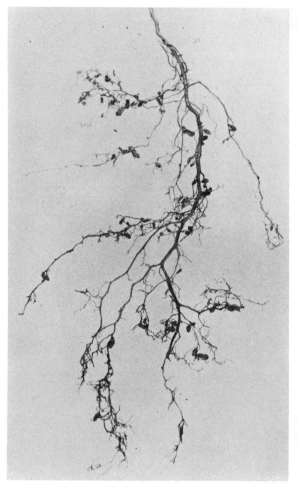

FIG. 5.9 Photograph of root system of broom seedling.

will be formed even if the plant is eaten by an animal. In this case some of the nitrogenous compounds are returned to the soil during the life-time of the animal by the excretion of urea.

Another group of bacteria in the soil obtains its supply of energy by the oxidation of these ammonium compounds. Some of these **nitrifying bacteria** oxidise the ammonium compounds to nitrites, others oxidise these to nitrates.

Thus, as a result of the activities of the putrefying, and nitrifying bacteria, the proteins and cytoplasm of the dead plants have been broken down to ammonia, and this has been oxidised to nitrates which other green plants will be able to use to synthesise more proteins and cytoplasm.

Unfortunately the cycle is not always as straightforward as this. A third group of bacteria—**denitrifying**—break down the nitrates and the resulting nitrogen escapes into the air. This takes place mainly in badly aerated soils.

This loss from the soil is made good by the activity of another group of bacteria. These **nitrogen-fixing bacteria** are able to use the nitrogen of the air to build up proteins if they are supplied

with carbohydrates. Some of these live freely in the soil and obtain their carbohydrate from the humus. Others live as symbionts (see sec. 8.5) in the roots of plants belonging to the leguminous family (e.g. clover, sainfoin, beans, vetches, lupins) and obtain their carbohydrate from the plant. In return, the plant obtains nitrogenous compounds from the bacteria and these will be returned to the soil when the plant dies. The roots of these plants bear many little swellings—**nodules**—which are packed with millions of nitrogen-fixing bacteria (see Fig. 5.9). It is because clover and sainfoin contain these nitrogen-fixing bacteria that they are used as 'green' manure and ploughed into the ground, thus increasing the supply of nitrates.

Thus if one could follow the career of a single atom of nitrogen it would be incorporated in many different compounds as the result of the activities of living organisms. The general circulation of nitrogen between compounds in the soil and proteins and cytoplasm of living organisms, which is often referred to as the nitrogen cycle, is shown in diagrammatic form in Fig. 5.10. It is

CIRCULATION OF NITROGEN

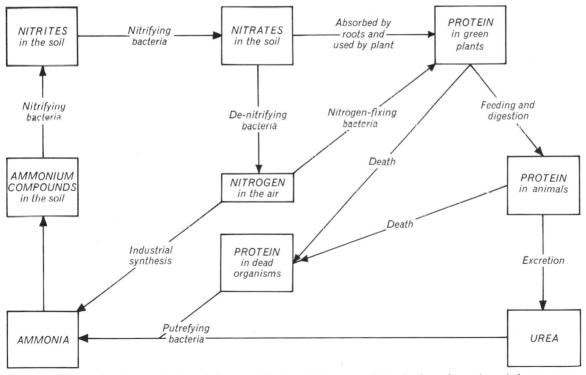

FIG. 5.10 The circulation of nitrogen. The chemical compounds involved are shown in capital letters and the names of the processes in smaller type.

impossible, however, for a state of equilibrium to be established in cultivated soils, and if man is to obtain sufficiently large crops to feed the world population, then large quantities of nitrates or ammonium salts have to be added to the soil (see sec. 5.4). The industrial 'fixation' of atmospheric nitrogen by the synthesis of ammonia from nitrogen and hydrogen is of great economic importance, since it produces cheap nitrogenous fertilisers and enables the farmer to make up for the annual drain on the nitrogen content of the soil caused by growing and removing crops.

5.6 Animals in the Soil

The soil has a very large population, but most of these living organisms are too small to be seen with the naked eye, though they play an important part in maintaining the fertility of the soil, as we have seen in the last section. Many, however, are larger. Foxes, rabbits, moles, and mice, for example, make their home in the soil; various insects, centipedes, millepedes, threadworms, and earthworms live in and find their food in the soil. Some of these affect the condition of the soil.

Darwin made a special study of the activities of earthworms and estimated that there might be as many as 50 000 in an acre of soil, and that as a result of their burrowing activities they bring up enough soil in five years to make a layer 25 mm deep on the surface. Much has been learnt about the earthworm since Darwin's day, and we now know that only two British species produce worm casts on the surface, so that we are less certain about the importance of earthworms in bringing fresh soil to the surface.

The earthworm's body is cylindrical and divided into about 150 segments. The anterior end is pointed and this helps the earthworm to push its way through the soil. The posterior end is flattened and can be used in the manner of a trowel to smooth the sides of the burrow. The skin is moist owing to the slimy secretion from the glands in the skin. This secretion keeps the skin clear of fungi and bacteria, acts as a lubricant when the earthworm moves through the soil, and helps to bind the soil particles together in the wall of the burrow.

The earthworm makes its burrows partly by forcing its way through the soil and partly by eating its way through the soil. It obtains some of its food from the humus in the soil, though it also drags leaves into its burrow to use as food. In this way the earthworms mix the dead surface litter with the main body of the soil, so making it more accessible to the attack of the micro-organisms in the soil and speeding up its rate of decomposition. The soil particles and the organic matter are ground up in the part of the alimentary canal known as the gizzard. This intimate mixture of soil particles and humus is passed out from the anus at the end of the body. In some species this takes place in the soil itself, but in others at the surface of the burrow and results in the formation of the familiar worm casts.

QUESTIONS 5

1. Name the constituents of a fertile soil. What is the importance in the soil of each constituent you mention? Describe the various ways in which soils may lose their fertility and the methods by which man tries to prevent such losses. [C]

2. What are the chief constituents of soil? Explain concisely the value of the following practices: (a) adding stable manure to the soil, (b) growing clover, and (c) weeding. [L]

3. Describe (a) one experiment by which you could find the water content of a sample of soil and (b) one experiment by which you could find the humus content of a sample of soil. Explain the part played by humus in soil fertility. [L]

4. An analysis of two soil samples, A and B, gave the following results:

	Soil A (%)	Soil B (%)
Coarse, medium and fine sand	75	22
Silt	7	20
Clay	15	48.5
Humus	3	9.5

(a) Using this information, briefly describe the general characteristics of each of these soils.

(*b*) State, giving reasons, in which of these two soils you would expect to find (i) the higher nitrogen content, (ii) the more vigorous root growth, (iii) the greater earthworm population.

(*c*) In what ways would you expect a large earthworm population to improve the quality of each of these soils?   [N]

5. What are the main differences between a sandy and a clay soil? If you were given a number of soil samples, describe how you would proceed to investigate them and sort them into various types.   [O & C]

6. Describe how the nitrogen in the atmosphere can become part of the protoplasm of a leguminous plant. How may the nitrogen in this plant become part of the protoplasm of a mammal?   [C]

7. Give an account of the nitrogen cycle in nature. How would you demonstrate that there are micro-organisms in soil?   [L]

8. How could you show that soil contains (*a*) air and (*b*) micro-organisms? Discuss the importance of the soil air and of soil micro-organisms to the plant.   [L]

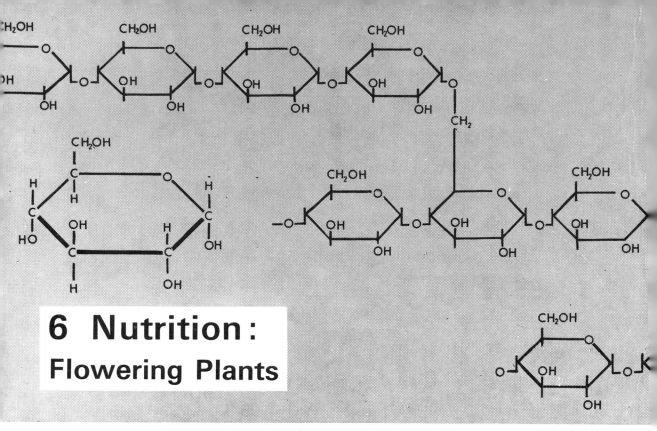

6 Nutrition:
Flowering Plants

Green plants are able to absorb inorganic materials such as water, carbon dioxide, and mineral salts from their surroundings and build them up into complex organic substances which can be used for growth, and for supplying energy. This building-up process goes on continuously in the plant, and is of great importance not only to the plant but to all animals, since they are unable to perform any similar process and are dependent on plants for their supply of organic foods.

The absorption of the raw materials takes place through the roots and the leaves, so we must now consider in detail the structure of these parts and of the stem also, since this is used to conduct the substances to the various parts of the plant.

THE ROOT SYSTEM AND ITS WORK

6.1 Internal Structure of the Young Root

If you study Fig. 6.1, which shows a transverse section of a young root as seen under the microscope, you will see two distinct regions—an outer **cortical region** and an inner **central core** or **stele**. The cortex consists mainly of large colourless thin-walled cells with air spaces between them. For a short distance behind the tip the root is bounded by a single layer of cells, the **piliferous layer** (*L. pilus*, hair; *ferre*, to carry), and many of its cells have grown long processes or root hairs. In this way the actual surface area available for absorption is greatly increased and the root hairs are able to penetrate between the particles of soil (see Figs. 5.6 and 5.7). In older parts of the root the piliferous layer has died away and the outer layer of the cortex—the exodermis—forms the boundary. The innermost layer of the cortex is the endodermis and this surrounds the central core or stele. The stele contains alternating groups of **xylem** and **phloem**. Most of the xylem cells have no living cytoplasm but their walls are strengthened with lignin (see sec. 6.6). Some form long tubes or vessels passing up through the root and stem to end in the veins of the leaf; they are the chief water-conducting tissue of the plant. Others are spindle-shaped with thick walls: these fibres help to make the stem rigid. The cells of the phloem contain abundant cytoplasm; they are not thickened and some form sieve-tubes which are used to transport the manufactured food, a process known as **translocation**. These sieve-tubes are made of cylindrical cells arranged end to end with their cross walls perforated by fine pores. Lying alongside each sieve-tube is a narrow companion cell,

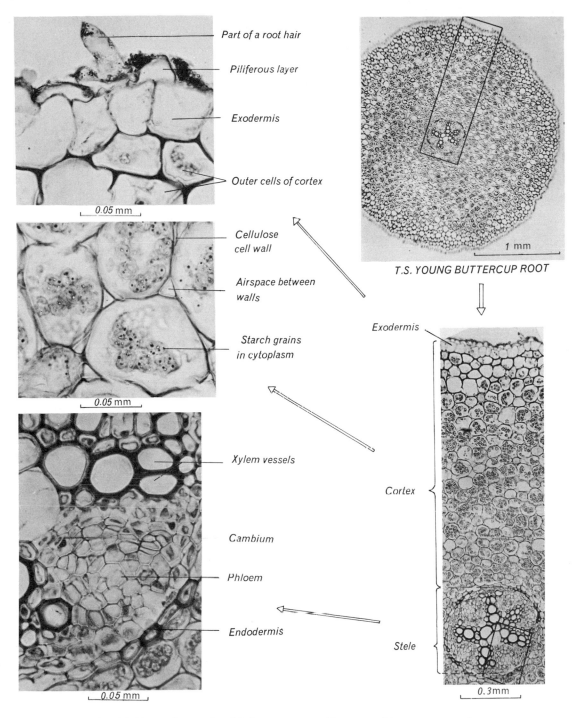

Part of a root hair

Piliferous layer

Exodermis

Outer cells of cortex

0.05 mm

Cellulose cell wall

Airspace between walls

Starch grains in cytoplasm

T.S. YOUNG BUTTERCUP ROOT

1 mm

Exodermis

0.05 mm

Xylem vessels

Cambium

Phloem

Endodermis

Cortex

Stele

0.05 mm

0.3mm

FIG. 6.1 Photomicrographs of T.S. young root of buttercup (*Ranunculus repens*).

densely filled with cytoplasm (see Figs. 6.6 and 6.7).

6.2 Water Uptake through the Root Hairs— Osmosis

The actual absorption of water from the films surrounding the soil particles is done mainly by the root hairs and the following experiment helps to explain the method. Cut a small circular piece from a pig's bladder, soak it in water, stretch it across the end of a thistle funnel, and tie it firmly in position. Fill the bulb of the funnel with golden syrup diluted with hot water until it pours easily. Finally, wash the outside of the funnel under the tap and then clamp it upright in a beaker of water as shown in Fig. 6.2. After a few hours you will find that there is liquid in the stem of the thistle funnel and that the level of the liquid continues to rise and eventually overflows from the top of the stem. This rise can only be the result of water passing through the pig's bladder, which is a semi-permeable membrane allowing water to pass through it but not sugar. Such a movement of water into a solution through a semi-permeable membrane is termed **osmosis** and it is by this process that the

root hair absorbs water from the soil. An apparatus employing visking tubing as a membrane, which allows osmosis to be demonstrated very rapidly, is shown in Fig. 6.2.

The cytoplasmic membrane of the root hair is semi-permeable, and is represented by the pig's bladder and visking tubing in these experiments, while the cell sap, represented by the golden syrup or sugar solution in the experiment, is a solution which is more concentrated than the soil-water films. The water drawn into the root hair by osmosis makes the cell wall stretch, but eventually an equilibrium is reached and no more water can pass into the cell, which is then described as **turgid.**

The cortex cells adjoining these turgid root hairs are able to draw water from them by osmosis, so that the root hairs have their water-absorbing power restored. Meanwhile the adjoining cortex cells have had water withdrawn from them by the inner cortex cells and so they can absorb more from the root hairs. Thus by the process of osmosis water is drawn from the soil into the root hairs and across the cortex towards the xylem. Such a movement of water across living cells can be shown by preparing three potatoes as in Fig. 6.3. There is no movement of water through the boiled potato since only living cytoplasmic membranes are semi-permeable.

6.3 Uptake of Mineral Salts

The pig's bladder and visking tubing are semi-permeable membranes which allow water but not sugar to pass through. Living membranes of cells, however, are surrounded on both sides by a great variety of dissolved substances and their degrees of permeability to these substances vary. Mineral salts, for example, potassium chloride (KCl) and magnesium sulphate ($MgSO_4$), exist in solution as electrically charged ions (K^+, Cl^-, Mg^{2+}, SO_4^{2-}) and these charged particles are invariably found in higher concentrations inside the root hairs and cells of the cortex of the root than in the soil water. Thus we would expect them to **diffuse** passively out of the cells down the concentration gradient into the soil water. In fact, however, the ions move in the opposite direction. How does this movement occur? It must be by some process involving the use of energy from respiration (see p. 3), since roots lose much of their ability to absorb mineral salts when deprived of oxygen. We say, therefore, that an **active** process is involved in the uptake of mineral salts from the soil, but the detailed mechanisms of the process are only imperfectly understood.

OSMOSIS

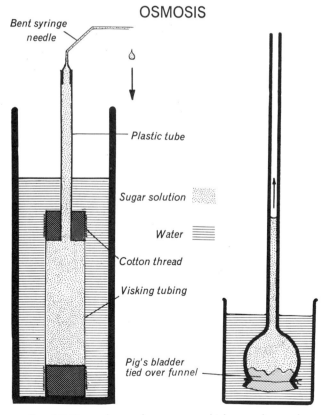

Bent syringe needle

Plastic tube

Sugar solution

Water

Cotton thread

Visking tubing

Pig's bladder tied over funnel

FIG. 6.2 Two pieces of apparatus which may be used to demonstrate osmosis.

OSMOSIS

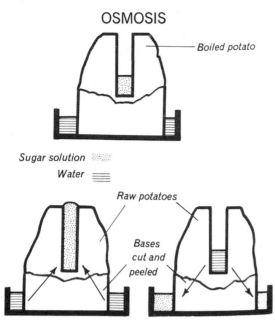

Sugar solution
Water

FIG. 6.3 The two raw potatoes show osmosis in living cells; the arrows show the direction in which the water moves. The boiled potato acts as a control.

The plant requires sulphates, phosphates, and nitrates as well as potassium, calcium, magnesium, and iron. These are obtained from the mineral salts in the soil, though they are actually absorbed in the form of their ions and not as molecules. (Ions are electrically charged particles derived from these salts when they are dissolved.) In addition to these, minute amounts of manganese, zinc, cobalt, and boron are needed; and because the amount required is so very small these are referred to as 'trace elements'. All these substances are dissolved in the films of water round the soil particles; some have been derived from the mineral matter of which the soil is composed, but others, like the nitrates, are formed from the humus (see sec. 5.5).

6.4 Water Culture Experiments

It is possible to show that these various elements are essential for plant growth by water culture ex-

periments. A plant is able to grow with its roots in water even though it is normally rooted in soil, and so we can grow a series of plants in different culture solutions. Litre gas jars fitted with a two-holed wooden stopper are ideal for the experiment, though large jam jars and two-holed corks will also serve (see Fig. 6.4). The jars must be absolutely clean and sterile before they are filled with the solutions. The recipes for some solutions are given in the table.

WATER CULTURE

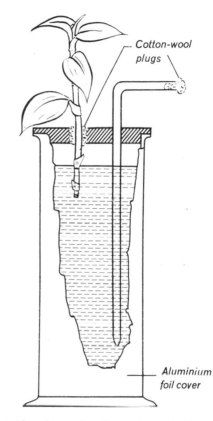

FIG. 6.4 The glass tube allows air to be blown through the culture solutions.

RECIPES FOR CULTURE SOLUTIONS

Substance	Complete Culture	Lacking Iron	Lacking Potassium	Lacking Nitrate
Distilled water	1 litre (10^{-3} m³)	1 litre (10^{-3} m³)	1 litre (10^{-3} m³)	1 litre (10^{-3} m³)
Potassium nitrate	1 g	1 g	—	—
Sodium nitrate	—	—	1 g	—
Potassium chloride	—	—	—	1 g
Potassium phosphate	0.5 g	0.5 g	—	0.5 g
Sodium phosphate	—	—	0.5 g	—
Calcium sulphate	0.5 g	0.5 g	0.5 g	0.5 g
Magnesium sulphate	0.5 g	0.5 g	0.5 g	0.5 g
Iron chloride	0.05 g	—	0.05 g	0.05 g

Notice that in the solution lacking iron it is sufficient to omit the iron chloride, since a supply of chloride is not essential, but that in the one lacking potassium, the sodium salts are substituted, since a supply of both nitrate and phosphate is required. Eight jars should be prepared, containing respectively: distilled water only, the complete culture, solutions lacking potassium, calcium, magnesium, iron, phosphate, and nitrate. The jars should be wrapped in black paper or aluminium foil (to keep the roots dark) and labelled clearly. Seedlings of oat or maize, and cuttings of Tradescantia, are all suitable for the experiment. They should be wedged with cotton-wool in one of the holes of the stopper; the other hole should contain a glass tube through which air can be blown from time to time. The experiment is best set up about April so that the results can be studied throughout May, June, and July, though it can be carried out at any time if a heated greenhouse is available. A summary of the results of such an experiment is given in the table and Fig. 6.5 is a photograph of an actual experiment.

WATER CULTURES

COMPLETE — MINUS Fe — MINUS K — MINUS N — DISTILL

Fig. 6.5 The maize grains were germinated in distilled water and then transferred to culture jars. The photograph shows the seedlings six weeks after germination. The aluminium foil has been removed so that the root system can be seen.

SUMMARY OF RESULTS FOR TRADESCANTIA

Solution	Result	Remarks
Complete culture	Normal healthy growth	
Distilled water	Hardly any growth	
Lacking potassium	Poor growth Leaves turn yellow	Essential for cell division
Lacking calcium	Stunted growth Short brown roots	Maintains selective permeability of cell membrane
Lacking iron	Poor growth Yellow leaves	Must be present before chlorophyll can be formed
Lacking magnesium	Poor growth Yellow leaves	Required for the formation of chlorophyll
Lacking phosphate	Thin weak shoots Poor roots	Required for the formation of nuclei
Lacking nitrate	Very little growth	Required for the synthesis of proteins

THE STEM AND ITS WORK

6.5 Internal Structure of Young Stem

The young stem is bounded by a single layered **epidermis** (*Gk. epi, on; derma, skin*) made up of colourless cells similar to those of the piliferous layer of the root, but very few are elongated to form hairs and all except the very young are covered on the outside by **cuticle** which is impermeable to both gases and water (see Fig. 6.6). Enclosed by the epidermis is the **cortex** which, unlike that of the root, contains a tissue whose cells have chloroplasts. Internal to the cortex is a ring of specialised strands called **vascular bundles.** These appear to be wedge-shaped in transverse section and consist of an outer mass of **phloem** and an inner larger mass of **xylem.** Between the xylem and phloem is a layer of actively growing cells called the **cambium.** The central row of cells is continually dividing, so that new layers of cells are formed on both the inside and the outside. Those on the outside gradually grow into phloem,

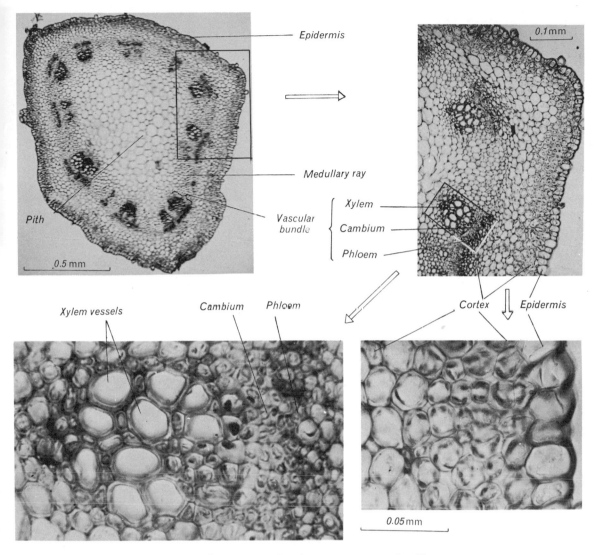

FIG. 6.6 Photomicrographs of T.S. young stem of wallflower.

while those on the inside develop into xylem. A large bundle enters the stem from the leaf stalk at each node and passes down to join the one entering from the leaf vertically below it. Thus in the wallflower there are five large bundles in a transverse section of an internode (see sec. 3.2). These bundles are continuous with the xylem and phloem of the root, and form the main conducting tissue. The middle of the stem is filled with **pith**, which is formed of cells similar to those found in the cortex but without chloroplasts. The regions between the vascular bundles are termed **medullary or vascular rays** and are made of cells similar to those of the pith but smaller.

6.6 Support and Conduction

The young stem has to support the leaves and the flowers, and is able to do this so long as all its cells remain turgid. Herbaceous plants often wilt on a hot summer's day when they are losing water rapidly but become erect again during the evening when their cells again become turgid. This stiffness due to turgidity is easily demonstrated. If you cut two cylinders out of a potato with a cork borer and put one in a beaker of water and the other in a beaker of concentrated salt solution you will find that the one in the water becomes stiff and hard, while the other becomes soft and flabby. Can you explain why?

Some extra support is also provided by the thick lignified walls of the xylem. Lignin is a material which differs from cellulose in that it is impermeable to water and does not stretch as much, though it is elastic and therefore returns to its normal length after stretching. These tough fibres in the vascular bundles are often discovered when we eat the stem and leaves of old cabbage, celery, rhubarb, etc. In a woody stem the number of these is greatly increased (see sec. 11.9) and there they provide the main means of support.

The stem also serves to conduct substances from the roots to the leaves, and vice versa. You can easily show that the vascular bundles are the main conducting channels by putting the cut end of a young cabbage or white dead-nettle plant in red ink (or 1 per cent eosin solution) for an hour and then cutting off the bottom inch of the stalk. Shoots of privet and St. John's wort are also useful. The newly cut surface then shows a ring of red dots when examined with a hand lens. These red dots mark the xylem of the vascular bundle and we may conclude that the xylem serves to transport the substances absorbed by the roots up to the leaves. It is not so easy to show that the

STEM STRUCTURE

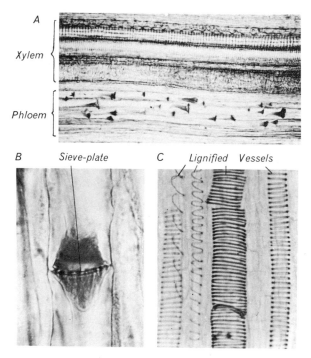

Fig. 6.7 A—Photomicrograph of part of L.S. Marrow stem. B—Part of two sieve tubes from the phloem more highly magnified. C—Part of the xylem more highly magnified.

phloem carries the manufactured food, but it can be done by a 'ringing' experiment. If a willow twig is 'ringed' by removing a complete circle of bark, including the phloem and cambium, about 0.05 m from the lower end and then stood in water so that the ring is immersed, it is found that adventitious roots (roots which do not grow from a primary root but from a stem or leaf) eventually appear above the ring. Unringed twigs form adventitious roots over the whole of the submerged area. This suggests that the lower part of the twig is unable to form roots as it is not receiving food, and that the phloem is probably the main food-conducting channel.

THE LEAVES AND THEIR WORK

6.7 Internal Structure of a Leaf

Fig. 6.8 shows the appearance of a transverse section of a leaf midrib and lamina. The leaf is bounded by a single layer of colourless epidermal cells, and this **epidermis** is continuous with that of the stem and like it is covered on the outside with cuticle. The lower epidermis (and in some leaves the upper epidermis also, though to a lesser extent) is pierced by numerous pores, each of which is bounded by two curved sausage-shaped guard cells. The two guard cells and the pore between them are called a **stoma** (*Gk. mouth*), and they provide openings through which gases can enter and leave the leaf. The size of the pore can be regulated by the guard cells by means of a mechanism which will be explained in the next section.

Between the upper and lower epidermis is a mass of cells which contain chloroplasts. This **mesophyll** is divided into two regions, an upper part or palisade mesophyll consisting of two or more rows of closely packed cylindrical cells (the long axes are at right-angles to the surface of the leaf) and the lower part or spongy mesophyll of irregular smaller cells which are loosely arranged so that there are large **intercellular spaces** between them.

The veins in the leaf are the vascular bundles which are continuous with those of the stem. The bundles gradually become smaller, losing their cambium and phloem so that the end consists of xylem tubes only.

6.8 Transpiration and the Movement of Water

You have probably noticed that the glass inside a greenhouse or garden frame is often covered with drops of water in the morning. This is due to condensation during the night of water vapour which

TRANSVERSE SECTIONS OF PRIVET LEAF

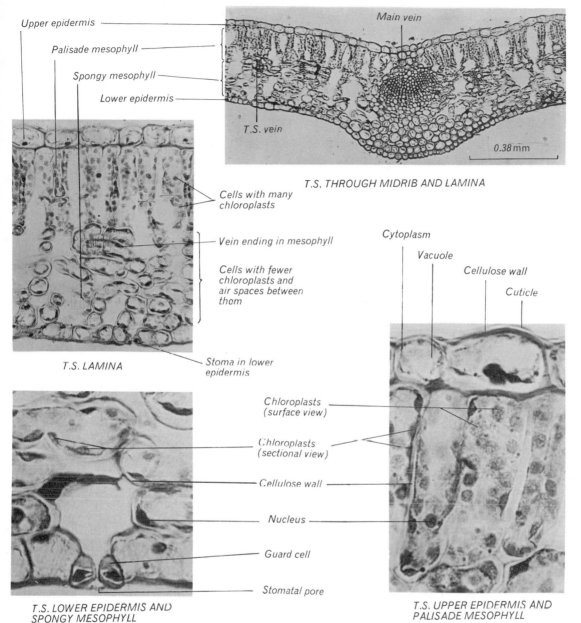

FIG. 6.8 Photomicrographs of parts of transverse sections of privet leaf.

has evaporated from the plants or soil. Evaporation of water from the plant is called **transpiration**. It may be demonstrated by putting a leafy shoot in a beaker of water under a bell jar; oil must be poured over the water in the beaker to prevent it from evaporating and a control experiment should be set up to show that it does so.

After the experiment has stood in the light for an hour or so you will find that tiny drops of moisture have condensed on the inside of the bell jar. The same type of apparatus covered with a black cloth can be used to show that little transpiration takes place in the dark, and by using red ink to colour the water in the beaker you can show that it is pure water and not a solution which is transpired.

The amount of water transpired by a plant is great and is controlled partly by the degree of opening of the stomata, since the rest of the leaf and stem is covered with cuticle. The under surface of most leaves contains more stomata than the upper surface and hence we should expect to find that the rate of transpiration of water is greater from the under surface. This can be shown by putting a lilac leaf covered on both surfaces with a piece of dry cobalt chloride paper between two pieces of glass. (Cobalt chloride paper, made by soaking filter paper in a 5 per cent solution of cobalt chloride, is normally pink, but when it is dry, the cobalt chloride is anhydrous and blue.) After a few minutes the paper is pink where it is in contact with the under surface, but still bluish on the upper surface, showing that more water has been lost from this surface than from the upper one.

If a geranium leaf is immersed in 1 per cent solution of gentian violet in absolute alcohol, the dye penetrates the open stomata and the leaf shows numerous black specks. If, however, this is repeated with a leaf which has been kept in the dark, no such specks appear, showing that the stomata are closed. The degree of opening of the stomata is controlled by the turgidity of the guard cells which, unlike the other epidermal cells, contain chloroplasts, and by the fact that the wall of the guard cell bordering the pore is much thicker than the other (see Fig. 6.9). When the guard cells are turgid they are very curved so that the pore between them is large, but when they lose water they become less curved so that the pore between them is nearly closed. The changes in turgidity are regulated by light. The guard cells contain chloroplasts and can photosynthesise, but the sugars they form accumulate within the vacuole and therefore increase its water-absorbing power; hence in the light the cells are fully turgid and the pore wide open. In the dark, however, the sugar is converted to starch so that the water-absorbing power of the cell is reduced, it becomes less turgid, and the pore closes. Unlike the starch grains in

STOMATA

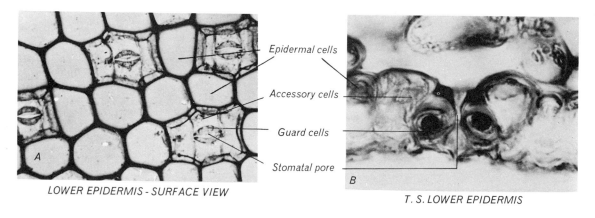

LOWER EPIDERMIS - SURFACE VIEW

T. S. LOWER EPIDERMIS

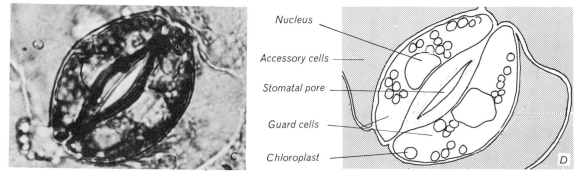

Fig. 6.9 A—*Tradescantia zebrina*. B—Cherry laurel. C—Stomata of Christmas rose in surface view. D—Key to C.

POTOMETERS

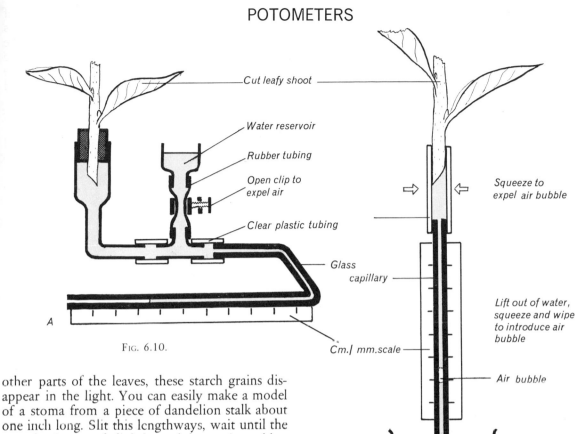

Fig. 6.10.

Labels in figure:
- Cut leafy shoot
- Water reservoir
- Rubber tubing
- Open clip to expel air
- Clear plastic tubing
- Glass capillary
- A
- Squeeze to expel air bubble
- Lift out of water, squeeze and wipe to introduce air bubble
- Cm.| mm.scale
- Air bubble
- Water
- B

other parts of the leaves, these starch grains disappear in the light. You can easily make a model of a stoma from a piece of dandelion stalk about one inch long. Slit this lengthways, wait until the parts begin to curl slightly, and then, using rubber bands at the ends, bind them together with their original outer surfaces in contact. The strips of dandelion stalk represent the guard cells. If you put the model in water you will see how the pore opens when the guard cells are turgid, and if you then transfer it to concentrated salt solution you will see how the pore closes as the guard cells lose their turgidity.

Transpiration is a process of evaporation and is therefore influenced by external conditions, as can readily be shown by using a piece of apparatus termed a **potometer**, two forms of which are shown in Fig. 6.10. The apparatus is most easily assembled under water, for it must be completely filled. As the leaves transpire, water is absorbed by the cut end of the shoot and the water is drawn up the capillary tube, dragging the air bubble with it. The potometer really measures the rate of water uptake rather than the rate of water loss, but in plants which are fully turgid the two rates will usually be equal. Thus the rate of transpiration in different atmospheres can be compared by noting the time taken for the bubble to move a given distance on the scale. It is found to move very quickly if the apparatus is put near an electric fan,

but slowly if the shoot is enclosed in a polythene bag.

The potometer also shows that transpiration causes a movement of water within the plant. Water evaporates from the mesophyll cells into the intercellular spaces and diffuses out through the stomata provided that the atmosphere is not already saturated with water vapour. This evaporation from the mesophyll cells uses latent heat and has, therefore, a cooling effect. As a result of the diffusion outwards of the water vapour more water vapour evaporates from the mesophyll cells into the intercellular spaces, resulting in a more concentrated cell sap and hence an increase in the water-absorbing power of these mesophyll cells. They absorb water from their neighbours and so, by osmosis, water is drawn from cell to cell until the xylem vessels at the ends of the veins in the leaf are reached.

The xylem vessels contain columns of water which are, in effect, unbroken, and the mesophyll cells, because their cytoplasm is a strong solution, can draw water osmotically from the top of this column. When water evaporates from the mesophyll cells a suction is applied to the column which is put in tension and water moves up the vessels without the column being broken by a gas bubble appearing. This movement of water is the transpiration stream, which is often loosely stated to be caused by leaf suction. Such a leaf suction can be admirably demonstrated by removing leaves from a shoot in a potometer and watching the rate of water uptake drop. The apparatus shown in Fig. 6.11 may also be used.

Water is also forced into the vessels from below by **root pressure**, a fact which may sometimes be demonstrated with a fuchsia or young sycamore plant. The plant should have been growing actively in a pot for some time before the experiment, which is most successful in the late spring. Cut off the shoot about 50 mm above soil level and bind on to the stump a piece of rubber tubing fitted to a glass tube about 0.3 m long. Pour a little water into the tube and a few drops of oil to prevent evaporation. You must keep the plant in a warm place and the soil very moist. The glass tube will gradually fill with liquid, owing to root pressure.

Thus, as a result of 'leaf suction', aided to a small extent by root pressure, a flow of water through the plant (often termed the transpiration stream) is achieved. Such a flow depends on the replacement of the transpired water by water absorbed from the soil, and if it is not replaced, then the plant wilts. Wilting often makes the work of a gardener difficult. Transplanted plants are very liable to wilt, as the uprooting process damages the root hairs and so greatly reduces the absorbing surface. Hence gardeners transplant in moist weather when transpiration is reduced, and water the transplants well to increase absorption.

6.9 Photosynthesis

In section 1.9 the basic nutritional differences between plants and animals were stressed. Green plants are able to build up complex organic substances by a process called photosynthesis. This process can take place only in the light and requires a supply of water and carbon dioxide. Such a method of nutrition, in which materials are absorbed in a fluid form and built up into complex organic compounds, is termed **holophytic**.

MOVEMENT OF WATER IN A PLANT

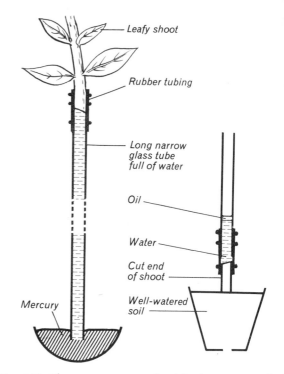

FIG. 6.11 The apparatus on the left demonstrates 'leaf suction' and that on the right root pressure.

The entry of water into the plant and its movement into the leaves were described in the last section. Carbon dioxide from the air diffuses into the leaf through the stomata, finally reaching the chloroplasts of the mesophyll cells. Water and carbon dioxide are the raw materials required by the green plant for photosynthesis, as a result of which carbohydrates are synthesised in the leaf.

We usually test for starch when trying to discover whether photosynthesis has taken place, though sugar is always formed first. Starch gives a blue-black colour with iodine solution, and it is easy to demonstrate its presence in a potato tuber by dipping a slice in iodine solution, but in a leaf the green colour masks the result so that a more complicated technique is required. Leaves such as nasturtium, lilac, and Tradescantia are very suitable for the test. Dip the leaf in boiling water for a few minutes to kill and soften it, then decolorise by soaking for several minutes in alcohol heated over a water bath. (Chlorophyll is soluble in alcohol.) This stiffens the leaf, which should therefore be re-dipped in boiling water before being immersed in dilute iodine solution. The resulting blue-black stain shows the presence of

PHOTOSYNTHESIS

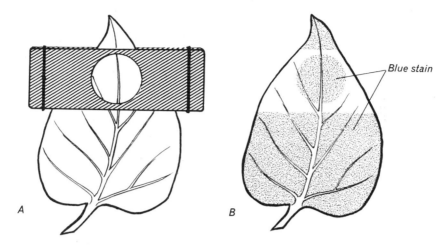

FIG. 6.12 A—Lilac leaf covered with a stencil made by pasting black paper over two microscope slides which are held together by rubber bands. B—The same leaf after exposure to light, followed by decolorisation and staining with iodine solution.

starch. The experiment can be completed more quickly if small circles of leaves, punched out with a cork-borer, are used instead of the whole leaf. In this case the test can be carried out in a test tube of alcohol in a beaker of hot water.

It is only the green parts of plants which are able to photosynthesise, as you can easily show by testing a variegated leaf such as that of geranium or Tradescantia for starch. If you make a drawing of the leaf first and shade in the green parts you will find that the leaf, after staining with iodine solution, shows exactly the same pattern in blue.

Photosynthesis requires light, and can be stopped by covering the leaf. First make two stencils by pasting pieces of black paper with a cut-out design on them on to two microscope slides. These should then be fixed one on either side of a lilac leaf, by rubber bands (see Fig. 6.12). Alternatively part of the leaf could be surrounded by aluminium foil. The leaf should be left attached to the tree and the stencils fixed in the evening. During the night starch will be removed from the leaf (i.e. it will be destarched), a fact which could be checked by detaching a leaf at dawn and testing it for starch. At the end of the next day the leaves should be picked and tested for starch. Those parts of the leaf which were exposed to the light will be found to stain blue, whereas the shaded parts will not, showing that light is necessary for the synthesis of starch. A mercury vapour lamp can be used instead of sunlight for this and other photosynthesis experiments.

Fig. 6.5 shows that the seedling in the complete culture solution flourished, though this solution contained no supply of carbon. The plant must have a supply of carbon and obtains it from the air; even though the air contains only about 0·04 per cent of carbon dioxide, this is the plant's sole source of supply. The apparatus shown in Fig. 6.13 can be used to show that a green leaf deprived of carbon dioxide is unable to photosynthesise. The plant should be destarched by keeping

PHOTOSYNTHESIS

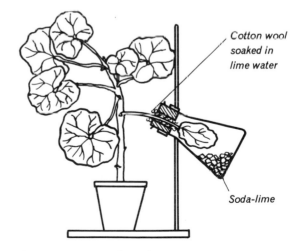

FIG. 6.13 Apparatus used to show that carbon dioxide is necessary for the formation of starch.

PHOTOSYNTHESIS

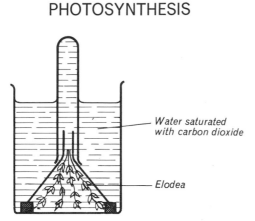

FIG. 6.14 Apparatus used to show that shoots of water plants give off oxygen in the light. The funnel is raised on rubber bungs. Why?

it in the dark for twenty-four hours before the experiment, and a leaf tested to confirm that no starch is present; after it has been exposed to the light for several hours the leaf from inside the flask, and one other to act as a control, should be picked and tested for starch, which will not be found in the leaf from the flask.

During photosynthesis oxygen is formed as a waste product; while some of this is used by the plant for respiration, the excess escapes from the plant into the air via the stomata. You can show this by putting a few short sprigs of mint in a little water in a gas jar inside which a tall candle has been fixed. Light the candle and cover the jar with a vaselined glass plate. The candle continues to burn until the oxygen supply is exhausted. Leave the apparatus in good light for several days and then test the air in the gas jar with a lighted splint. (Remember to set up a control.) It is possible to collect the oxygen freed during photosynthesis if a water plant such as Elodea is used. Fig. 6.14 shows how the apparatus should be arranged. Bubbles of gas escape from the stem and collect in the test-tube, so that it can later be tested with a glowing splint to show that it contains some oxygen. The bubbles will form more quickly if carbon dioxide is blown into the water through a glass tube. Why? On a dull day, however, the amount of oxygen in the gas which collects in the test-tube is too small to relight a glowing splint and a more convincing demonstration of the evolution of oxygen is given by the Kolkwitz method. Indigo carmine (sodium indigo sulphonate) when dissolved in water gives a bright blue solution

PHOTOSYNTHESIS

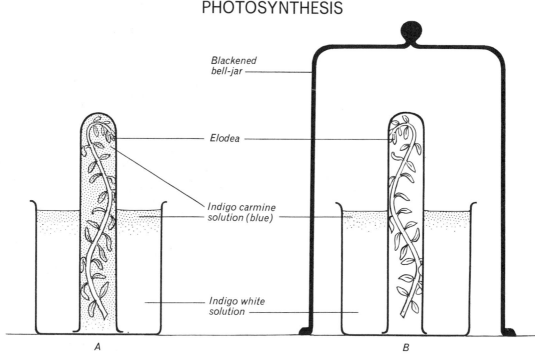

FIG. 6.15 Apparatus used to demonstrate the production of oxygen by a water plant in the light. Note that in the control experiment B, as well as in A, oxygen from the air turns the indigo white solution blue at the surface.

which is easily reduced to indigo white. The indigo carmine solution should be made by dissolving about 0.1 g in a litre of water and a 10 per cent solution of sodium hydrosulphite should be added to it drop by drop until the blue colour disappears. After reduction the indigo carmine solution should not be shaken. About 20 cm³ of the reduced indigo carmine solution should be carefully sucked into a pipette and used to fill a test-tube containing a young shoot of Elodea, taking care to avoid contact of the solution with the air as much as possible. If any blue colour appears, then another drop of sodium hydrosulphite must be added. The test-tube should then be inverted in a beaker of reduced indigo carmine solution. If the apparatus is left in bright light, clouds of blue will be seen around the leaves, showing that the indigo white is being re-oxidised to indigo carmine by the oxygen set free in photosynthesis. A control experiment should be set up in the dark or with roots in place of the green shoot.

By a series of experiments similar to those described, it is easy to show that a green leaf exposed to the light absorbs carbon dioxide, and evolves oxygen, and that carbohydrates such as starch or sugar accumulate within the leaf. The amounts of the gases concerned suggest that the chemical process involved is a reduction in which carbon dioxide is reduced by hydrogen obtained from the water.

$$CO_2 + H_2O \xrightarrow[\text{Chlorophyll}]{\text{Light}} CH_2O + O_2$$

Carbon   Water   Carbohydrates   Oxygen
dioxide

Experiments involving the use of radioactive tracer elements have shown that all the oxygen freed comes from the water, and hence the summary is better written as:

$$CO_2 + 2H_2O \xrightarrow[\text{Chlorophyll}]{\text{Light}} CH_2O + H_2O + O_2$$

Energy is required to bring about photosynthesis and is supplied by the light. The radiant energy is not lost to the plant during the process but is transformed into chemical potential energy in the carbohydrates which have been built up. This explains why photosynthesis is such a very important process, for all cytoplasm requires

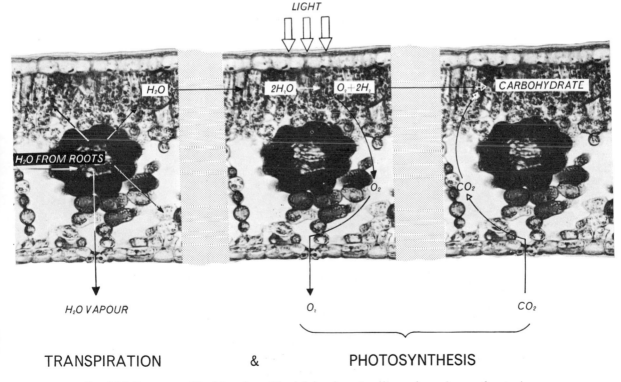

LIGHT

H_2O $2H_2O$ $O_2 + 2H_2$ CARBOHYDRATE

H_2O FROM ROOTS

O_2

CO_2

H_2O VAPOUR O_2 CO_2

TRANSPIRATION   &   PHOTOSYNTHESIS

Fig. 6.16 Summary of leaf functions. The left-hand section shows the pathway of water in a leaf during transpiration. The centre section shows the first stage and the right-hand section the second stage of photosynthesis.

a constant supply of energy (see sec. 1.3). This energy is released during respiration from these complex organic compounds manufactured by green plants during photosynthesis.

It is only recently that we have been able to discover something about the actual chemical reactions involved in photosynthesis, and the role played by chlorophyll and light. The process takes place in two main stages. The first stage involves the absorption of light by the chlorophyll, and the conversion of the radiant energy to chemical energy, with the result that the water is split up so that it can provide the hydrogen necessary for the second stage. This second stage can proceed in the dark, and during it the carbon dioxide is reduced to form simple organic compounds. This reduction, like most metabolic processes, involves many enzymes.

Whatever the reactions involved, green plants are able to synthesise sugars, and in the leaves of most plants these are immediately converted into the insoluble carbohydrate starch which accumulates during the day-time. at night it is reconverted to sugar and in this soluble form it is distributed to all parts of the plant through the phloem. This translocation (see sec. 6.1) enables all parts of the plant to obtain a continuous supply of oxidisable material for the release of energy while the excess material is carried to parts of the plant where it may be stored for later use. On arrival at these storage places (see sec. 16.8) the sugar is usually re-converted to the insoluble starch.

The final stages of photosynthesis involve the production of oils and proteins. During protein synthesis simple nitrogen-containing organic compounds called amino-acids are formed. To build up these amino-acids the plant also requires a supply of nitrogen, obtained from the soluble nitrates etc. in the soil, and a supply of energy.

QUESTIONS 6

1. (a) Describe an experiment to show that water vapour is given off by leaves while still attached to the tree on which they are growing.

(b) How would you show that water is taken up by the roots of a plant?   [C.]

2. Describe the conditions necessary for photosynthesis to take place. How would you demonstrate experimentally that a plant must have light in order to carry out this process?   [C]

3. Describe an experiment by which you could measure the uptake of water by a leafy shoot. Draw the apparatus you would use. How many environmental changes affect the rate of water loss from a plant?   [L]

4. In a climbing plant growing over both sides of a north-facing and a south-facing wall the sizes of many of its leaves from the sunny and shady sides were measured. The results were as follows:

	Sunny Side	Shady Side
Average length of 100 leaves	6 cm	10 cm
Average width of 100 leaves	4 cm	8 cm

(a) What is the effect of differing light intensities on the growth of these leaves on the sunny and shady sides of the wall?

(b) Name the nutritional process occurring within the plant which might be affected by the differences in light intensity on either side of the wall.

(c) Explain the effect on this process in well-illuminated leaves of (i) increasing the amount of carbon dioxide present in the atmosphere to 1 per cent, and (ii) greasing the lower surfaces of the leaves.

(d) Draw a simplified labelled diagram of the distribution of the tissues as seen in a section of a typical leaf, to show the structures involved in this process.

(e) Name the products of this process and briefly explain how they are removed from the leaves.   [N]

5. (*a*) (i) What is meant by transpiration and why must a plant transpire? (ii) Describe an experiment by which you would demonstrate that the rate of transpiration in a potted plant or a cut shoot varies with the external conditions. (iii) Through what leaf structures does transpiration occur?

(*b*) Explain how a green plant, planted in moist soil and sealed in an airtight glass jar on a window sill in a warm room, is able to survive in a healthy condition for many months. [N]

6. Define the process of osmosis. How would you demonstrate the occurrence of osmosis experimentally? Carefully explain the importance of this process in (*a*) root hairs, (*b*) guard cells. [O]

7. Give a concise account of the features of a typical land plant which tend to reduce loss of water.

Comment on the statement that extreme efficiency in reducing water loss is only achieved by seriously interfering with other functions of the plant. [O & C]

8. Make a large, labelled diagram of a mesophyll cell of a leaf. What changes would you expect in the cell and in the leaf as a whole when a leaf is placed in (*a*) a dry atmosphere and (*b*) a strong solution of sugar? [O & C]

9. Describe how water from the soil reaches the air spaces of the leaves on a herbaceous dicotyledon. Why is water important to this type of plant? [S]

10. (*a*) How would you demonstrate the process of osmosis, using non-living material to make your apparatus?

(*b*) Draw a labelled diagram to show the structure of the root of a herbaceous dicotyledon as seen in a transverse section. What role does osmosis play in drawing water from the soil into the xylem of such a root? [S]

11. Describe briefly experiments to demonstrate the following: (*a*) A potted plant loses water through its leaves. (*b*) Water passes up through the stem in the vascular bundles. (*c*) A herbaceous plant takes in water through its roots. [W]

12. In about ten lines, write an account to explain what you understand by photosynthesis. What part does this process play in maintaining the composition of the atmosphere? [A]

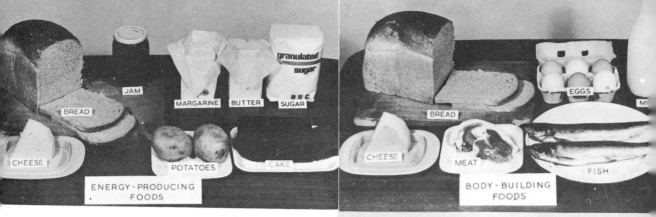

ENERGY-PRODUCING FOODS

BODY-BUILDING FOODS

PROTECTIVE FOODS

7 Nutrition :

Mammals

The bulk of a mammal's food is solid animal or plant material and requires to be broken up physically by the teeth into small pieces before it can be 'broken up' chemically and made soluble. The latter process, known as digestion, is carried out in the alimentary canal by means of digestive juices, and the soluble products from it are absorbed into the blood, passing through the intestinal wall. In the blood they are circulated to individual cells where they may be used in synthesising essential materials, or in energy production.

Such a nutrition is typical of animals, and is said to be **holozoic**, in contrast to the holophytic nutrition of green plants and the saprophytic nutrition of Mucor.

ESSENTIALS OF A HEALTHY DIET

Several different substances must be present in the food before a mammal can maintain health. They are:

1. Carbohydrates.	4. Water.
2. Fats.	5. Mineral Salts.
3. Proteins.	6. Vitamins.

The diet must also contain some 'roughage' (indigestible cellulose from plant cell walls) to add bulk so that the alimentary canal muscles can function easily.

It is possible to argue that water is not really a food, since it is removed from the body in sweat, urine, and expired air, but a supply taken by the mouth is clearly necessary to prevent desiccation of the body leading to death. Like water, oxygen is essential to a mammal, but it is excluded from the term 'food' since it is not swallowed.

7.1 Carbohydrates and Fats—Energy-producing Foods

Carbohydrates

Potatoes, bread, biscuits, and various cereal preparations contain a high proportion of starch, which is a useful energy-producing substance. Sweet foods containing relatively large quantities of sugar are also valuable in this way (see sec. 9.12). The word 'sugar', as used in everyday life, refers to cane-sugar or sucrose which is one member of a group of sweet-tasting compounds. In animals and plants several other members of the

group are found, for instance glucose, maltose (malt-sugar), and lactose (milk-sugar). Starch and the sugars are grouped together as carbohydrates since they all contain the elements carbon, hydrogen, and oxygen. Furthermore, the number of hydrogen and oxygen atoms in their molecules is in the same ratio as in the molecule of water. $C_x(H_2O)_y$ is their general formula,
e.g. glucose $(C_6H_{12}O_6)$ cane-sugar $(C_{12}H_{22}O_{11})$

One other carbohydrate remains to be mentioned—glycogen, which is insoluble and is stored in limited quantities in the liver of mammals and to a smaller extent in their muscles. Should any part of the body become short of energy-producing substances, it can rapidly be remedied by the conversion of glycogen to glucose, which is then carried in the blood to the part concerned.

The chemistry of carbohydrates is outside the scope of this book, but it is desirable to know certain chemical tests by which they can be recognised.

1. *The Naphthol Test*—given by all carbohydrates
Take a few cm³ of a solution or suspension of the substance to be tested in a test-tube or an evaporating dish, add a few drops of naphthol solution, and mix well. Tilt the vessel and carefully pour concentrated sulphuric acid down the side and leave for a minute. A purple ring at the boundary of the two liquids indicates the presence of carbohydrate in the original substance. Green and white rings often appear, but do not indicate the presence of carbohydrate.

2. *Benedict's Test*—given by some sugars, e.g. glucose, but not by sucrose
To a few cm³ of a solution or suspension of the substance to be tested add only enough Benedict's solution to give a faint blue colour (Benedict's solution contains copper sulphate). Boil carefully for one minute. A yellow, orange, or red precipitate of copper oxide indicates the presence of glucose. In this reaction glucose acts as a reducing agent and the orange precipitate is due to the formation of cuprous oxide from the cupric sulphate in the Benedict's solution. Sugars which bring about such a reaction are termed reducing sugars. Glucose, maltose, and lactose are all reducing sugars but sucrose is a non-reducing sugar. This test shows the presence of a reducing sugar but is not a specific one for glucose. Try this test on grape juice, honey and on milk.

3. *Starch Test*—given only by starch
To a solution or suspension of the substance to be tested add a few drops of dilute iodine solution. A blue-black colour indicates the presence of starch.

Fats

Fats are a group of organic substances which contain the same elements as carbohydrates but combined together in different proportions. They form an important part of the diet because of the energy which can be liberated from them. Fats are easily detected in foods on account of their greasy nature and their well-known ability to leave permanent translucent marks on paper and clothing. Butter, lard, and margarine are almost pure fat, and foods such as pork, bacon, milk, and nuts contain appreciable quantities. Carbohydrate is not stored in the body to the same extent as is fat, which is deposited round the kidneys and under the skin (see Fig. 10.5) where it forms a layer protecting the body from heat loss. In whales this layer is extremely thick and is known as blubber.

When 1 g of fat is fully oxidised it produces 38 kilojoules (or 9 kilocalories) of energy, but 1 g of carbohydrate produces only 17 kilojoules (or 4 kilocalories). In other words, 1 g fat supplies more than twice as much energy as 1 g carbohydrate and is therefore a more convenient reserve of energy to carry about.

How much energy-producing food must a man consume each day without having to draw on his reserves? This varies according to the type of occupation, as shown in the table.

DAILY ENERGY REQUIREMENTS

Occupation	Approximate Energy Requirement/Day	Equivalent to a Daily Intake of	
		Fat	Carbohydrates
Lumberman	27 200 kJ (6 500 kcal)	0.71 kg	1.59 kg
Navvy	18 850 kJ (4 500 kcal)	0.48 kg	1.11 kg
Average man	14 250 kJ (3 400 kcal)	0.37 kg	0.82 kg
Sedentary man	12 550 kJ (3 000 kcal)	0.34 kg	0.74 kg
Average woman	11 700 kJ (2 800 kcal)	0.31 kg	0.68 kg

It is, perhaps, disappointing to learn that the most arduous mental work continued for hours does not increase the energy requirement appreciably.

Both fat and carbohydrate are essential in the diet, but wide variations in the proportions of these two are possible without ill health. Where the energy requirement is high, most of the energy comes from fatty foods since these are usually more concentrated than carbohydrate foods (see Fig. 7.1). The latter, with their water and indigestible matter, would increase the total bulk of food to be consumed to an impossible size if they had to provide most of the energy.

PERCENTAGE COMPOSITION BY WEIGHT AND THE ENERGY CONTENT OF SOME COMMON FOODS

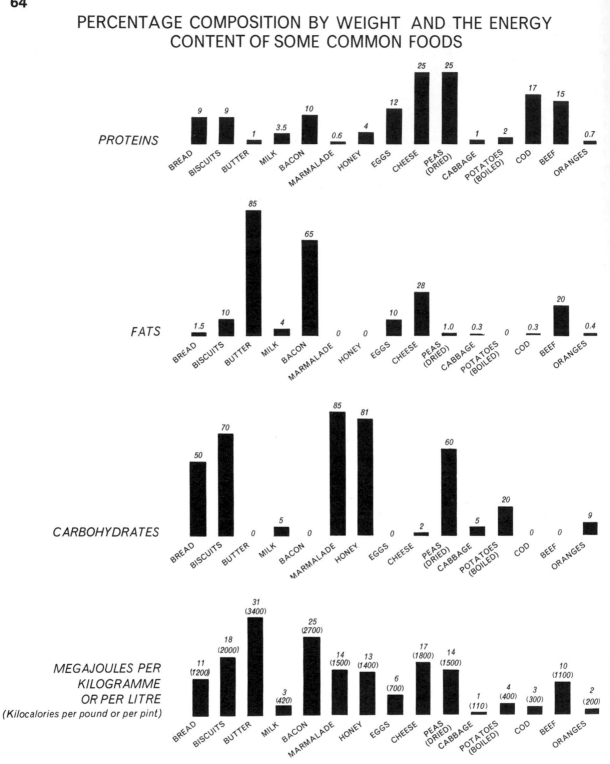

FIG. 7.1 Read this chart horizontally to find which foods are rich and which are poor in proteins, etc., then read it vertically. Notice that, in fresh foods such as milk, eggs, cabbage, potatoes, cod, beef, and oranges, the total percentage of solids is low because of the presence of much water. (1 kg = 2.2 lb, 1 kcal = 4.2 kJ = 0.0042 MJ, 1 pint = 568 cm³)

7.2 Proteins—Body-building Foods

Proteins were mentioned as an essential part of cytoplasm in section 1.1. Their main use in the body is in building the new cells involved in growth and tissue repair. In an average man the minimum daily intake required is 80 g but up to twice this amount is needed by pregnant and lactating women. Meat, fish, eggs, cheese, milk, bread, beans and ground-nuts are the commonest foods which are rich in protein. The presence of proteins in foods can be detected by the two tests given below.

1. *Biuret Test*—given by all soluble proteins

To a few cm³ of a solution or suspension of the substance to be tested add at least an equal volume of dilute caustic soda solution. Shake and then add 1 per cent copper sulphate solution drop by drop. A blue or violet coloration indicates that protein is present.

2. *Millon's Test*—given by all proteins

To a few cm³ of a solution or suspension of the substance to be tested add about 1 cm³ of Millon's reagent. A white precipitate may or may not occur. Boil carefully for half a minute. A brick-red coloration indicates the presence of proteins.

Proteins are built up by plants and animals from simpler nitrogenous substances called **amino-acids**, twenty-three of which are important biologically. These can be combined together in different ways and in different proportions, and, when one considers that more than a hundred amino-acid units are often required to make one protein molecule, it is clear that the number of possible proteins is almost infinitely large. Some idea of their complexity can be gathered from the following details of the molecule of insulin (see sec. 15.5), whose chemical structure is completely known. The insulin molecule ($C_{254}H_{377}O_{75}N_{65}S_6$) is composed of fifty-one amino-acid units and contains seventeen different amino-acids. It is probably true that every species of organism has its own particular types of protein which it makes from the food it takes in.

Man can synthesise thirteen of the amino-acids in the body, but he cannot make the other ten, which must, therefore, be present in his protein intake. Some proteins (first-class proteins) contain nearly all the twenty-three different amino-acids, but others (second-class proteins) have high proportions of one amino-acid to the complete exclusion of several others. In general, animal protein is more valuable in the diet than plant protein, a

difference which is exaggerated by the fact that much plant protein passes through the gut undigested, as it is protected from the action of digestive juices by cellulose cell walls. Kwashiorkor is a serious malnutrition of children found in tropical regions where animal protein is scarce or non-existent. In order to obtain enough of all the different amino-acids a mammal is bound to absorb an excess of some, and these are dealt with by the liver and kidneys, as explained in section 10.1.

7.3 Water

Water, which is present in the cells, the blood, and all other body fluids, has been found to make up 72 per cent of the body weight in rabbits. It is, of course, essential to life, for it is the medium in which all metabolic changes occur. The daily exchange of water in an average adult doing sedentary work is given in the following table.

Intake		Output	
Drinks	1 450 cm³	Urine	1 500 cm³
Food	800 cm³	Sweat	600 cm³
Respiratory oxidation	350 cm³	Evaporation from lungs	400 cm³
		Fæces	100 cm³
	2 600 cm³		2 600 cm³

7.4 Mineral Salts and Vitamins—Protective foods

It has been known since 1900 that a diet of pure fats, carbohydrates, and proteins with water is not sufficient to maintain health in mammals: in fact rats die more quickly on this diet than when completely starved. In addition, certain elements such as calcium, phosphorus, iron, sodium, potassium, iodine, and chlorine are necessary in the diet (they occur in foods chiefly as mineral salts) as well as vitamins. The table on p. 66 gives information about some of the essential elements obtained from the mineral salts in man's diet.

These elements are required in very small amounts—a few grammes of sodium, 12 milligrammes of iron, and minute traces of iodine are sufficient to supply the daily needs of a normal adult human. Iodine is an unusual case, because, in man's food, it is found only in marine fish and some drinking water. Before the nineteen-twenties cases of simple goitre (enlarged thyroid gland in the neck) were frequent in Derbyshire and parts of Switzerland, but the simple expedient of adding minute quantities of sodium iodide to either drinking water or table salt has eliminated this condition.

With the modern scope of advertising, few

NB—D

Element	Obtained from	Essential for	Deficiency Causes
Iron	Green vegetables, meat, potatoes	Making haemoglobin	Anaemia
Calcium	Milk, eggs, green vegetables	Making bones, blood clotting	Rickets
Phosphorus	Milk, meat, green vegetables	Making bones	Rickets
Iodine	Fish, iodised table salt	Making thyroxin (see sec. 15.6)	Simple goitre
Sodium and Chlorine	Table salt, green vegetables	Maintenance of correct composition of body fluids	Rarely deficient
Potassium	Green vegetables	Maintenance of correct composition of body fluids	Rarely deficient

people can be unaware of the need for vitamins, which are organic substances much simpler than proteins. They cannot usually be stored in the body but are needed in even smaller quantities than the mineral salts in order to prevent certain deficiency diseases. Scurvy is one of these, and was known amongst sailors in the sixteenth century, as also was the fact that eating fresh fruit and vegetables on reaching port caused the condition to clear up. It was not known that a few milligrammes of vitamin C in the food was the active agent in this recovery.

In every case, the existence of the vitamin was known before it was isolated from its food sources and identified chemically. Hence vitamins were given letters at first, but now that their chemical nature is known the chemical names, e.g. ascorbic acid for vitamin C, are coming into use.

What was originally called vitamin B is now known to contain several vitamins. One of these, vitamin B_1, is clearly distinct from the rest, which are collectively referred to as the vitamin B_2 complex. Although the actual number of vitamins known is a matter of opinion (there is no generally accepted definition of a vitamin at present) it can be stated that there are between fifteen and twenty at present recognised. Details of some of them are tabulated below and should be read carefully.

In some cases the mode of action of vitamins has been worked out. For instance, vitamin A, or carotene, is made into visual purple, an essential pigment in the retina of the eye without which a person cannot see in light of low intensity (see also sec. 14.3).

Mineral salts and vitamins are therefore highly potent in preventing deficiency diseases and are rightly called protective foods.

7.5 Analyses of Some Common Foods

Now that we have considered the six essentials of a human diet we are in a position to discuss the

VITAMINS

Vitamin	Solubility	Good Sources	Effect of Deficiency	Remarks
A Axerophthol $C_{20}H_{30}O$	Fat soluble	Fish-liver oils, milk, butter, eggs, green vegetables, carrots	Growth fails, reduced resistance to infection, night-blindness in some adults	Can be made in the body from carotene (provitamin A), which occurs in carrots and most green plants
D Cholecalciferol $C_{27}H_{44}O$	Fat soluble	Fish-liver oils, milk, butter, eggs, pig's liver	Malformation of bones, e.g. rickets in children	Made in the skin from ergosterol when exposed to sunlight
E α-tocopherol $C_{29}H_{50}O_2$		Wheat oil, eggs, lettuce, and other green vegetables	Sterility	Deficiency effect has been proved for rats, dogs, and rabbits, but not conclusively for human beings
B_1 Thiamine hydrochloride $C_{12}H_{17}ON_4SCl.HCl$	Water soluble	Yeast, Marmite, wheat embryo, pig's liver	Fatigue, muscular wasting, and nervous disorders known as beri-beri	Wholemeal bread contains more than twice as much as white bread
B_2 complex		Yeast, meat, milk, green vegetables	Dermatitis and nervous disorders including pellagra	At least nine substances are included in this complex
B_{12} Cyanocobalamin $C_{63}H_{90}O_{14}N_{14}PCo$		Liver	Pernicious anaemia and degeneration of the spinal cord	Contains cobalt (see p. 257). Human daily requirement one millionth of a gramme
C Ascorbic acid $C_6H_8O_6$		Oranges, lemons, blackcurrants, tomato, cabbage, other green vegetables	Scurvy	Destroyed by boiling, especially in alkaline solution. Rats can make it from other foods in the diet

SECRETION

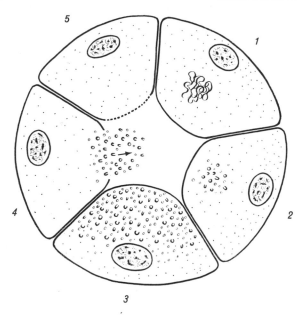

1 A network of secretory material
in the cytoplasm

2 Small granules of secretory
material appear in the network

3 Cell is full of secretory granules

4 Secretory granules extruded

5 Recovery

FIG. 7.2 A diagram of the cycle of changes which occurs within secretory cells.

diet as a whole. Each day an average adult person should eat about

0.08 kg (3 oz) protein,
0.099 kg (3.5 oz) fat,
and 0.51 kg (18 oz) carbohydrate,

but the amounts of the energy producers can be reduced a little in hot weather when the heat lost from the body surface is low. If these requirements are met with a **variety of foods** including some of each of the following—meat or fish, fresh fruit, milk, cheese, green vegetables, potatoes, bread, and butter or margarine—then there will usually be no deficiencies of protective foods, and there will be enough indigestible cellulose (roughage) to keep the muscles of the gut in good condition. Only when man is forced on to a restricted diet, through poverty or other causes, will symptoms of deficiency develop. There is certainly no need for the average person in Great Britain to administer expensive vitamin concentrates to himself if this advice is followed, except possibly in early childhood and during pregnancy. Variety in the diet is useful in another way; it promotes interest in the food, which has a stimulating effect on the flow of digestive juices.

Milk is sometimes called a complete diet in itself, which is, of course, true for the very early life of a mammal. It supplies a nicely balanced mixture of protein, fat, and carbohydrate, plenty of calcium and phosphorus for bone formation, and both fat-soluble and water-soluble vitamins. It does not, however, supply enough roughage for an active adult mammal and contains almost no vitamin C.

DIGESTION

Digestion includes the breaking up of lumps of food in the mouth, as well as the chemical changes carried out by the digestive juices with which the food is mixed during its passage down the alimentary canal. Only three of the six essentials of a healthy diet, carbohydrate, fats, and proteins, require to be chemically changed before they can be absorbed into the blood and distributed to the cells of the body; the others are readily soluble and diffusible.

The substances present in digestive juices are made within the cells of glands such as the pancreas and salivary glands. In producing these substances the cells of the glands undergo a cycle of changes which is known as **secretion** (see Fig. 7.2).

7.6 Teeth

The teeth of mammals are used to break the food into small pieces and exhibit interesting adaptations to deal with a certain type of diet. Although they vary tremendously in shape and size, they all possess the same essential structure (see Fig. 7.3). The greater part of mammalian teeth is **dentine** (ivory), a hard bone-like substance which surrounds the **pulp cavity**. Covering the part of the tooth which protrudes from its socket in the jaw is a layer of **enamel** which, although harder than dentine, is sometimes worn away with use. In the lower part of the tooth this is replaced by **cement** which, together with a **fibrous membrane** around it, fixes the tooth firmly to the jaw. At the base of the tooth is an aperture allowing the entry of nerves and blood vessels. In most teeth this aperture almost closes and prevents continued growth, but this is not the case in the cheek teeth of sheep, which are subject to a lot of wear.

Mammals possess two sets of teeth: the **milk dentition** of the young animal, and the **permanent dentition** of the adult, which replaces it. The former contains fewer smaller teeth than the latter. The four types of teeth present in the permanent dentition are listed below.

1. *Incisors*—more or less chisel-shaped teeth, used for cutting off pieces of food and cleaning the fur.
2. *Canines*—so called because of their prominence in dogs. Long, curved, pointed teeth posterior to the incisors. Never more than one on each side of each jaw.
3. *Premolars*⎫
4. *Molars*⎭ —the cheek teeth which are used for cutting or grinding the food. Premolars are represented in the milk dentition, but molars are not.

Example 1. The Dog—carnivorous dentition (see Fig. 7.4)

The number of the different types of teeth in a dog can be written briefly as a dental formula:

$$i\frac{3}{3}, \; c\frac{1}{1}, \; p\frac{4}{4}, \; m\frac{2}{3}. \qquad \text{Total 42}$$

If you compare this with Fig. 7.4 you should be in no doubt as to how dental formulae are made.

The twelve incisors of a dog are small, and probably as useful in cleaning the fur as in feeding. Behind them are four prominent, curved canines which serve to kill the prey and hold it, a job for which they are admirably fitted, since their points are backwardly directed. The cheek teeth are somewhat flattened from side to side and have

TOOTH STRUCTURE

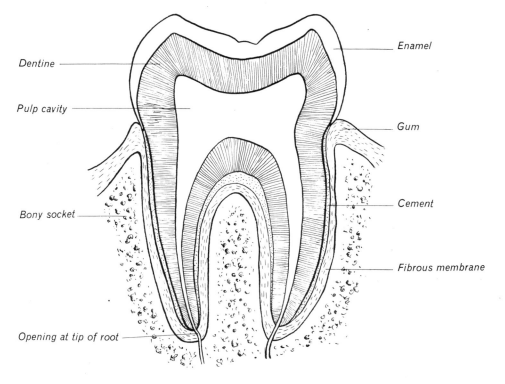

Dentine

Pulp cavity

Bony socket

Opening at tip of root

Enamel

Gum

Cement

Fibrous membrane

FIG. 7.3 Longitudinal section through a tooth with two roots.

CARNIVOROUS DENTITION

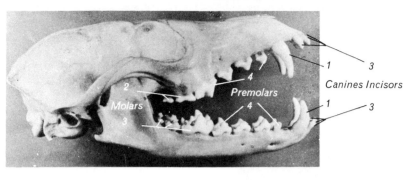

R. SIDE VIEW
OF DOG'S SKULL

R. SIDE VIEW
OF DOG'S JAWS

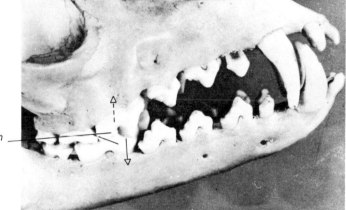

Note shearing action
of the carnassials

FIG. 7.4.

sharp ridges leading up to their points. Those of the lower jaw bite inside those of the upper jaw, resulting in a **shearing action** (rather like that of scissors) allowing the dog to cut up raw flesh efficiently. Particularly large and important in shearing are the fourth premolars of the upper jaw and the first molars of the lower jaw, which are know as carnassial teeth. If you examine the joint of the lower jaw with the skull you should be able to tell that the jaw's movement is restricted more or less to one plane; the joint is a precise one which positions the shearing surfaces of the carnassials close to one another.

Example 2. The Sheep—herbivorous dentition
(*see Fig. 7.5*)

The incisors and canines of a sheep are present only in the lower jaw. They are not used for cutting but for cropping grass when they bite against a horny pad on the upper jaw. Behind the canines is a long gap in the dentition, the **diastema**, which allows grass croppings to be manipulated with the tongue and cheeks so that they come to lie between the rough grinding surfaces of the cheek teeth. If

you examine these teeth you will see the obvious worn surfaces which fit exactly those of the opposite jaw, and have proud ridges of enamel raised above the dentine so that the surfaces are rough like a file. You will also realise that the lower jaw must be moved from side to side a great deal in order to produce the deep grinding valleys and ridges which are only shallow in a newly erupted tooth. Do sheep actually move their lower jaw in this way when grinding up grass? If you examine the cheek teeth carefully you will notice that the cement at first covers the whole of the crown of the tooth, and that this as well as the enamel and the dentine are in turn worn down to make the grinding surface. Enamel, being harder than cement or dentine, wears less rapidly and so remains raised above the general surface. The cheek teeth of sheep also differ from those of dogs in that they grow throughout life and their bases remain open, permitting the entry of blood vessels bringing nutrients to the tooth. Compare the precise jaw articulation of a dog with that of a sheep. The dentition and jaw articulation of a sheep are thus specialised for **cropping** and **grinding** vegetation.

HERBIVOROUS DENTITION

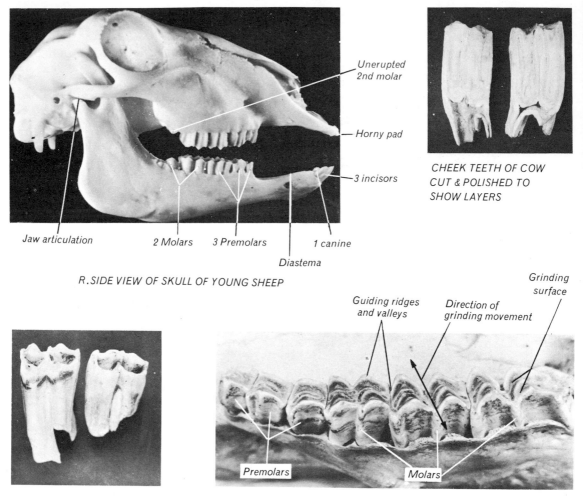

Unerupted 2nd molar

Horny pad

3 incisors

Jaw articulation

2 Molars

3 Premolars

1 canine

Diastema

R. SIDE VIEW OF SKULL OF YOUNG SHEEP

CHEEK TEETH OF COW CUT & POLISHED TO SHOW LAYERS

Guiding ridges and valleys

Direction of grinding movement

Grinding surface

Premolars

Molars

UNWORN CHEEK TEETH

WORN UPPER CHEEK TEETH OF OLDER SHEEP

Fig. 7.5.

You should also examine your own teeth (an omnivorous dentition) and those of any other available mammal, and try to relate the dentitions to the animals' habits and diet. Other dental formulae are given below:

Human	$i\frac{2}{2}, c\frac{1}{1}, p\frac{2}{2}, m\frac{3}{3}.$	Total 32
Rabbit	$i\frac{2}{1}, c\frac{0}{0}, p\frac{3}{2}, m\frac{3}{3}.$	Total 28
Horse	$i\frac{3}{3}, c\frac{1}{1}, p\frac{3}{3}, m\frac{3}{3}.$	Total 40

(canines usually absent from mares)

| Hedgehog | $i\frac{3}{2}, c\frac{1}{1}, p\frac{3}{2}, m\frac{3}{3}.$ | Total 36 |

7.7 Digestion in the mouth

During the chewing of food, **saliva**, the first digestive juice, is mixed with the food. The saliva is secreted by several pairs of glands in the head (four in the rabbit—see Fig. 7.8; three in man—see Fig. 7.9) and is conducted to the mouth through salivary ducts. The flow of saliva can be easily detected by curling the tip of the tongue down on to the floor of the mouth. A simple experiment will demonstrate the digestive action of saliva in man.

First wash out the mouth with lukewarm water and then obtain a copious secretion of saliva by a chewing action. Mix this in the mouth with a few cm³ of warm water and collect it in a test-tube.

SWALLOWING AND BREATHING

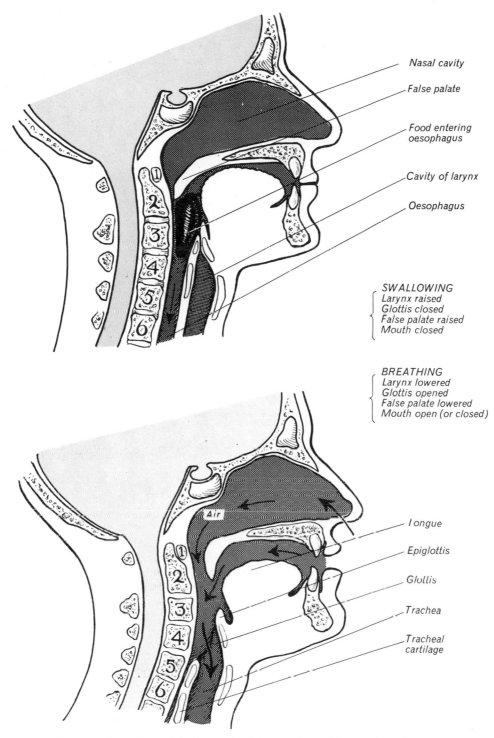

Nasal cavity

False palate

Food entering
oesophagus

Cavity of larynx

Oesophagus

SWALLOWING
Larynx raised
Glottis closed
False palate raised
Mouth closed

BREATHING
Larynx lowered
Glottis opened
False palate lowered
Mouth open (or closed)

Air

Tongue

Epiglottis

Glottis

Trachea

Tracheal
cartilage

FIG. 7.6 Diagrammatic sections of the human head close to the mid-line to show the intake of food (upper) and air (lower). Notice the different positions of the lips, the false palate, and the larynx during swallowing and at inspiration. The positions of the larynx should be compared with the neck vertebrae which are numbered.

Divide this mixture into two halves, label them both, boil one-half for one minute to act as a control and allow to cool. To each add equal quantities (about one-third of a test-tube is ample) of a dilute (0.5 per cent) starch solution, and agitate. At one-minute intervals take out sample drops from the experimental tube and the control tube, and add them to drops of dilute iodine

SALIVARY GLAND

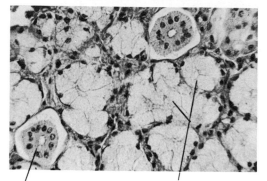

Branch of salivary duct   Groups of secretory cells

FIG. 7.7 Photomicrograph of mammalian salivary gland.

solution arranged in two rows on a white tile. At first the samples from both tubes give the blue-black colour of the starch test, but soon the colour of the samples from the experimental tube changes through purple to brown and then, as a rule, becomes colourless. Samples from the control tube retain the blue-black colour when added to the drops. The conclusion to be made from these results is that saliva contains some substance which, when unboiled, is able to change starch. That it is converted into sugar can be shown by applying a Benedict's test to a portion of the mixture left in each of the test-tubes. The result of this test is positive (i.e. a red precipitate) only in the case of the tube containing unboiled saliva.

The substance responsible for carrying out this change on starch is an enzyme called **ptyalin** (salivary amylase), which works best in a slightly acid medium, though its normal medium is slightly alkaline. **Enzymes** are complex catalysts which can be produced only by living cells, and they are destroyed by boiling. They are composed of proteins.

Apart from this digestive action, saliva enables dry food to be swallowed, facilitates speaking, and keeps the tongue moist so that the food can be tasted. For details of the mechanism of swallowing study Fig. 7.6.

7.8 Digestion in the Stomach

When the moistened food is pushed to the back of the mouth and swallowed it passes down a narrow tube, the **oesophagus**, into the large muscular **stomach** where it meets the **gastric juice** produced from tubular glands in the lining of the stomach. Because the gastric juice contains strong hydrochloric acid, salivary digestion is stopped in the stomach, for ptyalin cannot work in highly acid conditions. Food remains in the stomach for at least an hour and is thoroughly churned up with the gastric juice by muscular contractions of the stomach walls. The opening into the stomach and the opening from the stomach into the duodenum, the first part of the much-coiled small intestine, are each guarded by a ring of muscle known as a sphincter, which can close the opening effectively. These sphincters are kept closed during gastric digestion.

The protein of the food undergoes its first change in the stomach as a result of the action of an enzyme, **pepsin**, which 'splits' it into simpler substances called proteoses and peptones. In addition, another enzyme, **chymase** (rennin), is responsible for the clotting of the soluble protein of milk (casein), a change which helps its retention in the stomach where pepsin can act on it. Chymase is probably important only in young mammals which are on an exclusively milk diet.

Fluids or powders containing these enzymes can be prepared from the stomach wall of cattle and prove very useful when it is necessary to demonstrate the digestive changes which take place in the stomach. Many of you will have seen the making of junket from milk and a powder containing chymase, or a solution called rennet.

A convenient experiment demonstrating the action of pepsin on fibrin (a protein from blood) is described below. Fibrin which has been dyed with congo red is used, and the pepsin is made into a 1 per cent solution. Three test-tubes are filled as shown in the table below.

Tube	Distilled Water	Pepsin Solution	Congo-red Fibrin	Dilute Hydrochloric Acid
A	5 cm³	1 cm³	A few granules	1 cm³
B	5 cm³	1 cm³	A few granules	—
C	5 cm³	1 cm³ (previously boiled and cooled)	A few granules	1 cm³

In preparing these tubes you will notice that the congo-red dye acts as an indicator and turns blue when the acid is added. Label the tubes carefully and keep them at about 37°C in an oven or water bath. Shake each tube gently before examining at approximately ten-minute intervals. Digestion of the protein is indicated by the liberation of the dye into the solution and by the disappearance of the solid granules of protein. By this means a small amount of digestion can be easily detected and it is found to occur only in tube A. Think about the results and try to decide what they indicate.

7.9 Digestion in the Small Intestine (Duodenum and Ileum)

The food, which now has a soup-like consistency, is squirted into the **duodenum** by contraction of the stomach wall when the pyloric sphincter (see Fig. 7.8) is relaxed and open. Here it is mixed with three digestive juices:

1. Bile—a green alkaline fluid made by the liver and stored in the **gall bladder**, which lies embedded in the liver. A **bile duct** leading from the gall bladder opens into the duodenum in the rabbit, but in man it joins the pancreatic duct first.
2. Pancreatic juice—secreted by the **pancreas**, a gland which lies between the two limbs of the U-shaped duodenum.
3. Intestinal juice—produced by tubular glands in the walls of both the duodenum and the **ileum** which makes up the rest of the small intestine.

All these juices are alkaline and help to neutralise the acid stomach contents to provide the necessary alkaline medium for the activity of all the enzymes they contain. The bile does not contain enzymes but its bile salts have the property of breaking up oil droplets into very small ones, a process known as **emulsification**. This increases the surface area of the oil droplets tremendously and, because it is on this surface that fat-splitting enzymes must work, it speeds up the digestion of fats.

The pancreatic juice contains an enzyme, **trypsin**, which continues the protein digestion started in the stomach, but it is not active in this respect until it has combined with a substance known as **enterokinase** from the intestinal juice. Its **amylase** completes the starch digestion which was commenced in the mouth, and its **lipase** splits up the emulsified oils into glycerin and fatty acids. The latter are immediately converted to their sodium salts, which are known as soaps.

The intestinal juice contains several enzymes collectively known as **peptidases** which complete protein digestion by breaking down polypeptides to amino-acids. The sequence of changes, proteins → proteoses → peptones → polypeptides → amino-acids, represents a progressive reduction in the size of molecules by splitting them. **Maltase**, another enzyme of the intestinal juice, completes starch digestion, converting maltose into glucose.

Details of the chemical changes involved in making the complex organic compounds soluble and diffusible are summarised in the table.

THE CHEMISTRY OF DIGESTION

	Digestive Juice	From	Contains	Changes	Medium	Remarks
Mouth	Saliva	Salivary glands	Ptyalin	Starch → maltose	Slightly alkaline	Rarely completed
Stomach	Gastric juice	Stomach wall	Pepsin	Proteins → proteoses and peptones	Highly acid	Chymase only important in young mammals
			Chymase (rennin)	Clots protein of milk		
Small intestine	Bile	Liver	Bile salts	Emulsifies oils	Alkaline	Bile contains no enzymes
	Pancreatic juice	Pancreas	Trypsin	Proteoses + peptones → polypeptides + <u>amino-acids</u>	Alkaline	Secreted in an inactive form
			Amylase	Starch → maltose	Alkaline	
			Lipase	Emulsified oils → <u>glycerin + fatty acids</u>	Alkaline	Fatty acids absorbed as soaps; some fats are absorbed undigested
	Intestinal juice	Walls of duodenum and ileum	Enterokinase	Activates trypsin	Alkaline	
			Sucrase	Sucrose → <u>glucose + fructose</u>	Alkaline	
			Maltase	Maltose → <u>glucose</u>	Alkaline	
			Peptidases	Polypeptides → <u>amino-acids</u>	Alkaline	

Note: Final products of digestion which are absorbed are underlined.

ALIMENTARY CANAL OF A RABBIT

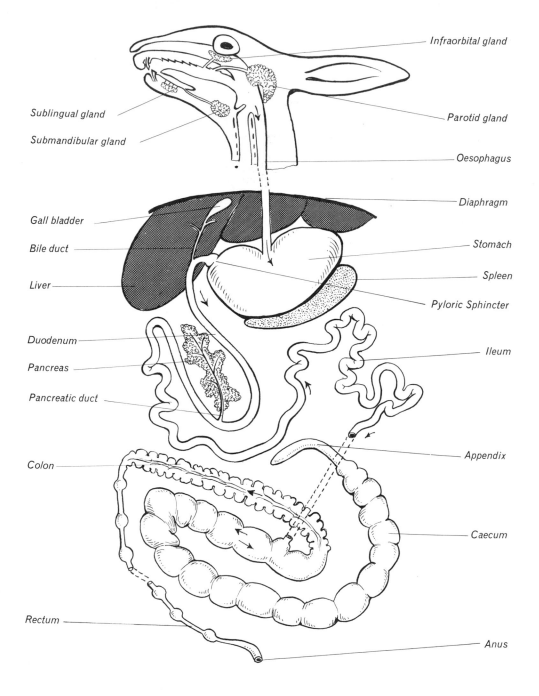

Infraorbital gland

Sublingual gland

Submandibular gland

Parotid gland

Oesophagus

Diaphragm

Gall bladder

Bile duct

Stomach

Liver

Spleen

Pyloric Sphincter

Duodenum

Pancreas

Ileum

Pancreatic duct

Appendix

Colon

Caecum

Rectum

Anus

FIG. 7.8 A diagram of the alimentary canal of the rabbit in ventral view. The thoracic part of the oesophagus, part of the ileum, and part of the rectum have been omitted. Compare this with Fig. 7.9.

HUMAN ALIMENTARY CANAL

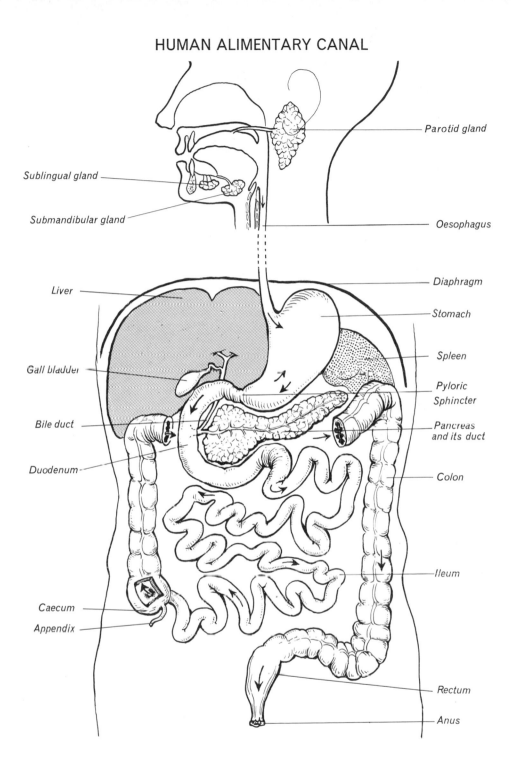

Parotid gland

Sublingual gland

Submandibular gland

Oesophagus

Liver

Diaphragm

Stomach

Spleen

Gall bladder

Pyloric Sphincter

Bile duct

Pancreas and its duct

Duodenum

Colon

Caecum

Ileum

Appendix

Rectum

Anus

Fɪɢ. 7.9 A diagram of the human alimentary canal in ventral view. The thoracic part of the oesophagus, and the transverse part of the colon which would otherwise obscure the pancreas, have been omitted. Compare this with Fig. 7.8 and note differences in number of salivary glands, shape of stomach, bile and pancreatic ducts, caecum and appendix.

STOMACH WALL OF RAT

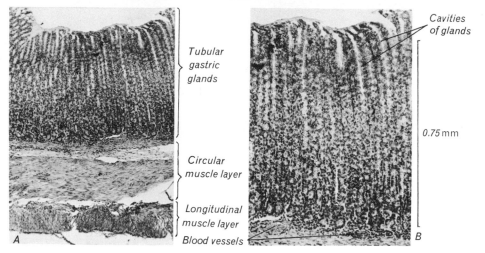

Tubular gastric glands

Circular muscle layer

Longitudinal muscle layer

Blood vessels

Cavities of glands

0.75 mm

A   B

FIG. 7.10 Two photomicrographs of stained sections of stomach wall of rat.

The muscular coat of the duodenum and ileum is constantly churning the food with the digestive juices by its muscular action, and occasionally it drives it down by a **peristaltic wave**. This is a muscular constriction which travels down many parts of the alimentary canal preceded by a dilation where the muscles are relaxed (see Fig. 7.11). The folded internal surface of the ileum is covered with finger-like projections or villi. If you study Fig. 7.12 you may appreciate the enormous surface area of the villi and microvilli—a surface through which the soluble products of digestion such as sugars, amino-acids, glycerin, and soaps can diffuse. The products of carbohydrate and protein digestion (sugars and amino-acids) diffuse into the blood in the villi and are transported to the liver by the portal vein. Fats are rarely digested completely, but there is evidence that some of the undigested fat is absorbed directly through being trapped in the mat of entangled cell filaments of the microvilli and there is no reason to suppose that the same is not true of the soluble products of digestion. The glycerin and soaps are reconstituted into fats and, together with the undigested

fat, collect in the lymph vessels of the villi from where they are carried in larger vessels to the neck region where they are emptied into the blood.

Thus the whole purpose of digestion is to change solid food containing complex substances with large molecules into simple, soluble, diffusible ones with much smaller molecules, so that the latter may be absorbed into the blood, either directly or indirectly, and distributed to all parts of the body where they may be used as a source of energy or incorporated in the cell structure. The latter is known as **assimilation**. Their fate is summarised in the table on p. 78. It is not surprising that the largest gland of the body, the liver, plays a major part in protein and carbohydrate metabolism, for the products of digestion are carried straight to this organ from the alimentary canal in the portal vein (see Fig. 9.14).

7.10 Digestion in the Large Intestine

As can be seen from Figs. 7.8 and 7.9, the large intestines of a rabbit and man show important differences. The long **caecum** and **appendix** of the rabbit provide a blind part of the gut where plant food stays for some time, during which bacteria are able to digest its cellulose. Other valuable products, e.g. vitamins of the B group, are produced here by bacterial action, but are not able to be absorbed into the blood. These substances are not lost at defaecation, however, for in their burrows during the day rabbits eat their own faecal pellets, a habit known as refection. In man the caecum and appendix are relatively short and there is no digestion of cellulose. This undigested cellulose with a large amount of dead bacteria, some

PERISTALTIC WAVE

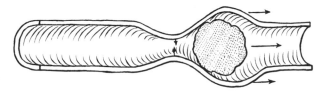

FIG. 7.11 A diagram of a bolus of food being driven down the oesophagus by a peristaltic wave.

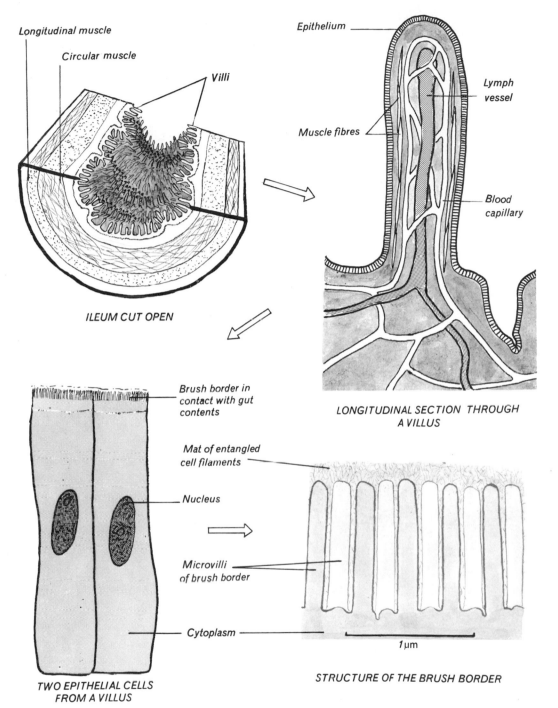

SMALL INTESTINE-VILLI AND MICROVILLI

Longitudinal muscle

Circular muscle

Villi

ILEUM CUT OPEN

Epithelium

Lymph vessel

Muscle fibres

Blood capillary

LONGITUDINAL SECTION THROUGH A VILLUS

Brush border in contact with gut contents

Mat of entangled cell filaments

Nucleus

Microvilli of brush border

Cytoplasm

TWO EPITHELIAL CELLS FROM A VILLUS

1 μm

STRUCTURE OF THE BRUSH BORDER

FIG. 7.12.

FATE OF PRODUCTS OF DIGESTION

Product	1st Stage	2nd Stage	3rd Stage	Remarks
Amino-acids	Carried in the blood to all parts of the body	Used in all tissues for protein synthesis	Excess of any amino-acid is converted in the liver to: 1. Urea, which is excreted 2. Glucose, which is used in respiration	The body does not accumulate reserve stores of proteins
Glucose	Carried in the blood to all parts of the body	Used by all cells to produce energy in respiration	Excess is usually stored as glycogen in the liver and muscles	Only limited amounts can be stored, more in the liver than muscles
Fatty acids and glycerin	Re-formed into fats, carried in lymph and then in blood to all parts of the body	Deposited as a reserve fat store under the skin and round the kidneys	Fat store mobilised and used to produce energy in respiration	Unlimited amounts of fat can be stored

living ones, cells from the gut lining, and a tiny amount of waste nitrogenous matter makes up the typical human faeces which are dried by water absorption in the **colon** and **rectum**, and evacuated at the anus each day by muscular action.

7.11 Digestion in Ruminants

Cattle, sheep, goats, and deer are mammals which chew the 'cud' and are, therefore, ruminants. Their special features include a four-chambered stomach (see Fig. 7.13) and crescent-shaped ridges of hard-wearing enamel on the grinding surfaces of the cheek teeth.

In sheep the cropped vegetation is mixed with saliva, and ground up between the upper and lower cheek teeth by a side-to-side movement of the lower jaw. It is then forced down the oesophagus into the reticulum and rumen of the stomach where it is kept for a short time. In these chambers the food is kept in motion by muscular

SHEEP'S STOMACH

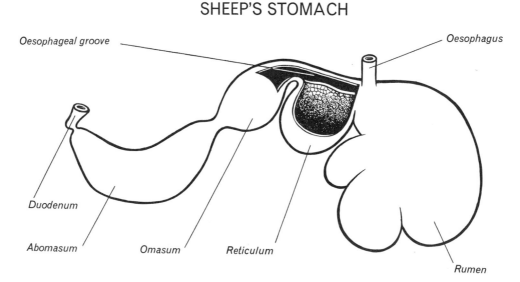

FIG. 7.13 A sheep's stomach displayed to show the four chambers. The reticulum is shown with part of its wall removed. Alternative names for the chambers are: rumen or paunch, reticulum or honeycomb-bag, omasum or psalterium, abomasum or true stomach or reed.

action and is acted on by various micro-organisms (bacteria and protozoa, see pp. 262 and 263). later, the mixture, known as 'cud', is passed back to the mouth where it is ground up more thoroughly with more saliva before being returned to the rumen. Now that the food is finely divided, the action of the micro-organisms is more efficient and converts much of the cellulose in the food to fatty acids. These products are absorbed directly into the blood and provide the sheep's main source of energy-producing food. The micro-organisms are useful in another way: they make vitamins of the B group which are also absorbed. The mixture of food and micro-organisms is kept alkaline in the rumen and reticulum by the large quantities of saliva which enter. It is then passed along the oesophageal groove, through the omasum where water is absorbed and into the abomasum, where it becomes acidic. The acidity kills the bacteria and protozoa, and normal gastric digestion of protein occurs. Fatty acids are absorbed in appreciable quantities by the blood flowing through the large intestine, which indicates that the bacteria present here are also breaking down cellulose.

It is important to realise that ruminants have themselves no enzymes capable of digesting cellulose; it is the micro-organisms which they harbour which are responsible for breaking down this resistant carbohydrate, and herein lies the value of sheep and cattle to man.

QUESTIONS 7

1. Why is the digestion of food essential? Name four digestive enzymes found in a mammal and state the part each plays in the process of digestion. Describe an experiment you could carry out to show the action of a named enzyme.   [A]

2. What are the essential constituents of the food of animals? How does the diet of a named herbivorous mammal differ from that of a named carnivorous mammal? In what ways do the teeth of these two types of mammal differ in relation to their diets?   [C]

3. Make a large labelled diagram of the alimentary canal and the associated glands in a named mammal. Indicate on this diagram where ptyalin, pepsin, and bile are produced. State briefly the functions of these juices. How are the contents of the gut moved along?   [L]

4. List the essential constituents of a human diet. Why does a mammal need to digest its food? Explain carefully what happens to the starch content of a meal from the time it enters the mouth until the products of digestion enter the liver.   [L]

5. Why is digestion necessary? Describe what happens to a piece of lean meat during its passage through the human alimentary canal. What use is made of this meat by the human body?   [O]

6. Make a labelled drawing to show the structure of a typical canine tooth. Compare the dentition of a named carnivore with that of a named herbivore, emphasising those features which you consider to be related to feeding habits. What differences are there between the digestive systems of these two types of mammal?   [O & C]

7. Describe what occurs in the various regions of the alimentary canal (including the mouth) when you eat bread and butter.   [S]

8. Fats, carbohydrates, and proteins are digested in the human alimentary canal. To what purpose does man put each of these foods? Give an account of the processes which the starch in a slice of bread undergoes until it is completely digested.   [A]

9. (i) Draw a very simple diagram of the digestive system of a named mammal. Label your diagram. (ii) What are the functions of the liver? (iii) Of what use are the villi of the small intestine?   [S]

10. (a) What three valuable ingredients of man's diet are found in wholemeal bread? (b) Briefly describe the value of each of these ingredients to man. (c) Outline what happens to bread as it passes through the digestive system until the blood-stream draws off its usable ingredients.   [M]

8 Other Types of Nutrition

SAPROPHYTISM

Some plants such as Mucor and yeast do not contain chlorophyll and are therefore unable to manufacture their own food by photosynthesis. They need dead organic matter, but it has to be in a dissolved form as plants cannot ingest solid particles. Such plants secrete enzymes which digest the food, which can then be absorbed. Plants feeding in this way are termed **saprophytes**; section 2.14 described how Mucor absorbs food from the bread on which it grows, causing it to decay, and mention has been made in section 5.5 of the numerous saprophytic bacteria which flourish in the soil.

8.1 Yeast

Yeast is another example of a saprophytic fungus. It consists of tiny oval cells, about $8\,\mu\text{m}$ ($\mu\text{m} = 1/1000$ mm) in length, with a thin wall and a relatively thick lining of cytoplasm surrounding a central vacuole (see Fig. 8.1). At one side of the vacuole is the nucleus with dark staining strands extending from it round the vacuole. The cells multiply by **budding**; a tiny area in the cell wall is pushed out into a bud by the cytoplasm. The bud increases in size and is finally nipped off. Under favourable conditions this goes on rapidly, and in a sugar solution the cells remain attached to one another in chains. When conditions are less favourable the cells form spores within themselves, and as spores the plant can pass through the winter.

Wild yeasts are found in the sugars which exude from plants, such as the nectar of flowers, or the broken surface of grapes and other fruits. Like many other plants, such as cereals, they are better known in the cultivated than in the wild state and are grown extensively for use in baking and brewing (see sec. 9.5).

Yeast feeds saprophytically on the sugar solution in which it grows. From these sugars and simple nitrogen compounds (e.g. ammonium tartrate, which is found in most fruit juices) it is able to synthesise proteins and obtains the energy required for this synthesis by respiration. If there is not enough oxygen present for the normal type of respiration, yeast is able to break down the sugar to alcohol and carbon dioxide, so releasing energy.

YEAST - A SAPROPHYTE

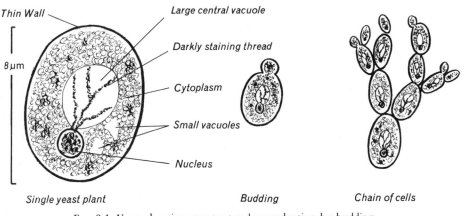

FIG. 8.1 Yeast showing structure and reproduction by budding.

This change is brought about by the group of enzymes known as zymase (see sec. *9.5*).

PARASITISM

An organism may also obtain its supply of food by living on or in another organism, which is called the **host**. Such an organism is termed a **parasite** and is completely dependent on its host for a supply of food. The host gains no benefit from the association, in fact it always sustains some loss or damage.

8.2 Dodder (*Cuscuta europaea*)

This is an example of a parasite which grows on its host. The nettle dodder is found in some parts of this country growing on clumps of nettles— especially those found on the banks of rivers. Its stem resembles a piece of fine pink cotton and it coils round the stem of its host plant, to which it is fixed at frequent intervals by small suckers or haustoria (see Fig. 8.2). These suckers penetrate the stem of the host and make contact with the vascular bundles so that the dodder can obtain a supply of water and complex organic compounds (see Fig. 8.3). The parasite has no roots, so that it cannot absorb materials direct from the soil nor can it photosynthesise as its leaves, which are small and scale-like, contain no chlorophyll. From July onwards the dodder bears clusters of tiny pink flowers in the axils of the scale leaves, and from these very numerous tiny seeds are formed. These are shed from the plant in the autumn and may be dispersed by wind and probably by flooding. Germination takes place in April or May and the tiny seedling shows no real differentiation into root and shoot. It is slightly green at the tip, which grows upwards, waving in circles as it does so. If a suitable host plant is touched, the seedling coils round it and forms suckers. Once it has become attached to its host the lower part of the seedling

DODDER - A PARASITE

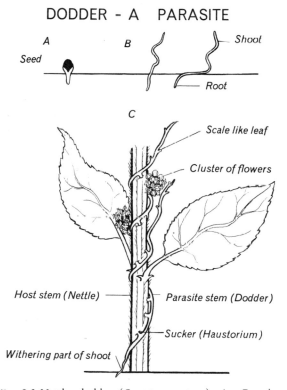

FIG. 8.2 Nettle dodder (*Cuscuta europaea*). A—Germinating seed. B—Young seedlings. C—Dodder plant growing on nettle.

DODDER ON NETTLE

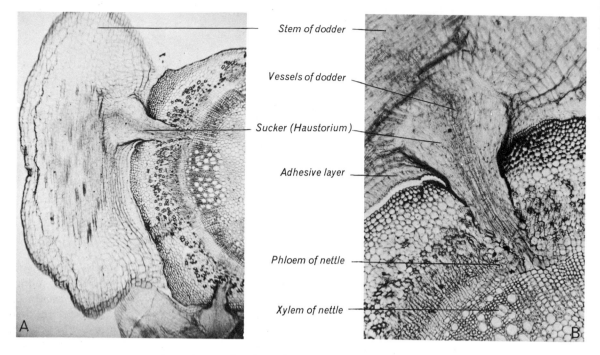

Stem of dodder

Vessels of dodder

Sucker (Haustorium)

Adhesive layer

Phloem of nettle

Xylem of nettle

FIG. 8.3 Photomicrographs of T.S. host stem. A—Low power. B—High power.

shrivels and dies. Other species of dodder are found growing on heather, gorse, and clover.

8.3 Mistletoe

Mistletoe (*Viscum album*) grows on trees, and in this country is found on apple and poplar. The plant (see Fig. 8.4) consists of a group of green twigs growing from a sucker embedded in the host plant. Unlike dodder, the mistletoe absorbs only water and mineral salts from its host, for it has green leaves which are able to photosynthesise. It is therefore said to be a partial or semi-parasite.

8.4 Tapeworms

Tapeworms are parasites which live inside their host. The adult animal is found in the intestines of the horse, dog, cat, and other mammals. It is attached to the intestine by suckers and hooks on its head, behind which is a long chain of segments (see Fig. 8.5). The segments are continually being formed in the region behind the head and the older segments gradually break off and are expelled from the host. The animal has no mouth or alimentary canal; it is surrounded by the digested food of its host, which it absorbs over the whole surface of its body. The segments are covered by

thick cuticle which protects them from the action of the host's digestive juices.

As it lives inside its host it is well protected, and consequently has little need of specialised locomotory or sense organs. It can, however, wriggle a little and has a very simple nervous system. Its main problem is to ensure that its eggs reach a suitable host. Each segment contains a complete

MISTLETOE — A PARTIAL PARASITE

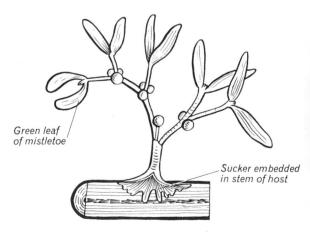

Green leaf of mistletoe

Sucker embedded in stem of host

FIG. 8.4 Mistletoe plant with twig of host in section.

TAPEWORM — A PARASITE

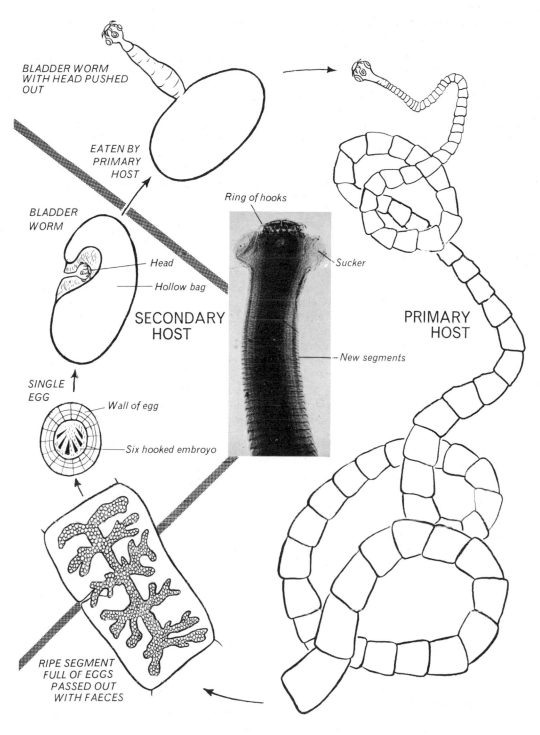

FIG. 8.5 Diagrams to show the life-history of a tapeworm.

set of male and female reproductive organs and the posterior segments are filled with eggs. These eggs pass out of the host with the faeces when the posterior segments break off. No further development takes place unless the eggs are swallowed by a secondary host. For the dog tapeworm—*Taenia serrata*—this is the rabbit. The gastric juices of the secondary host dissolve the wall of the egg and set free a six-hooked embryo which bores its way through the wall of the alimentary canal into the blood stream, which carries it to the muscles. There it loses its hooks and develops into a bladder worm, which consists of a hollow bag with the tapeworm head tucked inside it. The bladder worm cannot develop any further unless the flesh which contains it is eaten by the primary host—the dog. If this happens the head is pushed out of the bladder and becomes attached to the wall of the intestine where it starts to form new segments so that the life-cycle is complete.

Unlike its other systems the reproductive system is extremely complex, and a very large number of eggs is formed, for only a small proportion is likely to be eaten by the secondary host. The succession of stages in different hosts in the life-cycle is one way in which the tapeworm is adapted to ensure its survival and the passage from one host to another.

Man can acquire different tapeworms from various sources, but in civilised countries with meat inspection and controlled sewage disposal the chance of infection is slight. The pork tapeworm—*Taenia solium*—is probably extinct in this country now. The beef tapeworm—*Taenia saginata*—which spends its bladder-worm stage in cattle, and is acquired by man only through eating raw or undercooked infected beef, was relatively unknown in this country before World War II when some tapeworm carriers entered this country among the refugees and soldiers returning from the Continent and tropical areas. The tapeworm eggs pass into the sewage system and, because of their great resistance to external conditions, they are able to survive the various processes by which the sewage is treated, so that they may eventually be eaten by grazing cattle.

SYMBIOSIS

A partnership between two organisms is not always so one-sided as in the case of parasitism. There are many examples of two organisms living together in very close partnership from which both benefit; such an association is termed **symbiosis**.

The nitrogen-fixing bacteria which live as symbionts in the roots of plants belonging to the leguminous family were mentioned in section 5.5.

8.5 Chlorohydra, Lichens, and Mycorrhiza

In the nutritive cells (see sec. 2.11) of the green Hydra—Chlorohydra—there are small green algae called Zoochlorella. Each plant consists of a single nucleus and a spherical mass of cytoplasm containing chlorophyll surrounded by a cell wall. By living inside Chlorohydra, this plant gains protection and a supply of food materials (carbon dioxide from respiration and excretory nitrogenous compounds) which it requires to synthesise its own food. Chlorohydra also profits from the plant's presence by obtaining a supply of oxygen released during photosynthesis. The association, therefore, is of benefit to both partners and is a good example of symbiosis.

Lichens are peculiar compound plants which form incrustations on exposed wood and stones in country areas. Little structure can be seen with the naked eye, but examination with a microscope shows that the lichen is built up of two distinct partners, a fungus and an alga. In this case the association between the two partners is very close and both plants benefit, for each assists the nutrition of the other. Because of their association they can survive and grow slowly in the most difficult conditions, such as occur on bare rock faces.

There are some fungal saprophytes which live in very close association with flowering plants, and apparently both benefit from the association. The roots of most forest trees which grow in rich humus are surrounded by a mass of fungal hyphae. Such trees have few root hairs and depend on the fungus for a supply of water, giving the fungus in exchange a supply of food. A fungus which associates symbiotically with a flowering plant is called a mycorrhiza. Sometimes the mycorrhiza is internal as in some orchids. In the common heather (ling) the association is ensured by the fact that the fungus is transmitted in the seed. This fungus is known to fix atmospheric nitrogen.

INSECTIVOROUS PLANTS

Some plants find difficulty in obtaining an adequate supply of nitrates from the soil in which they grow. One group of such plants obtains its nitrogenous supply by catching insects in various ways and afterwards absorbing parts of the insect's body.

INSECTIVOROUS PLANTS

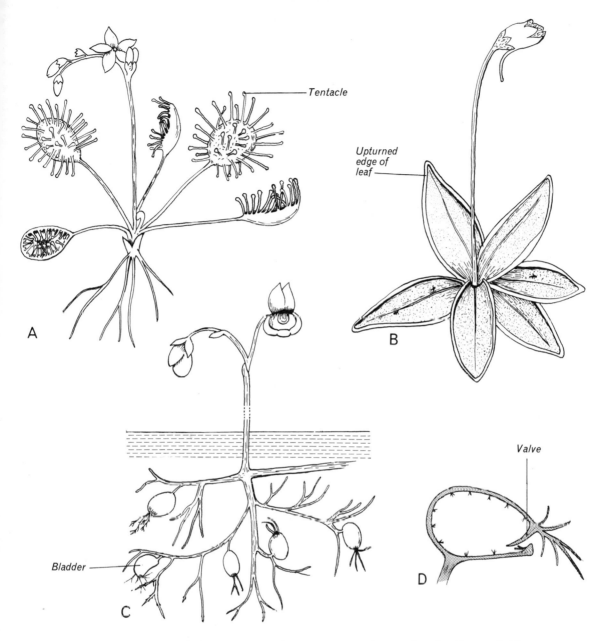

FᵢG. 8.6 A—Sundew (*Drosera rotundifolia*). B—Butterwort (*Pinguicula vulgaris*). C—Bladderwort (*Utricularia minor*). D—Section through a bladder of bladderwort.

8.6 Sundews, Butterworts, and Bladderworts

The British insectivorous plants are the sundews,
the butterworts, and the bladderworts and all are
found in wet areas (see Fig. 8.6).

Sundew is a low-growing plant found amongst
the bog mosses. It has a flat rosette of leaves which
are covered with delicate red tentacles, each ending in a glistening knob so that the whole leaf
resembles a pin-cushion. The tentacles secrete a
sticky fluid—the 'dew' which glistens in the sun.
The 'dew' traps the insects, but as soon as the
tentacles are touched they curl over so that the

FOOD WEB

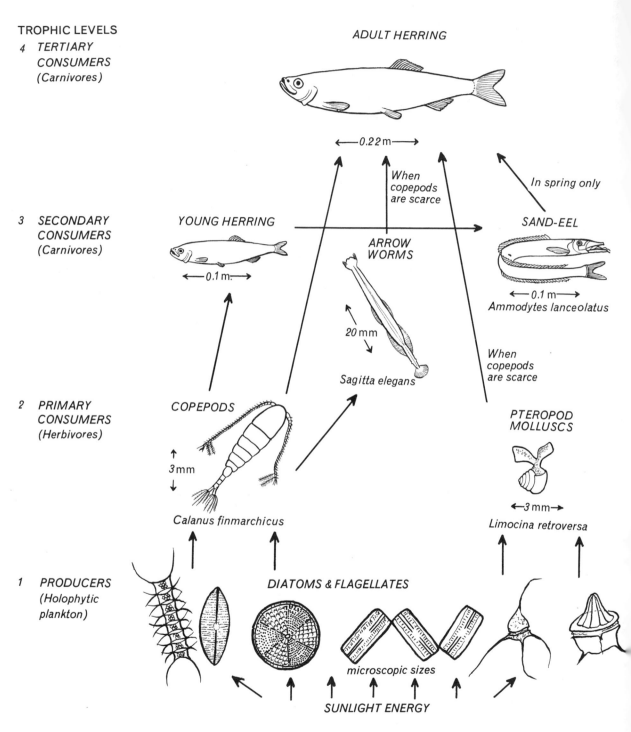

TROPHIC LEVELS

*4 TERTIARY
    CONSUMERS
    (Carnivores)*

ADULT HERRING

←—0.22 m—→

*When
copepods
are scarce*

In spring only

*3 SECONDARY
    CONSUMERS
    (Carnivores)*

YOUNG HERRING

←0.1 m→

ARROW
WORMS

SAND-EEL

Ammodytes lanceolatus
←—0.1 m—→

20 mm

Sagitta elegans

*When
copepods
are scarce*

*2 PRIMARY
    CONSUMERS
    (Herbivores)*

COPEPODS

PTEROPOD
MOLLUSCS

3 mm

Calanus finmarchicus

←3 mm→
Limocina retroversa

*1 PRODUCERS
    (Holophytic
    plankton)*

DIATOMS & FLAGELLATES

microscopic sizes

SUNLIGHT ENERGY

FIG. 8.7 Simplified food web of the North Sea herring (*Clupea harengus*). (*From information supplied by D. G. Lambert.*)

insect is held securely. The contact of the insect also stimulates glands in the tentacles to secrete juices containing enzymes, which digest the body of the insect, changing it to a form which the leaf can absorb. Later, the tentacles uncurl and dry, and the undigested parts of the insect may be blown away.

The butterworts also grow in the damp soil of moors and they too have a flat rosette of leaves, but in this case each leaf has an upturned edge so that a shallow trough is formed. The surface of the leaf contains glands, some of which secrete a liquid that attracts the insects, and others secrete the digestive fluid.

The bladderworts are rootless plants which grow submerged in the acid pools of moorland ponds. They have finely divided leaves covered with tiny bladders which trap small water animals. These bladders are pear shaped and have a small opening guarded by a valve which can be pushed inwards but not outwards. When the water animal touches the valve, it opens, and the inrush of water carries the animal with it and the valve shuts. The trapped animal eventually dies and decays, forming soluble substances which can be absorbed by the plant.

All these plants contain chlorophyll, though they supplement their nitrate supply by catching insects.

INTERDEPENDENCY

8.7

Since green plants are the only organisms able to synthesise carbohydrates and proteins from inorganic substances, it is obvious that all animals and non-green plants are dependent on them for their food supplies. Some of the foods we eat, like lettuce and cabbage, are green leaves themselves; others, like potatoes and apples, are the food stores which the green plants have built up. When we eat mutton we are eating the flesh of the sheep which has been built up from the grass it has eaten. Such **food chains** all start from green plants. You have probably seen thrushes smashing snails on a large stone. Snails feed on green plants. Other examples of food chains are:

It is not so easy at first sight to trace the food chain back to green plants when we think of a meal including herrings or cod. These fish live in the sea away from the shore, and appear to have few plants available for food. There are, however, an immense number of small floating plants and animals, together known as **plankton**. Many of these small organisms are **diatoms**, which are minute non-cellular algae. They are found in both salt and fresh water and vary considerably in their shape. Diatoms can manufacture their own food by photosynthesis (the products of photosynthesis are stored in the form of oil) and they multiply very rapidly by a simple form of fission. They are eaten by small animals, which are in turn eaten by the larger animals on which the herrings feed, so that once again we have a food chain beginning from holophytic plants (see Fig. 8.6).

OWL PELLETS

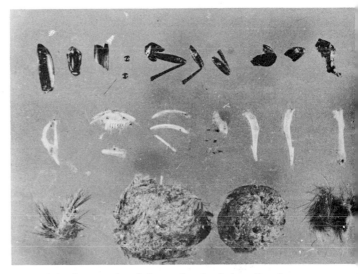

Fig. 8.8 Photographs of dissected and whole pellets from a little owl. On what has it been feeding?

It is possible to discern a pattern in these food chains. They are all based on one or more photosynthetic plants, **the producers**, which acting as a solar energy trap, make organic material for the next level, the herbivores or **primary consumers**. The herbivores in turn provide food and energy

Habitat	Primary producers	→	Herbivores	→	Carnivores					
Suburban garden	Rose trees	→	Greenfly (aphids)	→	Ladybirds	→	Blue tits	→	Cats
Pasture land	Ryegrass	→	Short-tailed vole	→	Weasel	→	Barn owl		
North Sea	Diatoms	→	Calanus (a copepod)	→	Herring	→	Cormorant		

for the next level or levels which includes the carnivores. Such feeding levels in a community of organisms are often referred to as **trophic levels**. Most commonly the number of organisms decreases the further one moves from the producers: thus we can envisage the feeding relationships in a community as a pyramid of numbers with the top carnivores (though the largest organisms), being the least numerous.

The food and feeding relationships of plants and animals are rarely, if ever, as simple as the examples of food chains quoted. Fig. 8.7 shows some of the complexity of the feeding relationships of an adult herring. There are several food chains which are interconnected and these interconnections lead to the food relationships of an organism being described as a **food web** rather than a food chain. A complete food web for a

community is probably impossible to establish, so great is the variety of feeders and their foods. It is, however, important to realise that in all communities, whether they are marine, fresh-water or soil-based, there are detritus feeders which receive food in the form of dead plant and animal fragments or excreta from all trophic levels. Bacteria and fungi are important microscopic examples of detritus feeders. Water fleas, mosquito larvae, mussels, barnacles, mites, and springtails are among the larger members of these scavengers which clear up the habitat.

We must remember also that living organisms require a supply of food so that they can get energy for their various metabolic processes. This energy is stored in the food by the green plant during photosynthesis, and hence all non-green plants and animals are dependent upon green

CIRCULATION OF CARBON

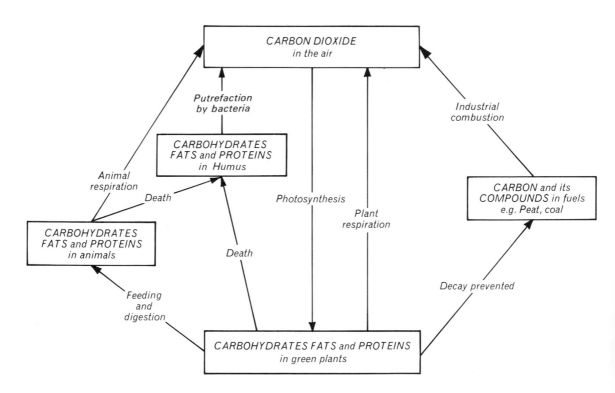

Fig. 8.9 The circulation of carbon between living organisms and the carbon dioxide of the air. The chemical compounds involved are shown in capital letters and the names of the processes in smaller type.

plants for their supply of energy. This energy is released from the food during respiration, and for this a supply of oxygen is also required. Once again we find that animals are dependent upon green plants for maintaining a constant supply of oxygen in the atmosphere, for they release it as a waste product during photosynthesis.

The process of photosynthesis removes carbon dioxide from the air, but the amount does not decrease. This is partly because some is released from the plant in respiration, but most of it is built up into the tissues of the plant. These may be eaten by animals and the carbon returned to the air once more as carbon dioxide during respiration. There may be considerable delay before the carbon dioxide is returned to the air, as very large quantities of carbon are 'locked up' in the huge forests of two hundred to three hundred million years ago, which have been buried and slowly converted to coal. This carbon is returned to the air when the coal is burnt. When plants and animals die their bodies decay owing to the activities of the numerous saprophytic organisms, such as bacteria and fungi, which feed on them, and in so doing release carbon dioxide to the air. This circulation of carbon between the carbon dioxide of the air and the tissues of living organisms is known as the **carbon cycle** and can be summarised as a diagram shown in Fig. 8.9.

QUESTIONS 8

1. Distinguish between the parasitic and the saprophytic modes of life. Give an illustrated account of the life-history of either a named parasite or a named saprophyte.
[L]

2. (a) Describe the life-history of a named animal parasite such as the liver fluke or tapeworm. (b) What features enable this animal to live successfully as a parasite? (c) How may the spread of this animal be prevented? [S]

3. (a) What is meant by (i) symbiosis and by (ii) parasitism? (b) Give an example of each. (c) Select a parasitic plant or animal, and explain in words, or suitably labelled drawings, the ways in which the organism is specially adapted for a parasitic life. (Describe its shape, how it attaches itself to its host, how it feeds, and its method of reproduction.) [S]

4. Give concise accounts of the modes of nutrition of a green plant, a saprophytic fungus, and a named mammal. How are the three methods interdependent? (Details of the alimentary canal of the mammal are not required.) [L]

5. Show how a tapeworm appears to be suited to its mode of life. Explain how a knowledge of the life-history of such a parasite often makes it possible to limit its distribution. [L]

9 Respiration

Respiration, in its unrestricted sense, occurs in the cytoplasm of every living thing, and includes all the processes involved in liberating energy for the use of the organism. The energy, which has been stored in substances such as starch, glycogen, fats, and proteins present in the cytoplasm, is liberated by their chemical breakdown, a process which usually requires free oxygen. It is convenient to consider respiration in two parts:

1. The chemical breakdown or oxidation of sugar, in which energy is liberated. This is sometimes known as internal respiration.

$$C_6H_{12}O_6 + 6O_2 \xrightarrow{\text{Enzymes}} 6CO_2 + 6H_2O + \text{Energy}$$

Sugar   Oxygen   Carbon   Water   Energy
dioxide

The breakdown of sugar, or any other energy-producing substance, in respiration takes place in several successive changes, each one carried out by an enzyme. Thus the summary above indicates only the 'raw materials' used and the end products; it gives no information about the intermediate stages in the breakdown of sugar. It is only within the cytoplasm of living cells that this type of oxidation occurs.

2. The part known as the gaseous exchange, whereby the respiring cells of an organism obtain oxygen and get rid of carbon dioxide. This is sometimes known as external respiration.

RESPIRATION IN PLANTS

9.1 Detection

One method of demonstrating plant respiration by detecting the carbon dioxide produced was described in chapter 1 (see Fig. 1.4). Another quicker method is to pump or draw a stream of carbon-dioxide-free air through a chamber containing the plant, and then through a U-tube of lime-water (see Fig. 9.1). The soda-lime tower is an efficient absorber of carbon dioxide, and the U-tube next to it containing lime-water acts as a check that the soda-lime is doing its job. If the plant material is green it is necessary to cover the bell jar with a lightproof cloth so that photosynthesis will not interfere with the respiratory exchange. For the purpose of this experiment it is necessary to ignore the small amount of carbon dioxide which was in the bell jar when set up, unless one takes the special precaution, before connecting the second U-tube, of pumping air through the apparatus until it is completely flushed out with air which has been through the soda-lime tower. If a potted plant is used, a control experiment should be set up without the plant, but with a pot of soil similar to that which contains the plant. The results of both experiments should be considered before reaching any conclusion.

DETECTION OF CARBON DIOXIDE

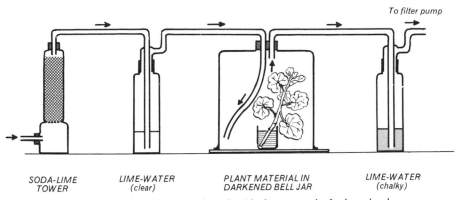

SODA-LIME TOWER   LIME-WATER (clear)   PLANT MATERIAL IN DARKENED BELL JAR   LIME-WATER (chalky)

FIG. 9.1 Apparatus for the detection of carbon dioxide from a cut leafy shoot by the gas-stream method.

Some of the energy set free in respiration appears as heat, which suggests another way of detecting it. If the organism is kept in a heat-insulated chamber, such as a vacuum flask, a rise in temperature is noticed. Fig. 9.2 shows how this can be done. Pea seeds are suitable material for this experiment, since they are easily introduced into the flask without damage. Vacuum flasks are not perfect heat insulators, so it is essential to carry out a control experiment with the same number of seeds which have previously been killed by being soaked in 10 per cent formalin solution (Flask C). The dilute solution of Milton or Chloros does not kill the seeds but it does destroy the micro-organisms on their surface (Flask B). Fig. 9.3 shows the result of such an experiment. Why is the temperature rise in flask B less than that in flask A?

HEAT PRODUCTION

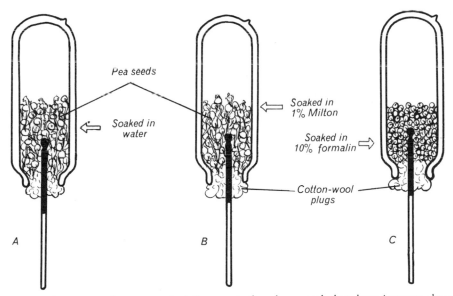

FIG. 9.2 The pea seeds were soaked for twenty-four hours and then kept between damp blotting paper for twenty-four hours to start germination before being placed in the flasks.

HEAT PRODUCTION IN SEEDS

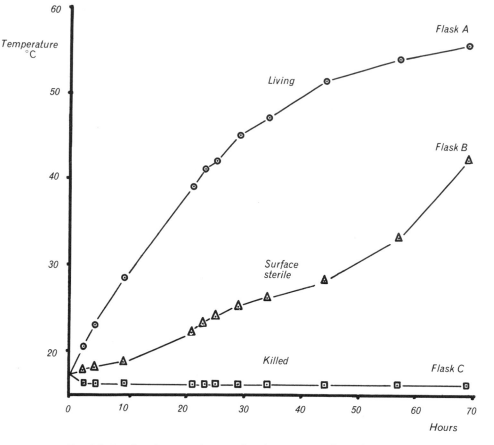

FIG. 9.3 Results of an experiment using the apparatus shown in Fig. 9.2.

A familiar example of organisms producing heat is a heap of grass mowings which, after a few days, is found to be quite warm at the centre. This is due to the heat given off in the respiration of certain bacteria which thrive under these conditions.

9.2 Exchange of Gases

How do all the cells of a flowering plant, especially those in the centre of thick roots and stems, get an adequate supply of oxygen; and how is carbon dioxide prevented from accumulating in these places? The answer to these questions lies in the presence of a continuous network of tiny air spaces which pervades the whole plant, and is particularly well developed in the leaves. This system of air spaces, and the way in which the gases tend to distribute themselves evenly in it by diffusion, was mentioned in section 6.9, but then we were concerned only with the gaseous exchange of photosynthesis.

In the dark, when respiration alone is taking place, there will be less oxygen in the air spaces than outside the plant, and, as a result of the concentration gradient set up, the oxygen will diffuse inwards through lenticels (see sec. 11.10), and even through the walls of the epidermal cells to a very small extent. (The stomata, which are usually closed at night, will not allow gases to pass as they do when open.) Conversely, carbon dioxide, which is being produced in the plant, will diffuse outwards.

In the light, however, oxygen will be produced in the green cells of the plant by photosynthesis, and this gas will be used in respiration. In the same way, the carbon dioxide from respiration will be used in photosynthesis to build up carbohydrates. Which gas is taken in by the plant, and which is given off, will depend on the relative rates of the two processes. For instance, in bright sunlight there will not usually be enough respiratory carbon dioxide in the leaves to supply the needs of photosynthesis, and there will be an excess of oxygen above that required for respiration. Oxygen will, therefore, be given out and carbon dioxide absorbed. Fig. 9.4 summarises these gas exchanges.

9.3 Energy Production and Use

Many flowering plants store food in the form of insoluble starch and, as we have seen, this appears in the leaves during periods of sunlight, later to be transported elsewhere for more permanent storage, e.g. to the tubers of a potato plant, or to the developing seeds of wheat. Starch is often used as an energy-producing fuel, but first it has to be converted into soluble sugar. In addition to starch, fats, stored in the seeds of the sunflower and castor-oil plants, are broken down to provide energy, but it is doubtful whether proteins are used in this way, except during periods of starvation.

The uses to which the energy is put are not at all obvious, since they are mostly included in the building-up processes of growth. Thus a plant, in order to make more cellulose for cell walls, and more proteins for cell contents, requires a source of energy. Plant roots also require energy to absorb mineral salts from the soil. It is from respiratory breakdown that both these needs are met.

9.4 Respiration and Photosynthesis compared

Both respiration and photosynthesis are vital complex metabolic changes which occur within plant cells, but they are often confused because both involve an exchange of gases. The most important differences are set out below:

Photosynthesis	Respiration
Takes place only in light in green cells of plants	Occurs at all times in all living cells
Carbon dioxide used, oxygen produced	Oxygen used, carbon dioxide produced
Light energy 'stored' in carbohydrates built up	Energy liberated from the breakdown of carbohydrates
Plant gains weight	Plant loses weight
Chlorophyll essential	Chlorophyll plays no part

ANAEROBIC RESPIRATION

9.5

Not all organisms require a source of free oxygen to support life, in fact some are unable to live in the presence of the gas. Such organisms are said to respire anaerobically. Pea seeds show some

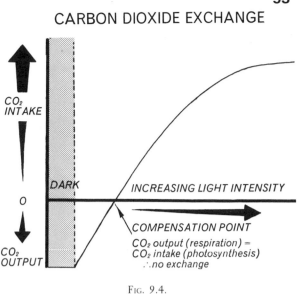

CARBON DIOXIDE EXCHANGE

FIG. 9.4.

capacity for existence without oxygen, as can be demonstrated by the following experiment (see Fig. 9.5).

Fill a small test-tube with mercury and invert it in a shallow dish also filled with mercury, making sure not to include any bubbles of air. Then remove the testas from two soaked pea seeds, and introduce the latter into the tube by means of a pair of curved forceps. They come to rest at the top of the tube, completely immersed in mercury, and therefore without any gas supply around them. The removal of the testas ensures that no air is trapped between them and the cotyledons of the seeds. After one or two days in the dark a small volume of gas appears at the top of the tube, having displaced the mercury. That this is carbon dioxide can be shown by inserting a moist pellet of caustic potash into the tube, when the mercury will rise owing to absorption of the gas. It is reasonable to deduce from this that pea seeds can respire without oxygen, but they do not grow normally under these conditions and would eventually die.

Yeast (see sec. 8.1) is also capable of anaerobic respiration even though it can, and usually does, live where oxygen is present. When little or no oxygen is present it is able to break down sugar to alcohol and carbon dioxide with the release of some energy, a process known as **fermentation**.

$$C_6H_{12}O_6 \xrightarrow{\text{Enzymes}} 2C_2H_5OH + 2CO_2 + \text{Energy}$$

Sugar · · · · · · · · Alcohol · · · · Carbon dioxide · · · · Energy

This process is catalysed by zymase, which is now

ANAEROBIC RESPIRATION

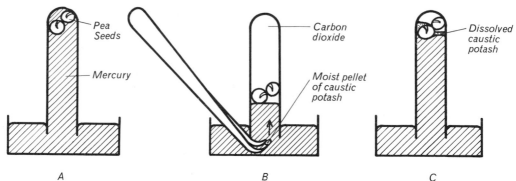

Fig. 9:5 A—Soaked pea seeds are stripped of their testas and introduced into the inverted tube of mercury. B—After two days a gas liberated from the peas has pushed the mercury down. C—After introducing a moist pellet of caustic potash as in B, the mercury level rises slowly almost to the top of the tube. This shows that the gas produced is mostly carbon dioxide.

known to be a complex of several enzymes. A simple demonstration of fermentation is illustrated in Fig. 9.6. The suspension of yeast in 10 per cent sugar solution, when kept warm over a radiator, soon froths as bubbles of gas are given off. The first bubbles to issue from the delivery tube cause no chalkiness in the lime-water, since they are mostly air which was included in the flask and tube when the apparatus was set up. Soon, however, a chalky precipitate in the lime-water indicates that carbon dioxide has been produced. Although yeast can respire in the complete absence of oxygen, it is not well fitted for these conditions, since in them it fails to bud and grow.

Fermentation by yeast is made use of commercially in wine making, brewing, and baking. In

FERMENTATION BY YEAST

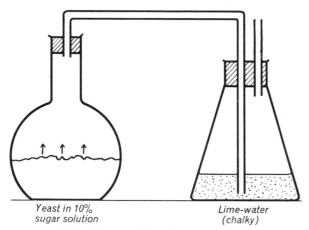

Yeast in 10%
sugar solution

Lime-water
(chalky)

Fig. 9.6 If the flask containing the yeast is put in a water bath at 25°C a result will be obtained more quickly.

brewing, a liquid obtained from germinating barley grains is boiled with hops, cooled, and then fermented by yeasts. The resulting solution contains alcohol formed from the sugar dissolved out of the barley grains; beer is produced from this by clearing the solution of yeast and other suspended particles. To make it appear attractive when poured, carbon dioxide is dissolved in it under pressure whilst it is being bottled. The same principle is used in making wines, but here wild yeasts from the skin of the fruits are usually used in the fermentation instead of adding cultivated yeasts, as in brewing. In baking, yeast is mixed with a little sugar and added to the flour and water to make a heavy dough, which, when kept warm, increases in volume owing to bubbles of carbon dioxide being produced by the yeast's fermentation. These bubbles persist in the bread after baking, which causes it to be light and palatable.

There is increasing evidence that many small worms and crustaceans can live in the mud of ponds and lakes, where they may be without oxygen for long periods. Certainly there is no doubt that individual animal tissues such as muscle can respire anaerobically. This happens when an oxygen debt is incurred in strenuous exercise, sugar this time being broken down, not to alcohol as in yeast, but to lactic acid (see sec. 13.7).

RESPIRATION IN MAMMALS

When the thorax of a mammal is expanding and contracting rhythmically, the organism is taking into its lungs air which contains the normal percentage of oxygen, and then breathing out air which has less oxygen and more carbon dioxide than that which was taken in. The blood in the lungs absorbs some of the oxygen in the air there, in exchange for carbon dioxide which it gives up. The oxygen it carries to all other parts of the body, where it loses it and picks up more carbon dioxide, only to reverse the process again on its return to the lungs. There is obvious evidence that this gaseous exchange is connected with energy production, since, during violent exercise, when much energy is being expended, breathing is much faster and deeper, and the body temperature may rise a little. These facts we must examine in greater detail.

9.6 Breathing Mechanism

The lungs of mammals are paired spongy structures, protected by the bony cage made by the ribs, sternum, and vertebrae (see sec. 13.2). They are composed of millions of microscopic air sacs or **alveoli** whose walls are thin and moist, and cover a huge area. It is through these walls that the oxygen and carbon dioxide pass between the air in the lungs and the blood. Tiny elastic tubes or **bronchioles** lead from the alveoli and unite to form larger and larger tubes, until one large tube, a **bronchus**, issues from each lung. The **trachea** connects both bronchi with the back of the mouth cavity (see Fig. 7.6), and is supported by incomplete hoops of cartilage, which help to prevent its closure under external pressure. Nearest the mouth are four special cartilages which form the **larynx**, stretched across the cavity of which are the vocal cords.

The lungs do not lie freely in the thorax, but each is covered by two thin, transparent elastic sheets known as **pleurae** or pleural membranes (see Fig. 9.7). The outer one on each side lines the inside of the thorax, and the inner one covers the lungs. Thus each of the two lungs is in a separate pleural cavity; but these cavities do not usually contain any air. The inner and outer pleurae are thus in contact, and, being moist, glide over one another as the lungs fill and empty of air.

The mammalian thorax is completely air-tight, being closed posteriorly by the muscular **diaphragm**, and thus, if its volume is increased the internal pressure will be reduced. However, the lungs are open to the air outside, so atmospheric pressure will cause air to fill the lungs until the pressures inside and outside become equal. Careful

THORAX

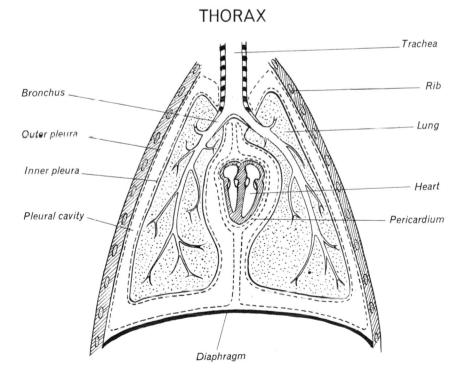

Fig. 9.7 A diagrammatic section through the thorax of a mammal. The pleural cavities are shown much enlarged and the heart much reduced.

use of the thorax models (see Fig. 9.8) will make this quite clear. When the volume of the thorax is decreased, the reverse process occurs and air is expelled from the lungs.

There are two ways in which the volume of the thorax is increased, thus allowing the lungs to be filled with air:

1. By contraction of the muscles of the diaphragm, which causes it to flatten, so pressing the stomach and liver further down into the abdomen.

2. By contraction of the inspiratory intercostal (*L. inter, between; costa, rib*) muscles pulling the lower ribs, which pivot at the vertebrae, upward and outward in man (outward and forward in the rabbit).

The thorax model A can imitate only the first of these movements. Fig. 9.9 shows the position of the ribs and diaphragm after expiration and inspiration.

Breathing out is normally carried out by:

1. Relaxation of the inspiratory intercostal muscles, which allows the ribs and sternum to fall under gravity and reduce the volume of the thorax.

2. Relaxation of the diaphragm muscles whilst

THORAX MODELS

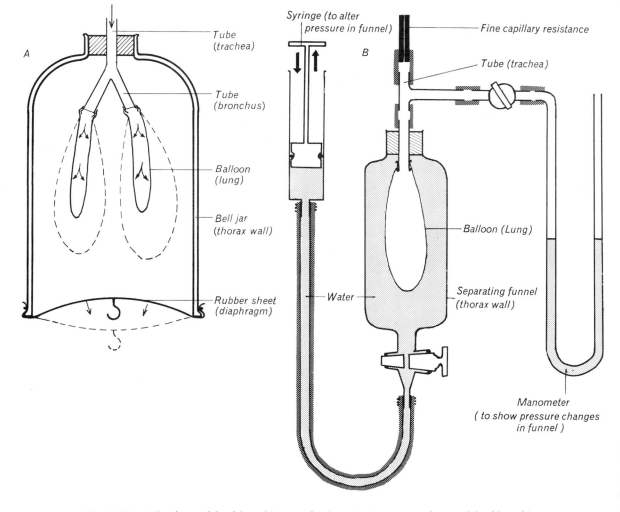

FIG. 9.8 A—Simple model of breathing mechanism. B—More complex model of breathing mechanism. Model A can be used to imitate the action of the diaphragm. In model B a syringe is used to change the pressure in the 'thorax'. Which is the more accurate model of the breathing mechanism?

BREATHING MECHANISM

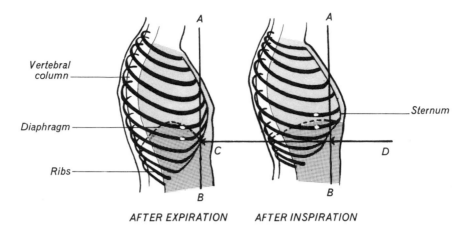

AFTER EXPIRATION   AFTER INSPIRATION

Fig. 9.9 Side views of human thorax at the end of expiration and inspiration. Compare the positions of the ribs, sternum, and diaphragm with the reference lines AB and CD. A white dot is placed on ribs 6 and 7 in each diagram to make comparison easier.

those of the abdominal wall are contracting and pushing the stomach and liver up into the diaphragm, which arches into the thorax so reducing its volume. (It is possible to breathe in when these abdominal muscles are contracting.)

3. The elasticity of the lungs which causes them to expel air.

During forced expirations, such as happen when a person coughs, another set of intercostal muscles contracts, pulling the ribs downward and inward, thus helping to expel air from the lungs.

A resting person may pass 15 000 litres of air in and out of the lungs during one day. It is not surprising, therefore, that the body has several mechanisms which prevent air-borne dust and

GASEOUS EXCHANGE

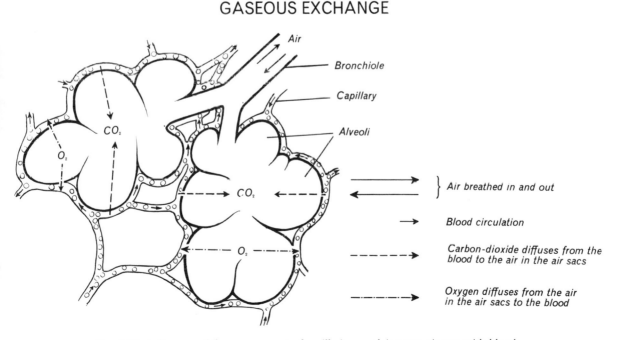

Fig. 9.10 A diagram of the arrangement of capillaries supplying two air sacs with blood.

other solid particles from accumulating in the air sacs.

1. The tortuous nasal passages are lined with a layer of mucus (slime), which collects dust particles and bacteria from the air taken in, and is propelled back by cilia into the mouth, where it is swallowed. In cats, dogs, and rabbits, this mucus is propelled in the opposite direction, making the nose cold and wet to touch. The very large surface of the nasal passages is well supplied with blood, and has the important functions of warming and moistening inspired air.
2. The trachea, bronchi, and bronchioles are also lined with this layer of mucus, which is propelled up into the mouth and works in the same way as that in the nose.
3. The **epiglottis**, a cartilaginous flap, closes off the trachea from the back of the mouth during swallowing, a movement that usually prevents any food from entering this tube (see Fig. 7.6).
4. If large particles do somehow enter the trachea, they cause coughing, which may be successful in removing them.

If analyses of expired and inspired air are compared, it will be seen that, in passing through the lungs, the air has lost about a third of its oxygen and picked up carbon dioxide and water vapour. The oxygen has been absorbed into the blood running through the innumerable tiny vessels or capillaries on the walls of the air sacs (see Fig. 9.10), whilst the carbon dioxide has passed in the reverse

| | Composition in Volumes per cent | |
	Inspired Air	Expired Air
Oxygen	20.7	14.6
Carbon dioxide	0.04	3.8
Water vapour	1.25	6.2
Nitrogen	78.0	75.4

direction. You may be surprised to know that the internal surface of the alveoli if opened and flattened would form an area about the size of a tennis court.

This gaseous exchange, involving, as it does, the blood and its circulation, leads us to consider these two topics in detail before we can understand respiration as a whole in mammals.

9.7 Structure and Composition of the Blood

If fresh blood, to which a little sodium citrate solution has been added, is allowed to stand for an hour or more, it will separate into a red lower

layer and a pale straw-coloured upper layer. (This separation can be speeded up by spinning for a short time in a centrifuge.) In the red layer are cells or **corpuscles** of two types, the red ones and the white ones, the former being much more numerous than the latter. The upper layer is the **plasma**, or fluid part of the blood, in which the corpuscles are suspended. The composition of the plasma and some facts about the corpuscles are given on p. 99.

When a drop of blood is examined microscopically, it is an easy matter, given a little practice, to distinguish between the two types of cell. Most obvious are the red cells (see Fig. 9.11), which appear faint yellow when seen separately; they

BLOOD CORPUSCLES

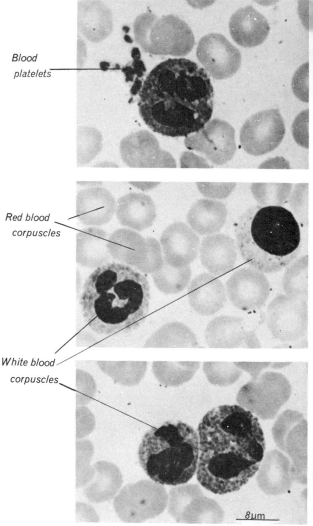

Blood platelets

Red blood corpuscles

White blood corpuscles

8μm

FIG. 9.11 Photomicrograph of stained human-blood smears.

COMPOSITION OF BLOOD

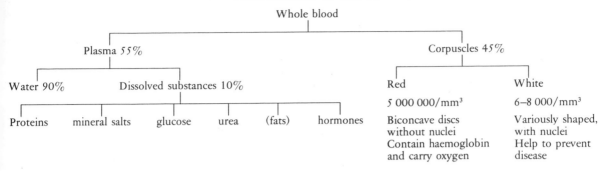

are disc-shaped and are slightly concave on both sides, which causes the centre of the cell to appear paler than the outer region. The red cells are without a nucleus and have, in their cytoplasm, the iron-containing pigment **haemoglobin**, which gives the blood its red colour. It is calculated that they last for only 120 days in the blood before they are replaced by new ones made in the bone marrow. Their short life is, perhaps, not surprising when one remembers that they are repeatedly shot out of the heart at high speeds, and then squeezed through the tiny capillary blood vessels.

The white corpuscles have to be searched for on account of their relatively small numbers. Three facts help in finding them; firstly their shape is irregular, secondly they possess nuclei of various shapes, and thirdly they tend to stick to the glass on which they are mounted. They contain no haemoglobin, and some are able to ingest foreign matter such as bacteria in the same way as Amoeba feeds. In section 19.4 the way in which they protect the body from disease is described.

Blood platelets are the smallest bodies suspended in the plasma. They play a part in the clotting of blood over a wound, but their origin and precise functions are uncertain.

The composition and functions of blood are summarised in the tables.

9.8 Carriage of Oxygen and Carbon Dioxide

Blood reaching a lung capillary on the wall of an alveolus has its lowest concentration of oxygen. The oxygen concentration in the air of the alveolus is greater than this, so that an oxygen **concentration gradient** is set up through the very thin walls of the blood capillary and the alveolus. This causes oxygen to diffuse 'down the gradient' and so pass into the blood. The reverse is true for carbon dioxide. Blood reaching a lung capillary has previously absorbed carbon dioxide from

FUNCTIONS OF THE BLOOD

1. TRANSPORT

Transport of	From	To	How carried	See Sec.
Oxygen	Lungs	All tissues	As oxyhaemoglobin in red corpuscles	9.8
Carbon dioxide	All tissues	Lungs	As bicarbonates in solution in plasma	9.8
Urea	Liver	Kidney	In solution in plasma	10.1
Products of digestion	Small intestine	All tissues	In solution in plasma (fat droplets in suspension)	7.9
Hormones	Ductless glands	All tissues	In solution in plasma	15.6
Heat	Produced in all tissues (especially in liver and muscle) and distributed evenly		In all parts of the blood	10.2

2. DEFENCE AGAINST DISEASE (see sec. 19.4)

Production of antitoxins by white corpuscles.
Ingestion of bacteria by white corpuscles.
Formation of blood clots, which seal wounds against bacteria.

respiring cells somewhere in the body, and it has a higher concentration of carbon dioxide than the air taken into an alveolus. The gas therefore diffuses 'down the concentration gradient' into the air in the alveolus.

In all tissues other than the lungs, the opposite exchange of gases to those described above occurs. These are summarised below.

In tissues other than lungs:

Concn. O_2 in blood > Concn. O_2 in tissue cells

$$\therefore \quad O_2 \xrightarrow{\text{diffuses}} \text{Tissue cells}$$

Concn. CO_2 in tissue cells > Concn. CO_2 in blood

$$\therefore \quad CO_2 \xrightarrow{\text{diffuses}} \text{Blood}$$

The blood transports oxygen and carbon dioxide efficiently because it carries them not only as gases dissolved in the plasma, but also in the form of chemical compounds. Oxygen combines with the haemoglobin in the red corpuscles to form **oxyhaemoglobin**, and carbon dioxide combines with water in the plasma to produce **bicarbonates** as shown below:

$$\underset{\text{(purplish red)}}{\text{Haemoglobin}} \quad \underset{-O_2}{\overset{+O_2}{\rightleftharpoons}} \quad \underset{\text{(bright red)}}{\text{Oxyhaemoglobin}}$$

$$\underset{\substack{\text{Water Carbon} \\ \text{dioxide}}}{H_2O + CO_2} \rightleftharpoons \underset{\substack{\text{Carbonic} \\ \text{acid}}}{H_2CO_3} \rightleftharpoons \underset{\substack{\text{Hydrogen} \\ \text{ion}}}{H^+} + \underset{\substack{\text{Bicarbonate} \\ \text{ion}}}{HCO_3^-}$$

When haemoglobin takes up oxygen, it changes from a purplish red colour to the bright red of oxyhaemoglobin, and vice versa. This colour difference is noticeable when the arteries and veins of a freshly killed mammal are examined. It can be demonstrated with fresh diluted blood, which when shaken with air shows the bright red colour of oxyhaemoglobin. If a few minute crystals of sodium hydrosulphite are added, the oxygen is removed from the oxyhaemoglobin and the solution turns to a purplish red colour. Provided not too many crystals were added, the bright red colour will return on shaking with air.

The volumes of the two gases carried by human blood are given in the table below.

Apart from the carriage of respiratory gases, the blood has many functions which are summarised in the table.

It is interesting to note that haemoglobin is not confined to vertebrates: it is present in water-fleas, earthworms, insect larvae known as 'blood-worms', and some pond snails, but in these invertebrates it is not present in corpuscles.

9.9 The Circulation of the Blood

Without an efficient method of circulating the blood round a mammal's body the power to carry the gases would be of little use to the animal. In section 4.2, the heart, arteries, veins, and capillaries were mentioned as a closed system of tubes making up the **double circulatory system** of a mammal.

The heart, which is covered by a thin, transparent membrane called the **pericardium**, is a muscular pump, inside which are four separate cavities and a series of valves permitting blood to pass in one direction only. The two anterior chambers of the heart (**auricles**) receive blood conveyed to them in veins, which are thin-walled tubular vessels that show clearly the dark red colour of the blood they carry. The two posterior chambers (**ventricles**) are bigger and have more muscle in their walls than the auricles. They pump the blood out along thick-walled arteries which have a pale pink appearance because the bright red colour of the blood they carry does not show clearly through the walls. The **pulmonary arteries and veins**, which convey blood to and from the lungs, are exceptions to the colour difference between the two sets of vessels.

Whilst reading the following description of the circulatory system you should make frequent reference to Figs. 9.12, 9.13, and 9.14.

The paired pulmonary veins convey oxygenated blood from the lungs back to the left auricle of the heart (on the right of the diagram), which contracts when full, forcing its blood through the bicuspid valve (mitral valve) into the large left ventricle, whose muscular wall relaxes accommodating the blood. The bicuspid valve is composed of two flaps attached by string-like tendons to the muscle of the ventricle wall. When the ventricle of this side contracts, the valve flaps are forced upward and together, thereby preventing

Types of Blood	Total CO_2 per 100 cm³ Blood	Total O_2 per 100 cm³ Blood	Colour
Oxygenated blood (blood leaving the lungs)	53 cm³	19 cm³	Bright red
Deoxygenated blood (blood leaving parts other than the lungs)	58 cm³	13 cm³	Purplish red

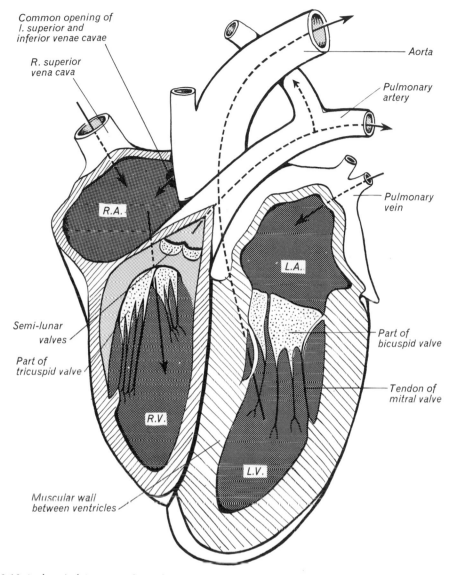

Common opening of
l. superior and
inferior venae cavae

R. superior
vena cava

Aorta

Pulmonary
artery

R.A.

Pulmonary
vein

L.A.

Semi-lunar
valves

Part of
bicuspid valve

Part of
tricuspid valve

Tendon of
mitral valve

R.V.

L.V.

Muscular wall
between ventricles

FIG. 9.12 A sheep's heart seen from the ventral side after cutting away the ventral walls of all four chambers. Part of the tricuspid valve is hidden by the base of the common pulmonary artery.

any backflow into the auricle, and the blood is pumped out through the biggest artery of the body (the **aorta**). Branches from this lead to the head, limbs, alimentary canal, kidneys, and the muscular wall of the heart; in fact, to all parts except the lungs. From all these parts deoxygenated blood is collected into three main veins, two **superior** (anterior) **venae cavae**, and one **inferior** (posterior) **vena cava**, which all open into the right auricle. This chamber contracts in unison

with the left auricle, driving blood into the right ventricle through the **tricuspid valve**, the latter being similar to the bicuspid valve except for the possession of an extra flap. When the right ventricle contracts in unison with its 'partner', blood is pumped along the two pulmonary arteries to the lungs where it is once again oxygenated. Thus the blood goes through the heart twice before it can return to any particular spot, an arrangement which is simply described by the term 'double

DOUBLE CIRCULATION

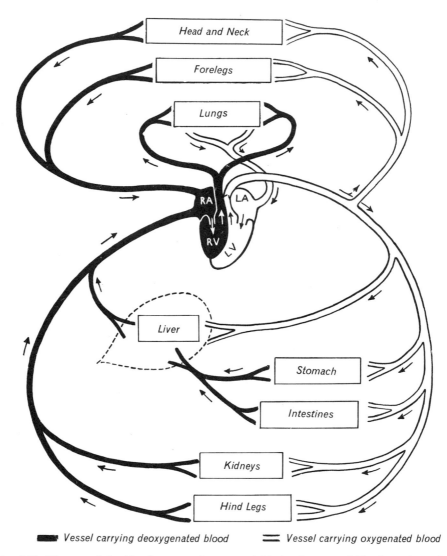

▬ *Vessel carrying deoxygenated blood*	═ *Vessel carrying oxygenated blood*

FIG. 9.13 Diagram of the blood system of a mammal. Notice the unusual blood supply to the liver, which comes in an artery and a vein. Note that the right side of the heart is on the left since this is a ventral view. Compare this with Fig. 9.14.

circulation'. Valves are present both in the veins (see Fig. 9.15) and where the two big arteries leave the ventricles. They resemble pockets, with the openings directed away from the heart in the case of the arteries, and towards the heart in the case of the veins.

Capillaries, the tiniest of blood vessels, form a complicated network in all parts of the body, and provide a pathway for the blood from the smallest arteries to the smallest veins. It is during passage through these vessels that the blood comes in intimate contact with the respiring cells, and is

able to give up its oxygen and absorb carbon dioxide. The reverse is, of course, true for the lung capillaries, a photomicrograph of which is at the head of this chapter. Capillaries are contractile, that is they are able to close and cut off the blood flow in a localised region, or open widely and increase it. This provides some explanation of such common everyday experiences as blushing, and facial pallor due to fear or shock.

William Harvey (1578–1657), the first man to realise that blood moves 'as it were in a circle', published his work in 1628, and it had the pro-

BLOOD VESSELS OF A RABBIT

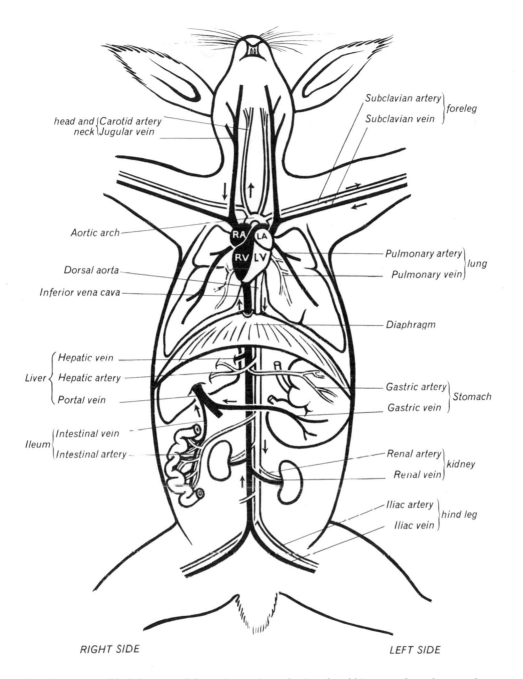

head and {Carotid artery
neck {Jugular vein

Subclavian artery ⎫
                  ⎬ foreleg
Subclavian vein   ⎭

Aortic arch

RA LA
RV LV

Pulmonary artery ⎫
                 ⎬ lung
Pulmonary vein   ⎭

Dorsal aorta

Inferior vena cava

Diaphragm

Liver { Hepatic vein
Liver { Hepatic artery
      { Portal vein

Gastric artery ⎫
               ⎬ Stomach
Gastric vein   ⎭

Ileum { Intestinal vein
      { Intestinal artery

Renal artery ⎫
             ⎬ kidney
Renal vein   ⎭

Iliac artery ⎫
             ⎬ hind leg
Iliac vein   ⎭

RIGHT SIDE

LEFT SIDE

FIG. 9.14 A simplified drawing of the main arteries and veins of a rabbit as seen from the ventral side. Vessels carrying deoxygenated blood are shown black, those carrying oxygenated blood are shown white. Liver lobes on the animal's left side have been removed and only two portions of the gut are shown. The common pulmonary artery disappears dorsally to the heart after leaving the right ventricle.

VALVES

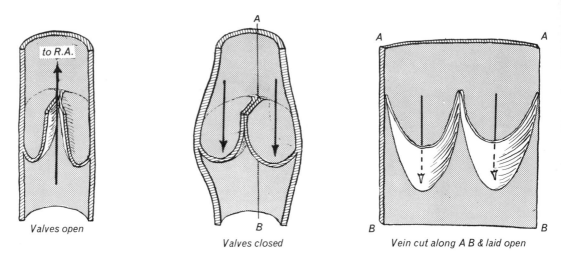

to R.A.

Valves open

A

B

Valves closed

A

B

A

B

Vein cut along A B & laid open

FIG. 9.15 Diagram to show the movement of valves in a vein.

found effect of stimulating a trend of biological thought which is still profitable today, namely to interpret biological activities in terms of physics and chemistry. Although Harvey never saw blood passing from the small arteries to the small veins, the presence of capillaries was inferred by him. Nowadays, the capillary circulation can be seen with a microscope, and what a beautiful sight it is! It is easy to demonstrate in embryo guppies or zebra fish, in the web of a frog's foot, and in the external gills or tail fin of a tadpole, especially if one uses an anaesthetic (1 per cent urethane for the frog, 1 per cent chlorbutol solution for the tadpole) to help keep them still.

Thus the arrangement of the blood vessels, the pumping action of the heart, and the ability of the blood to carry and give up oxygen and carbon dioxide, provide an efficient means of supplying every cell of a bulky mammalian body with oxygen, and removing its carbon dioxide.

9.10 Energy Production and Use

Once oxygen has diffused into the living cells from the blood it can be used to break down the complex organic compounds (fats, carbohydrates, and proteins) which contain far more energy than the carbon dioxide and water produced from them. Thus, when oxidation occurs energy will be set free. Fats and sugars are most commonly employed as energy sources by man, but if there is an excess of protein in the diet, or if all the fat and carbohydrate reserves are used up as in starvation, proteins can be used. This energy freed by oxida-

tion is probably not used immediately by the body. Instead it may be used to change ADP (adenosine diphosphate) to ATP (adenosine triphosphate), which is an 'energy-rich' compound that can provide the body with a source from which energy can be obtained quickly. The store of ATP is replenished by the oxidation of foods.

The energy liberated is used in the following ways:

1. To maintain a constant body temperature (37°C in man) higher than that of the environment.
2. To build up proteins required for making new cells in growth and tissue repair.
3. To produce the muscular movements of vital activities such as breathing, the pumping action of the heart, and the propulsion of food along the alimentary canal.
4. To convey 'messages' along nerve cells from one part of the body to another.
5. To produce non-vital muscular movements involved in running, walking, bending down, etc.

In the vital processes alone, an adult man uses up 7.5 MJ (1 800 kcal) per day, and so, if he is not to lose weight, food which can provide this amount of energy must be eaten by him. This minimum energy requirement is reached only when the person is at rest in bed all day, and any external work done will make energy demands in addition to this minimum.

A normal person carries quite a lot of reserve energy in the fat laid down in the cells under the skin and round the kidneys, and in the form of glycogen in the liver and muscle cells. It is the fat stores which are usually the more extensive, which is not surprising, since one ounce of fat can, when oxidised, produce over twice as much energy as the same weight of a carbohydrate such as glycogen. The proteins of living cells are best considered as part of the essential structure of the body, and not as reserves of energy, even though they can be used as such in the event of starvation.

RESPIRATION IN OTHER ANIMALS

The energy-liberating processes which take place within living cells are believed to be fundamentally similar in all organisms, but the methods of gaseous exchange vary a great deal.

9.11 Invertebrates

In Amoeba (see sec. 2.1) no part of the cytoplasm is so far from the pond water that it cannot get oxygen fast enough by direct diffusion from there; neither does it produce carbon dioxide so quickly that diffusion cannot keep its concentration down to a tolerable level in the animal. Its gaseous exchange thus goes on through the whole body surface.

Hydra (see sec. 2.9), although much bigger than Amoeba, relies on diffusion for its gaseous exchange. This is facilitated by the arrangement of cells in two layers round a central body cavity, so that no cell is far from the pond water where there will usually be inexhaustible supplies of oxygen.

It is noteworthy that in the higher animals, which are generally bulkier than Amoeba and Hydra, there exist special organs or systems, some of which help in the exchange of gases with the surroundings by providing a large wet respiratory surface for diffusion, and some of which are concerned in the transport of the gases about the animal body.

Earthworms show this sort of specialisation, since they possess haemoglobin in their blood, which circulates in closed vessels propelled by five pairs of pseudohearts (see Fig. 13.16). The moist skin contains a well-developed network of capillaries where the exchange with the exterior occurs.

Insects are remarkable in that they have a series of openings along the sides of the body, from which lead a branching system of tubes (**tracheal tubes**). A pair of these openings, or **spiracles** (see Fig. 18.7), is commonly found in most segments of an insect, and they are often equipped with a closing device. Inside the insect these tubes branch extensively until the finest endings, which contain

GILLS

A

Water in

Water out

Operculum

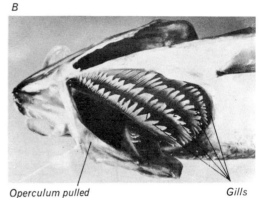

B

Operculum pulled away to expose the gills

Gills

Fig. 9.16 A—Photograph of the head of a herring from the left side. B—Photograph of the ventral view of the head of a herring.

a liquid, are amongst, or even within, the respiring cells. Oxygen and carbon dioxide diffuse along the tracheal tubes to and from the body cells, although in some insects, particularly after severe exercise, a rhythmic 'bellows' type of movement may be observed in the abdomen. This movement promotes rapid ventilation of the tracheal system.

9.12 Fish

Fish carry out their gaseous exchange by means of **gills** (see Fig. 9.16) which lie between gill slits. The gill slits, of which there are five pairs as a rule, are paired lateral clefts which perforate the body right through to the cavity of the alimentary canal in the pharyngeal region (see Fig. 9.17). The arch of the body between adjacent gill slits is slender in bony fish, and is produced into a large number of fine projections (gill filaments), which are red because of their copious blood supply. There are four pairs of gills (one between each pair

FISH — BREATHING MECHANISM

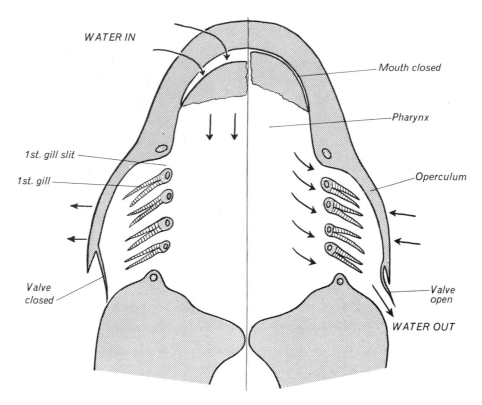

WATER IN

Mouth closed

Pharynx

1st. gill slit

1st. gill

Operculum

Valve
closed

Valve
open

WATER OUT

Fig. 9.17 Diagrammatic section through the head of a bony fish to show the breathing mechanism. The inhalant current of water entering the mouth is shown on the left, and the exhalant current coming out of the branchial aperture is shown on the right.

of slits) in bony fish, each made up of filaments which present a very large surface over which oxygen can diffuse into the blood from the water, and over which carbon dioxide can diffuse in the opposite direction. The gills are extremely delicate and are protected externally by a hard plate, known as the **operculum**, which extends posteriorly at the side of the body from in front of the first gill slit, thus hiding the gills (see Fig. 9.16). You should examine a fairly large fish (herrings are quite suitable) by lifting the operculum and separating the gills, and then open the mouth and pass a bristle through a gill slit from inside.

The water in·contact with the gills is constantly being renewed by the breathing mechanism of the fish, thus providing a continuous supply of oxygen. In breathing (see Fig. 9.17), the fish opens its mouth, closes the valve of its operculum, and enlarges its mouth and pharyngeal cavities by muscular action. This causes water to enter at the mouth. The mouth is then closed, and muscles constrict the mouth and pharyngeal cavities, so forcing water past the gills and out under the operculum of each side. If you watch a fish which has been very active, the movements of the operculum and mouth are easily observed.

9.13 Frog

Frogs exhibit several gaseous exchange mechanisms during their life-history. Before hatching, the embryos obtain oxygen and get rid of carbon dioxide by diffusion through the whole body surface. As they grow and become more active this method alone is apparently insufficient to supply their needs, for three pairs of branched finger-like processes, or **external gills**, are visible soon after hatching. These gills have very thin walls and allow gases to diffuse freely in and out of the blood which is pumped through them. External gills are soon replaced by **internal gills** which closely resemble those of a fish, already described. The operculum of a tadpole, however, differs from that of a fish and is described in section 17.7

where there is a complete account of the life-history.

Adult frogs carry out their gaseous exchange by three methods. The **skin** is well supplied with blood and, provided it remains moist, offers no barrier to the diffusion of oxygen and carbon dioxide. Roughly half the respiratory exchange when on land, and the whole of it when in water, takes place through the skin.

The lining of the **mouth cavity** also has a copious blood supply, to which the air in the mouth cavity gives up its oxygen. Carbon dioxide, of course, diffuses in the opposite direction. This air is constantly renewed through the nostrils (nares) by muscles which raise and lower the cartilaginous hyoid plate in the floor of the mouth cavity (see Fig. 9.18). You can hardly fail to notice this movement if you observe a live frog.

By a slight modification of this 'mouth-breathing' mechanism the **lungs** are ventilated. With the mouth closed and the nares open, the hyoid plate is pulled down, causing air to enter the mouth cavity. The nares are then closed and the hyoid plate raised, which forces air through the glottis into the lungs where gaseous exchange occurs. With the nares open, the elasticity of the lungs together with pressure from surrounding viscera cause them to empty, and the cycle is complete. It is possible to detect 'lung breathing' from outside, for the eyeballs are pulled in, so helping to increase the pressure in the mouth cavity when the lungs are being inflated. Whilst at rest, an adult frog may be moving its hyoid plate up and down rapidly, but it may fill its lungs only about once every minute. The lungs have relatively a much smaller internal surface than those of a mammal and are important respiratory organs only during active movement.

We have considered sufficient examples to realise that there is a wide variety of gaseous exchange mechanisms in animals, and that the exchange must take place through a **moist surface**, whether this be an internal one, as in lungs, or an external one, as in the skins of the frog and earthworm.

FROG — BREATHING MECHANISM

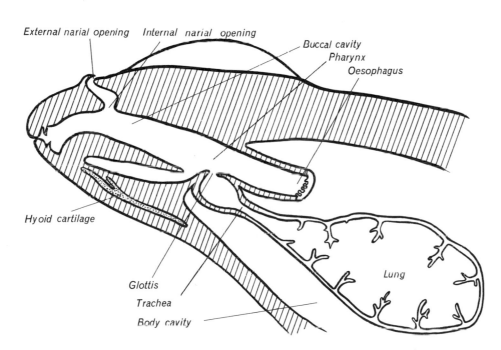

Fig. 9.18 Diagram of a median longitudinal section through the anterior region of an adult frog to illustrate the breathing mechanism.

QUESTIONS 9

1. State the functions of the blood. Show how the structure and composition of the blood are related to the functions you have mentioned. [C]

2. Make a large labelled diagram to show the internal structure of the mammalian heart. Give a concise description of the heart and explain precisely how the circulation of the blood is maintained in a mammal. [L]

3. (*a*) Make large labelled diagrams to show the structure of red and white cells (corpuscles) from mammalian blood, and list the essential differences between them.

(*b*) Trace the path of a blood cell leaving the small intestine, passing by the shortest possible route, until it returns to the same organ. Give any changes in composition of the blood along this route and say where each change occurs. [N]

4. (*a*) (i) Make a labelled drawing of the mammalian breathing system. (ii) What features of this system enable it to permit gaseous exchange to occur? (iii) Name two factors which could cause a decrease in the rate of gaseous exchange and two which could increase the rate.

(*b*) How is gaseous exchange brought about in (i) an insect, (ii) a fish? [N]

5. What are the functions of (*a*) plasma, (*b*) lymph? With the aid of a large-scale labelled diagram, carefully describe how the heart of a mammal functions. [C]

6. Why is blood necessary in a mammal, but not in Hydra? Describe the structure of mammalian blood and the functions of its several components. [O]

7. What is understood in biology by the term respiration? Describe fully *either* how you would show that germinating seeds respire, *or* how gaseous exchange is effected in a mammal. [O & C]

8. What do you understand by the term 'respiration'? How does oxygen reach the respiring tissues in (*a*) a terrestrial flowering plant and (*b*) a mammal? [S]

9. A mammal was observed for a period of a week and recordings were made of pulse and breathing under the conditions shown. The average rates are given in the table below.

Condition of Mammal	Body Temperature	Pulse Rate per Minute	Breathing Rate per Minute
Sleeping	36.1°C	60	10
Eating	36.7°C	72	18
Moving	37.1°C	120	25

(*a*) Explain the differences in pulse and breathing rates, under these different conditions.
(*b*) Immediately after exercise what other changes would you notice? [*M*]

10. (*a*) Why is oxygen necessary to all living things? (*b*) Describe how a named fish obtains its oxygen. (*c*) How does a frog obtain its oxygen when swimming in water? (*d*) How does a plant leaf cell obtain its oxygen? [*M*]

10 Excretion

Organisms are constantly taking in food materials, from which they are able to make vital substances such as the proteins in both cytoplasm and nuclei, enzymes, reserve foods such as fats, starch, and glycogen; structural materials including bone, cellulose, and lignin; haemoglobin, chlorophyll, and a host of others. In the tremendous number of chemical changes involved, waste substances which would hamper the normal functioning of an organism are bound to be produced, and their complete removal from the body or their deposition as insoluble substances within the body is known as excretion. Excretion can therefore be defined as the methods by which an organism prevents the waste products of metabolism from accumulating in solution in its cells.

EXCRETION IN MAMMALS

Mammals possess two main excretory organs:

1. Lungs—which excrete carbon dioxide in expired air.
2. Kidneys—which excrete urea and mineral salts in urine.

Some excretion also takes place through two other organs:

3. Skin—which excretes salt in the sweat, though its main function is temperature regulation.

Like the kidney it is concerned with maintaining steady conditions inside the body.
4. Liver—which excretes some water and salts in the bile, though its main function is concerned with carbohydrate distribution.

The removal of carbon dioxide, a waste product of respiration, has been dealt with in the previous chapter and will not be considered further. Although water is produced in respiration, and is expelled by all three organs listed above, it can hardly be classed as excretory, since its supply is vital to the body and it is the medium in which all metabolism occurs.

10.1 The Urinary System

The gross structure of the urinary system of a mammal is shown in Fig. 10.1. The kidneys are a pair of dark red bodies closely attached to the dorsal part of the abdominal wall, one on each side of the vertebral column. They are shaped like beans, with their indentations toward the midline, and are usually embedded in fatty tissue. A **renal artery** and **renal vein** carrying blood to and from each kidney enter and leave at the indentations. A narrow tube known as the **ureter** also leaves from this point and carries the urine made by that kidney down to a median muscular sac, the **bladder**. As the urine drips in, the muscles of the bladder relax, thus accommodating it without

URINARY SYSTEM

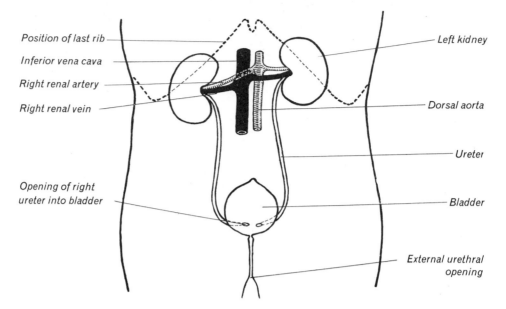

Position of last rib
Inferior vena cava
Right renal artery
Right renal vein
Opening of right ureter into bladder

Left kidney
Dorsal aorta
Ureter
Bladder
External urethral opening

FIG. 10.1 A diagram of human urinary system from the ventral side.

any increase in pressure. However, the pressure does eventually rise, whereupon the muscle contracts, forcing the urine out through a tube known as the **urethra**. This tube opens at the end of the penis in the male, and at the vulva in the female (see Fig. 17.6).

Before the way in which the kidneys make urine can be understood their internal structure must be known. They consist of a mass of microscopic **tubules**, blind at one end, and joining together at their other ends to open into a large space, the pelvis of the kidney, from which leads the ureter of that side. In a section of the kidney (see Fig. 10.2) two distinct layers, the medulla and cortex, can be distinguished, with the naked eye, round the pelvis. The blind ends of the tubules lie in the cortex and are expanded to form a cup-shaped **capsule** which encloses a small knot of approximately two hundred capillary loops known as the **glomerulus** (see Figs. 10.3 and 10.4). The remainder of the tubule is composed of two twisted portions lying in the cortex amongst the capsules, and separated from each other by a U-shaped loop which dips deeply into the medulla. The second twisted portion gives way to a collecting duct which joins others as it runs through the medulla to open into the pelvis.

In the plasma of the blood reaching the kidneys is a small quantity of **urea** (CON_2H_4), a waste nitrogenous substance which is formed in the

liver. It is made from carbon dioxide and ammonia, the latter being derived from excess amino-acids not required for protein formation (see sec. 7.2). The formation of ammonia in this

KIDNEY

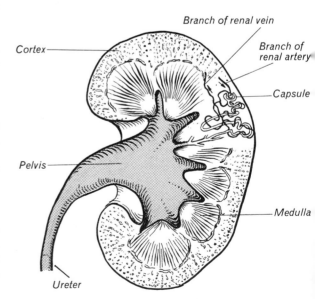

Branch of renal vein
Branch of renal artery
Cortex
Capsule
Pelvis
Medulla
Ureter

FIG. 10.2 Section through a mammalian kidney. The position of one kidney tubule (much enlarged) is shown.

KIDNEY SECTIONS

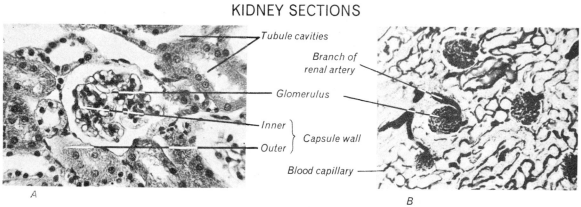

Tubule cavities

Branch of
renal artery

Glomerulus

Inner }
Outer } Capsule wall

Blood capillary

A

B

FIG. 10.3 Photomicrographs. A—Stained preparation. B—Preparation injected through the renal artery before sectioning (blood vessels show dark).

manner is known as **deamination.** Urea is a convenient form in which to transport waste nitrogen, since it is far less poisonous than ammonia and can be tolerated in much higher concentrations. The urea in the plasma, together with other nitrogenous wastes, glucose, mineral salts, and water, is forced by the blood pressure through the walls of the glomerular blood vessels and the capsule into the tubule cavity. Blood corpuscles and the plasma proteins are not able to get through. The **filtrate** which collects in the capsules is not urine; it has to pass down the tubules, and is modified on its way by the **re-absorption** of 99 per cent of its water, all of its glucose, and some of its mineral salts. Re-absorption here refers to the activity of the tubule cells passing water, etc., from the filtrate back to the blood in their capillaries. The fluid which collects in the pelvis of the kidney and

KIDNEY TUBULE

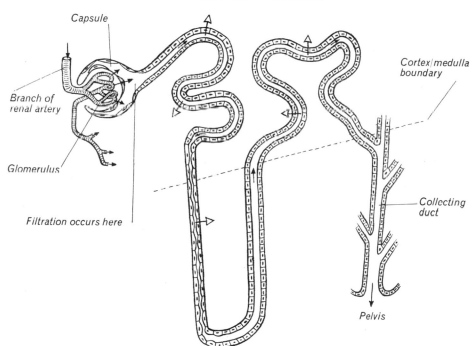

Capsule

Branch of
renal artery

Glomerulus

Filtration occurs here

Cortex/medulla
boundary

Collecting
duct

Pelvis

FIG. 10.4 Diagram of a mammalian kidney tubule. For simplicity its blood supply has been omitted with the exception of that to the capsule. Open arrows indicate reabsorption.

passes to the exterior is urine and has the following composition in a normal person.

Average composition of human urine in g per 100 cm³.

Urea	2.0
Other nitrogenous substances	0.2
Sodium chloride	1.0
Other mineral salts	0.8

In humans the average quantity of urine passed each day is 1 500 cm³, but this figure varies widely according to conditions. For instance, after copious draughts of water, or on cold, windy days when sweating is reduced to a minimum, large volumes of dilute urine are produced. On the other hand, if a lot of water is lost in sweat the volume of urine will be reduced. Thus the kidneys are not only important in excretion, they also serve to regulate the water content of the body.

10.2 The Skin

The skin makes up one-sixth of the body weight and covers an area of 1.6 to 1.8 m² in a human adult. It performs several functions and, although excretion is not the most important, it is convenient to deal with them here.

The skin is composed of two layers, the outer **epidermis** and the inner dermis (see Fig. 10.5). The outer part of the epidermis is dead and serves as a protection to the delicate underlying tissues. It is known as the **stratum corneum** and its cells are no longer distinct as their cytoplasm has been converted into **keratin**, a tough protein. The inner part of the epidermis contains living cells which are dividing and becoming cornified to replace the layers of stratum corneum which are constantly being rubbed off. Between these two layers can be seen the stages in the development of cornified cells from the soft inner cytoplasmic cells. The epidermis is thickest on the palms of the hands and the soles of the feet, these being parts of the body which are constantly in contact with other objects and are therefore most liable to wear.

The **dermis**, usually a thicker layer, contains blood vessels, fatty tissue, nerves, the sense organs of touch, hair follicles, sweat glands, and elastic fibres which made the whole skin supple. The hairs are made of keratin, and are formed in follicles, which are deep pockets of living epidermal cells projecting into the dermis. Associated with

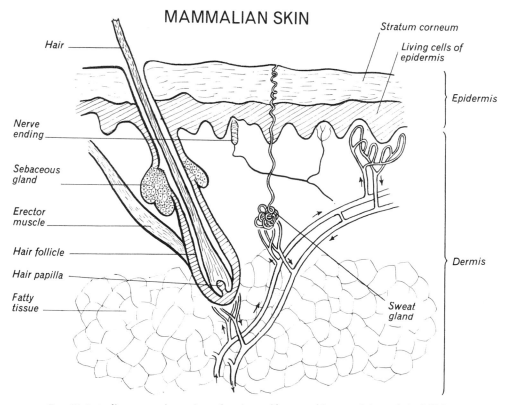

MAMMALIAN SKIN

Hair — Stratum corneum — Living cells of epidermis — Epidermis — Nerve ending — Sebaceous gland — Erector muscle — Hair follicle — Hair papilla — Fatty tissue — Sweat gland — Dermis

Fig. 10.5 A diagrammatic section of a piece of human skin containing a hair follicle.

these are small sac-shaped **sebaceous glands** whose ducts open into the follicles. Their fatty secretion, known as **sebum**, keeps the hairs oiled, prevents the skin from becoming too dry, and has a destructive action on some bacteria and fungi. Attached to each hair follicle is a tiny muscle which, when contracted, erects the hair, at the same time causing the familiar 'goose-skin'.

The **sweat glands** are tubular and lie coiled in the dermis with their spiral ducts opening on to the skin surface. There are about three million of these distributed over the whole surface of the body in a human being. During exertion or hot weather these glands produce a copious watery secretion whose main function is to keep the body temperature down to its normal level. (Mammals and birds keep their body temperatures fairly constant, no matter what the temperature of the environment.) The sweat is the most dilute body fluid, containing about 99.5 per cent water, 0.3 per cent sodium chloride, 0.03 urea, and traces of other mineral salts. For this reason, and because it is not produced regularly in quantity, it cannot be considered as an important excretory product.

The cooling effect of sweat is the result of heat being taken from the skin to evaporate its water, and it is often intensified by the dilation of all skin capillaries, which increases the blood flow near the surface of the body and makes it look red. In an atmosphere saturated with water-vapour, sweating does not cool the body, since its water cannot evaporate and merely drips off the skin. Close-fitting clothing or a dense layer of hair holds a moist atmosphere close to the skin and prevents the cooling effect being felt. Such coverings also restrict heat loss due to radiation, conduction, and convection. Dogs, having sweat glands only on their feet, resort to rapid shallow breathing when they are hot, which has a similar effect to sweating, since it causes cooling by the rapid evaporation of water from the mouth, trachea, and lungs.

The body temperature is regulated not only by varying the heat loss but also by varying the heat production. A familiar example of the latter is shivering—a rapid series of muscle contractions which generates heat.

The skin, then, protects the underlying tissues, houses important sense organs, and plays a large part in regulating the body temperature, but only a minor part in excretion.

10.3 The Liver

The liver is a large red organ which is divided into lobes; it lies below the diaphragm overlapping the stomach (see Fig. 4.2). It receives a double blood supply, (see Fig. 9.14) via the hepatic artery directly from the heart and via the portal vein from the alimentary canal. It performs many important functions, including:

1. Storage of reserve carbohydrate in the form of glycogen (see p. 63).
2. Regulation of blood sugar (see p. 172).
3. Formation of bile, which is really an excretory product (see sec. 7.9). Besides the salts, bile contains green pigments formed during the decomposition of haemoglobin from the worn-out red corpuscles (see sec. 9.7).
4. Storage of iron. The decomposition of red corpuscles is completed in the liver, and iron from haemoglobin is stored.
5. Deamination of proteins (see p. 110).

10.4 Summary of Excretion

To make the following table complete, the large intestine has been included, but it should not be regarded as equal in importance to the other three excretory organs since the bulk of the faecal matter (undigested food and bacteria) is not excretory, never having been involved in metabolism.

Excretory Organ	Excretion	Excretory Substances	Remarks
Lungs	Expired air	Carbon dioxide	Product of respiration
Kidneys	Urine	Urea, mineral salts	Also regulates the water content and salt content of the body
Skin	Sweat	Sodium chloride, urea	More important in temperature regulation and protection than in excretion
Large intestine	Faecal matter	Organic nitrogenous substances	Not more than 0.1 g per day
Liver	Bile	Mineral salts	More important in digestion and carbohydrate distribution

NB—F

EXCRETION IN FLOWERING PLANTS

10.5

Probably because they are far less active than animals, flowering plants produce little that can be classed as excretory, and they have no system of organs whereby excretory substances could be continuously passed out of the body. A further difference is that they have not the problem of getting rid of the nitrogen of excess amino-acids, for they make proteins as they require them from amino-acids they themselves have synthesised.

Perhaps the most obvious case of excretion in plants is the respiratory carbon dioxide passed out in the dark, but the same view cannot be taken of this as in an animal because, when illuminated, the plant uses this gas as food material in photosynthesis. The oxygen liberated during photosynthesis is a similar case, since this is used in respiration. In this case, however, unlike the previous one, more gas is produced than is needed and the excess passes out of the plant.

Crystalline deposits of calcium oxalate are common in the leaves and flowers of many plants, and particularly in the leaves of deciduous trees just before they fall. Organic substances, called tannins, are deposited in the bark cells of many trees, but these may serve a useful purpose in making the bark unpalatable to rodents. It is difficult to be sure that any permanent deposit in plants does not serve some useful purpose, and therefore it is always doubtful whether or not it should be called excretory.

QUESTIONS 10

1. State what is meant by excretion. Describe the structure and function of the mammalian kidney. [L]

2. (a) (i) What is meant by excretion and why is it necessary? (ii) Name the principal excretory products of the mammal and state where they are formed.
   (b) Give an illustrated account of the elimination of waste nitrogenous material by the mammal. [N]

3. Describe the appearance and position of the liver in any named mammal. Give an account of the functions performed by the liver. [O & C]

4. Make a large labelled diagram to show the structure of human (mammalian) skin as seen in vertical section. What are the functions of the human skin? [S]

5. By means of a labelled diagram, show the gross structure and position of the nitrogenous excretory system in a named mammal.
   Describe briefly the functions performed by the kidneys. Explain concisely the changes that take place in the blood as it passes through the kidneys. [W]

11 Growth

Growth is a characteristic activity of living things, and involves an increase in size and an increase in complexity. Increase in size—e.g. the development of a plant from the seed or the growth of an adult man from a baby—is easily seen, but a more careful study is needed to appreciate the increase in complexity. Growth is possible only if conditions are suitable and if sufficient food is available. The best criterion to use in measurements of growth is change in dry weight, since absorption of water often takes place without genuine growth occurring.

In plants, growth is a continuous process and occurs mainly at the tips of the root and shoot systems, whereas in an animal growth often proceeds for a limited period only and usually occurs throughout the whole body.

GERMINATION OF SEEDS

11.1 Broad Bean (*Vicia faba*)

The seed is kidney-shaped and consists of the embryo protected by a leathery skin or **testa**. If the seed is soaked in water for several hours it swells and is easier to examine. One end of the seed is much broader, and has a dark coloured scar, the **hilum**, showing where it was attached to the pod by the **funicle**. Close to one end of the hilum there is a triangular swelling, and between this and the hilum is a tiny hole, the **micropyle**

(*Gk. mikros, little; pyle, gate*). In an immature seed this is quite clear, but in the mature seed it is too small to be seen, though if the soaked seed is squeezed at the sides, water can be seen as it is forced out through the micropyle. It was through the micropyle that a tube from the pollen grain grew preparatory to fertilisation of the ovule, and prior to germination some water enters the embryo by the same route.

The testa, which is very tough, can be slipped off the soaked seed if you make a shallow cut right round the seed from one end of the hilum to the other. If you look inside the testa you will see a tiny pocket. The rest of the seed, which is cream-coloured, has a white conical peg projecting from it. This is the **radicle** or young root which fitted into the pocket of the testa. The seed can now be separated into two halves; each half, which is a seed leaf or **cotyledon**, is really a food store (starch and protein). Lying between the two cotyledons is a small curved structure continuous with the radicle. This is the **plumule** or young shoot, and you should use a hand lens to examine it carefully. You will then see that each cotyledon is joined by a short stalk to the region between the radicle and plumule.

Under suitable conditions (see sec. 11.5) the seed will germinate and grow into a plant (see Fig. 11.1). It first absorbs water, mainly through the micropyle, and this makes the cotyledons swell

GERMINATION OF BROAD BEAN

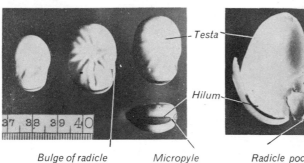

Entrance to radicle pocket

Cotyledon   Cotyledon stalk

Testa

Hilum

Bulge of radicle   Micropyle   Radicle pocket   Plumule   Radicle

DRY, PART-SOAKED & FULLY-SOAKED SEEDS   *TESTA SLIT & OPENED, EMBRYO REMOVED*   *EMBRYO WITH COTYLEDONS SEPARATED*

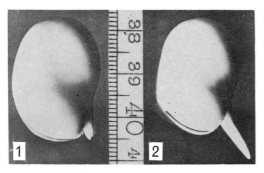

1   **2**

RADICLE BURSTS TESTA AT POCKET

Cotyledon stalk

Hooked plumule

3   **4**

1st lateral root

Plumule straight

Shoots growing from the axils of the cotyledons

Swellings of lateral roots inside main root

6

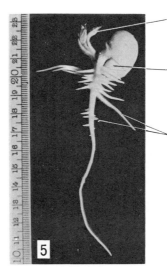

Plumule almost straight

Cotyledon seen through ruptured testa

Lateral roots

5

Fɪɢ. 11.1.

EPIGEAL GERMINATION OF FRENCH BEAN SEED

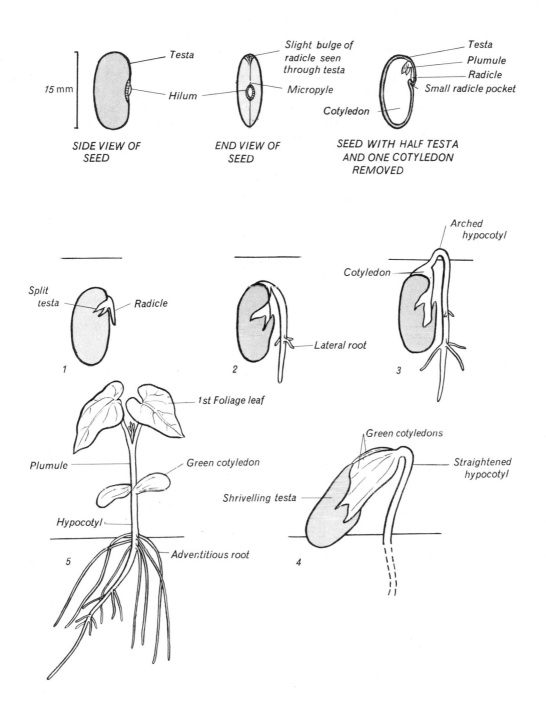

FIG. 11.2 French-bean seed and its germination. Stages 1 to 4 are natural size but stage 5 is a half natural size. The horizontal lines represent the soil surface.

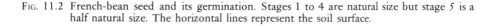

and the testa splits. The water also enables the enzymes in the cotyledons to make the stored food soluble so that it can be passed from the cotyledons to the radicle and plumule and used for growth. First the radicle grows downwards, then the cotyledon stalks grow longer so that the cotyledons are forced a little way apart and the plumule starts to grow upwards. As its tip is bent over, the delicate, pale yellow leaves are protected from damage by movement through the soil. When the plumule reaches the surface it straightens and the green leaves start to develop. The first two leaves are quite simple in structure, but the third is more complex and resembles those of the mature plant.

The growth of the radicle and plumule involves the formation of new tissues and requires a supply of energy. The materials for this are supplied by the food stored in the cotyledons, so that as growth proceeds they shrivel and the testa becomes wrinkled. This type of germination, in which the cotyledons remain in the soil, is termed **hypogeal** (*Gk. hypo, under; ge, earth*). The plumule is carried upwards by the growth of its lowest internode, i.e. the part of the stem just above the cotyledon stalks. Runner bean seeds are also hypogeal.

11.2 French Bean (*Phaseolus vulgaris*)

The french-bean seed (see Fig. 11.2) is similar in structure to that of the broad bean, but its method of germination is quite different. After the radicle has grown downwards, the **hypocotyl**, the part of the stem between the radicle and the cotyledon stalks, starts to grow, so that the cotyledons and the testa are carried upwards. As it lengthens, it bends over and forms a hook which pushes through the soil. Once above ground this hook straightens; the cotyledons then grow apart, freeing themselves from the testa, expand, and become green. By this time most of the food from the cotyledons has been used up, but they are now able to manufacture more by photosynthesis. This type of germination, in which the cotyledons come above ground and form the first leaves, is termed **epigeal** (*Gk. epi, on; ge, earth*).

11.3 Sunflower (*Helianthus annuus*)

The so-called 'seeds' of the sunflower are really one-seeded fruits, and the outer skin is therefore the pericarp and not the testa as in the broad and french bean. The fruit (see Fig. 11.3) is triangular and was attached to the receptacle at its narrower end; the scar at the broader end was left when the corolla and style fell off. If the pericarp, which is

ribbed and brittle, is removed carefully, the single seed, covered by a thin yellowish-brown testa can be seen joined to it by a short stalk—the funicle. The testa can be peeled off to reveal, at the narrower end, a short pointed radicle joined to two flat cotyledons between which there is a small plumule. The cotyledons act as a food-store and contain protein and oil.

The germination of the sunflower, like that of the french bean, is epigeal. The pericarp and testa both split at the narrow end and the radicle grows downwards. Next, the hypocotyl elongates and forms a hook which pushes its way through the soil, carrying with it the two cotyledons still within the pericarp. As the cotyledons open out, the pericarp is thrown off and the plumule begins to grow. The cotyledons turn green and start to synthesise food to supply the plumule in its early stages of growth, for by this time the food store within them is practically used up.

11.4 Maize (*Zea mais*)

The maize 'seed' is really a complete fruit, but the pericarp (fruit wall) and testa (seed coat) are so closely united that they cannot be separated. The grain is yellow except for a light-coloured oval patch on one side; this is the embryo (see Figs. 11.4 and 11.5). If you remove the skin from this region the radicle, which points toward the apex of the grain, and the plumule, which points in the opposite direction, can be levered up. They are attached to a single cotyledon which partly surrounds them. Behind the cotyledon is the hard yellow **endosperm** which forms the food-store.

When the grain germinates the radicle appears first, bursting through its protective sheath. It does not form a conspicuous taproot but soon gives rise to numerous lateral roots so that a fibrous root system is formed. In the mature plant adventitious roots arise from the nodes. The plumule too is protected by a sheath, or **coleoptile**, which lengthens as the plumule grows so that it gives protection until the surface is reached. The cotyledon remains within the grain, changing the starch of the endosperm to sugar and passing it to the developing embryo (see Fig. 11.4).

11.5 Conditions necessary for Germination

When the seed is dispersed from the plant it is usually very dry; in this condition growth cannot take place, so the embryo remains dormant within the seed until conditions are suitable for growth. This renewal of growth is germination, for which **moisture, oxygen,** and a **suitable temperature** are all essential. The necessity for these conditions

GERMINATION OF SUNFLOWER FRUIT

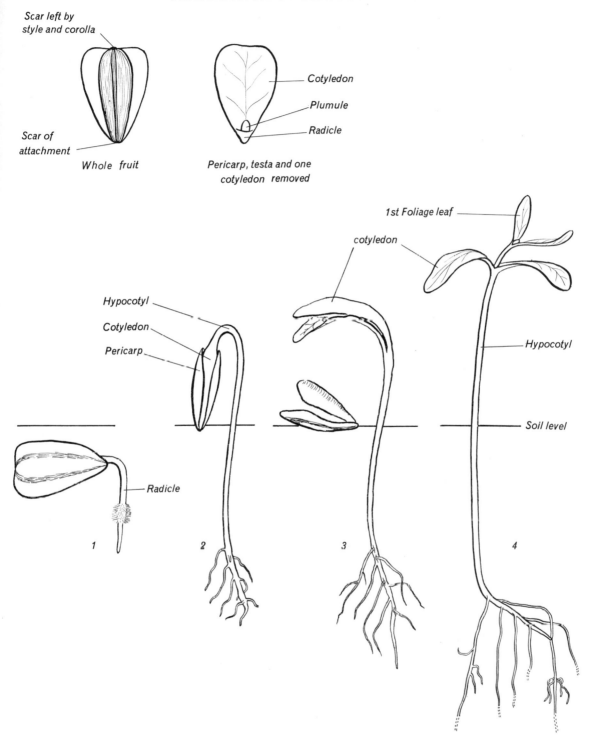

Scar left by style and corolla

Scar of attachment

Whole fruit

Cotyledon

Plumule

Radicle

Pericarp, testa and one cotyledon removed

1st Foliage leaf

cotyledon

Hypocotyl

Cotyledon

Pericarp

Hypocotyl

Soil level

Radicle

1   2   3   4

FIG. 11.3 The drawings on the sunflower fruit and its germination are all twice natural size. Only half the total root length has been drawn in 3. The horizontal lines represent the soil surface.

GERMINATION OF MAIZE GRAIN

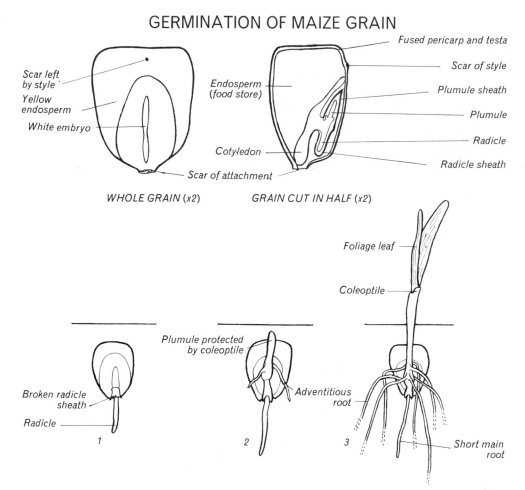

Scar left
by style

Yellow
endosperm

White embryo

Scar of attachment

Endosperm
(food store)

Cotyledon

Fused pericarp and testa

Scar of style

Plumule sheath

Plumule

Radicle

Radicle sheath

WHOLE GRAIN (x2)

GRAIN CUT IN HALF (x2)

Foliage leaf

Coleoptile

Plumule protected
by coleoptile

Broken radicle
sheath

Radicle

Adventitious
root

Short main
root

1

2

3

Fig. 11.4 Maize grain and its germination. The germination stages are natural size. In stage 3 only half the total root length has been drawn. The horizontal lines represent the soil surface.

CONDITIONS FOR GERMINATION

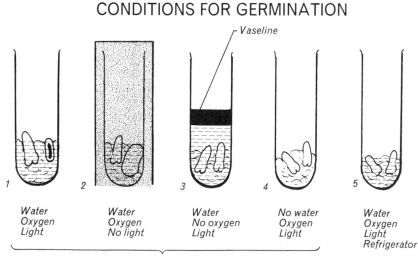

Vaseline

1

2

3

4

5

Water
Oxygen
Light

Water
Oxygen
No light

Water
No oxygen
Light

No water
Oxygen
Light

Water
Oxygen
Light
Refrigerator

In a warm place

Fig. 11.5

L. S. MAIZE GRAIN

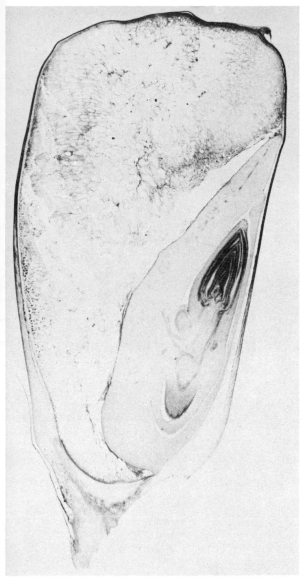

Fig. 11.6 Try to identify the parts of this photomicrograph of a section of a maize grain by reference to Fig. 11.4. (*Photomicrograph by M. I. Walker.*)

can be shown by arranging five boiling tubes as shown in Fig. 11.5 and watching the behaviour of the seeds. Broad bean, pea, and mustard seeds are all suitable for the experiment. The water in tube 3 should be boiled to expel the dissolved air, and cooled before the seeds are put in and the molten vaseline or oil poured on. Tube 2 is covered with a black cloth or aluminium foil to exclude light. Examine the tubes daily and consider your results carefully. The seeds in tube 2 should germinate,

but the seedlings will have lanky stems and yellow leaves—i.e. they will be etiolated (see sec. 12.1). Such seedlings are unable to make food by photosynthesis and eventually die.

GROWTH IN LENGTH

11.6 Region of Growth

After germination the root continues to grow downward, and the shoot upward. That growth in length takes place at the apex in both cases is very easily shown in a broad-bean seedling with a straight radicle about 40 mm long (see Fig. 11.7). Mark the radicle at intervals of about 1–2 mm by means of a piece of cotton dipped in indian or cyclostyle ink. The seedling should then be suspended in a moist atmosphere (e.g. a jam jar lined with moist blotting paper). In both the root and the shoot the actual region of growth is just behind the tip.

In the apex of the root is a group of cells which are capable of division. These cells form a tissue known as a **meristem** and they are protected by a conical root cap. The outer layer of the meristem produces the root hairs, and the lateral roots develop from the internal tissue and grow out through the cortex.

The apex of the young stem is protected by overlapping, undeveloped foliage leaves. In the axils of the larger leaves are slight swellings, formed from the outer tissues of the meristem; these develop into the axillary buds.

GROWING REGION OF A ROOT

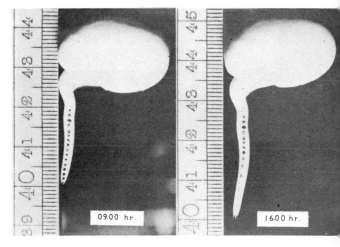

Fig. 11.7.

LONGITUDINAL SECTION OF A ROOT TIP

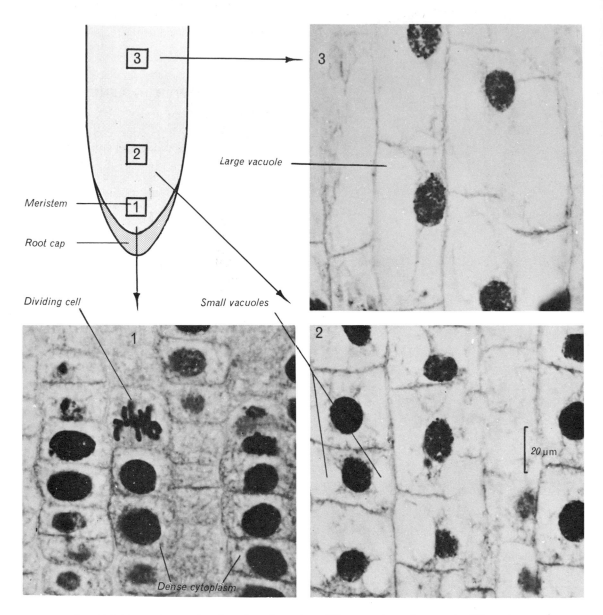

FIG. 11.8 Photomicrographs of sections from three different regions of a root tip. (*Photomicrographs by M. I. Walker.*)

11.7 Stages in Growth

Growth in length is the result of two processes: cell division and cell elongation. The division of the cells takes place in the meristems and after this there is a period of growth caused by the extension of the cells. At first the cells have thin walls and are filled with dense cytoplasm, but as more water is absorbed vacuoles appear in the cyto-

plasm and cause cell extension. The actual growth in length takes place during this **vacuolation** and **extension** of the individual cells (see Fig. 11.8).

Cell division is always preceded by nuclear division (mitosis), which takes place in a special way. At certain stages in cell growth there can be seen with the aid of a microscope one or more pairs of threads or **chromosomes** in the nucleus

MITOSIS AND CELL DIVISION

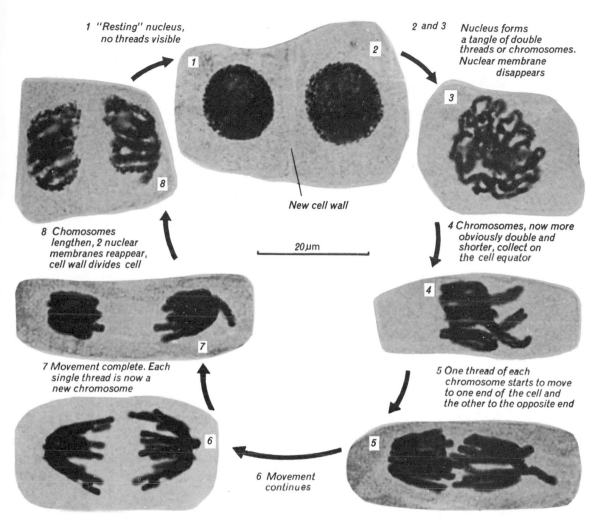

1 "Resting" nucleus, no threads visible

2 and 3 Nucleus forms a tangle of double threads or chromosomes. Nuclear membrane disappears

New cell wall

8 Chomosomes lengthen, 2 nuclear membranes reappear, cell wall divides cell

20 μm

4 Chromosomes, now more obviously double and shorter, collect on the cell equator

7 Movement complete. Each single thread is now a new chromosome

5 One thread of each chromosome starts to move to one end of the cell and the other to the opposite end

6 Movement continues

Part of a stained root tip squash

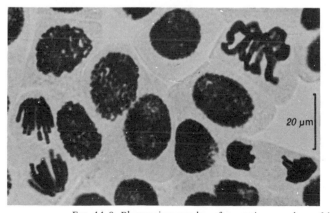

20 μm

Stage 4 with cell squashed to spread chromosomes

FIG. 11.9 Photomicrographs of root-tip squashes of broad bean (*Vicia faba*), showing mitosis.

HORSE — CHESTNUT

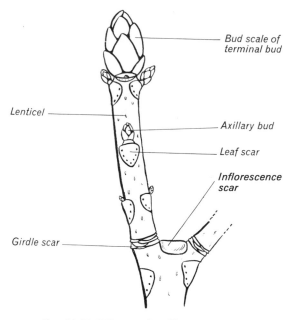

Labels on figure:
- Bud scale of terminal bud
- Lenticel
- Axillary bud
- Leaf scar
- Inflorescence scar
- Girdle scar

FIG. 11.10 Winter twig of horse-chestnut.

(see Fig. 20.10). The sequence of events leading to the formation of two cells is shown in Fig. 11.9: the whole process and its significance will be referred to again in chapter 20.

11.8 Growth of Woody Twigs

Plants, though they continue to grow throughout their whole life, do not grow at a uniform rate. Deciduous trees, for example, remain dormant during the winter months and then have an active period of growth during the spring and early summer. This periodicity of growth is easily followed in the horse-chestnut.

In winter the tree is devoid of leaves and each twig is curved upwards at the tip. The twig (see Fig. 11.10) is stout, brown, and covered with slit-like scars or **lenticels** (see sec. 11.10). At the tip of the twig there is a large cone-shaped **terminal bud** beneath which is a pair of small **axillary buds**. The rest of the twig also bears axillary buds arranged in pairs, each pair being at right angles to those above and below it. Below each axillary bud there is a shield-shaped **leaf scar** and on it are raised dots which mark the ends of the veins.

The buds are covered with reddish-brown sticky **bud scales**. If the bud is soaked in methylated spirit you can wipe off the stickiness and examine the bud more carefully. Remove the bud scales and arrange them in pairs (see Fig. 11.11).

Notice how they are arranged and how their shape, size, and colour alter as you get nearer the inner fluffy mass of tightly folded foliage leaves. Remove one of these leaves and try to separate the tiny leaflets.

In the spring, the apex begins to grow again (see Figs. 1.6 and 11.12), and as the foliage leaves swell the bud scales are forced apart and eventually drop off, leaving a **ring of scars** or **girdle scar** round the stem, so that it is possible to discover the age of a twig by counting these girdle scars.

In winter the soil becomes cold, and though it still contains plenty of water the absorptive power of the roots decreases because of the lower temperature. Since the intake of water by the roots is reduced, the evaporation of water from the leaves must be cut down (see sec. 6.8) and in deciduous trees this is accomplished by leaf fall. In autumn, the food remaining in the leaves passes to the stem where it is stored, the chlorophyll is broken down, and the familiar autumn tints develop. Meanwhile, the cells at the base of the petiole separate and become rounded, so that the leaf easily falls off. The exposed surface is protected by the development of a layer of cork (see p. 126) so that a leaf scar is formed.

GROWTH IN GIRTH

11.9 Cambial Activity

The aerial stems of trees and shrubs, unlike those of most perennial herbs, do not die down at the beginning of winter but persist. During the lifetime of the plant these stems increase in length and form new branches. They also increase in thickness. This is necessary, not only so that the stem shall be strong enough to bear the weight of the new branches but also so that there will be an increased amount of conducting tissue to supply materials to the increasing number of leaves, and to transport to other parts the increased amount of food manufactured by the leaves.

Growth in girth begins with the formation of a band of cambial cells across each medullary ray, so that a continuous ring of cambium is formed. Like the cambium within the original vascular bundles, this forms new cells both internally and externally. The inner cells are transformed into xylem and the outer cells into phloem. In this way compact rings of xylem and phloem are formed and the thickness of these rings increases as growth proceeds. The lignified walls of the xylem form the main supporting tissue of the woody stem (see sec. 6.6).

HORSE — CHESTNUT

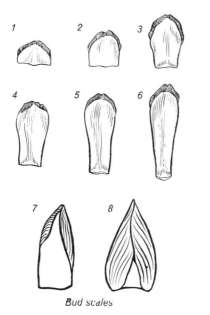

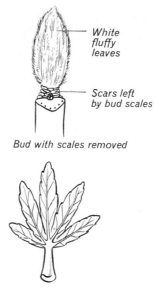

White
fluffy
leaves

Scars left
by bud scales

Bud with scales removed

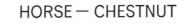

Bud scales

Foliage leaf

FIG. 11.11 Dissected horse-chestnut bud. The inner surface of the bud scales is drawn and only one of each pair is shown.

HORSE — CHESTNUT

FIG. 11.12 Stages in the opening of the horse-chestnut bud.

ANNUAL RINGS

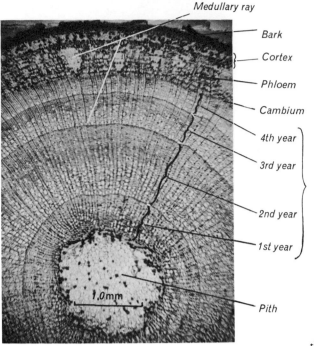

FIG 11.13 Photomicrograph of part of a transverse section of a four-year-old lime twig (*Tilia europaea*).

LENTICELS

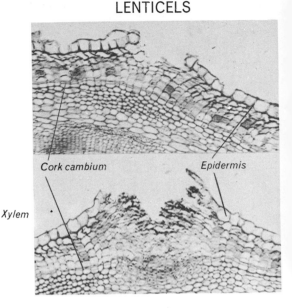

FIG. 11.14 Photomicrographs of transverse sections through two lenticels of elder (*Sambucus nigra*). The upper one shows an early developmental stage and the lower one a later stage.

The formation of new xylem and phloem continues throughout the life of the tree. It does not, however, take place at a uniform rate throughout the whole year, probably because of the varying amounts of water required in the different seasons. In autumn, the formation of xylem is very slow and only a few small vessels are formed, but in the spring it is rapid and a large number of wider vessels is formed. This different texture of the autumn and spring xylem forms **annual rings** and by counting the rings of autumn or spring xylem the age of the tree can be established (see Fig. 11.13).

It is only the newly formed secondary xylem that is able to conduct water and this is termed **sap wood**. In the older secondary xylem or **heart wood** the cells become blocked and no longer able to conduct water. Some waste substances may be stored here so that the heart wood is often darker than the sap wood.

11.10 Bark Formation and Lenticels

As the stem increases in thickness owing to cambial activity, the epidermis becomes stretched and finally splits. The cortex does not, however, become exposed, for a layer of cork develops below the epidermis. The cork cells are formed by a **cork cambium** which usually arises in the cortex. This divides to form an outer layer of closely fitting cells whose walls become impregnated with suberin, which is impermeable to water, and this layer of cork, together with the outer dead cells, forms the **bark**. It is the suberin which gives this layer its corky texture.

When the epidermis has been replaced by bark, gases are no longer able to enter or leave the stem by stomata. At intervals the cork cambium forms groups of loosely fitting cells, and gases are able to pass between them. Such a group of cells is termed a **lenticel** (see Fig. 11.14).

QUESTIONS 11

1. Draw a large labelled section of a named seed. What conditions are necessary for its germination? What are the food reserves in this seed, where are they stored, how are they moved from the place of storage to the places where they are used, and what use will be made of them? How will the seedling be nourished when its initial food reserves have been used up? [O]

2. Draw and label the twig of a named deciduous tree in its winter condition. How would you determine its age? How is the twig adapted to survive the winter and what happens to it when growth is resumed in the spring? [O]

3. What is meant by (*a*) epigeal and (*b*) hypogeal germination? Name one seed for each type of germination and with the help of diagrams describe the structure and germination of one of these. [O]

4. Make a large labelled diagram of a two-year-old woody stem seen in transverse section. Label five tissues and give a brief account of the function each performs in the living tree. [O & C}

5. Write an illustrated account of how a tree increases in girth (thickness). What are the functions of the bark of a tree? [S]

6. What do you understand by the term *growth* as applied to plants? Where does growth take place in (*a*) a broad-bean seedling, (*b*) a woody twig? [O]

12 Plant Movement

Movement is one of the characteristic activities of living materials, though in plants the movement is usually restricted to organs such as stem tips and leaves. One example of plant movement, the attachment of leaf tendrils of pea plants to supports (see Fig. 1.8), was described in section 1.6. This movement is the plant's response to the contact stimulus of the support. In section 1.7 the bending of stems in response to the stimulus of light coming from one side was described. This sensitivity (ability to respond to stimuli) is another characteristic of living cytoplasm. The stimuli which produce responses in plants include light, gravity, and water.

GROWTH MOVEMENTS—TROPISMS

12.1 Phototropism

We have all noted how an indoor plant, such as the geranium, turns its leaves toward the window. The shoots of plants are particularly sensitive to light and bend their stems or petioles so that the arrangement of the leaves is such that they do not overlap one another, and each one receives the maximum amount of light.

This response to light is termed phototropism. A **tropism** (*Gk. tropos, a turning*) is the response of the growing region of a plant to an external stimulus and the direction of the growth is controlled by the direction of the stimulus.

You can easily show how shoots respond to light by germinating three sets of pea seeds under different conditions (see Fig. 12.1). One set should be left to grow in the full light; another set should be covered by a box painted black on the inside; and the third set should be covered by a similar box but with a slit cut in one side. After a few days you will find the first set has erect stems with short internodes and spreading leaves; the set in the dark box has elongated internodes making the stems straggling, and small pale yellow leaves (etiolated); the set with light falling on one side has stems bent towards the light and leaves at right-angles to the rays of light. We may conclude then, that light retards the rate of growth. Darkness, however, prevents the development of leaves and chlorophyll, so that growth can continue in the dark for only a short time. The extent of growth under conditions of total darkness is limited by the food supply available, since no more can be made by photosynthesis. The shoot is said to be **positively phototropic** because it bends towards the light. The actual bending toward the light always takes place in the growing region of the stem and you should be able to devise an experiment to show this, using the technique described in section 11.6.

It is really the tip of the shoot, however, which is sensitive to light. An experiment to illustrate this can be carried out with oat seedlings, the coleoptiles (see Fig. 12.2) of which are very sensitive to light. Sow about a dozen oat seeds in each of three small cactus pots, moisten the soil, enclose

128

EFFECT OF LIGHT ON PLANT GROWTH

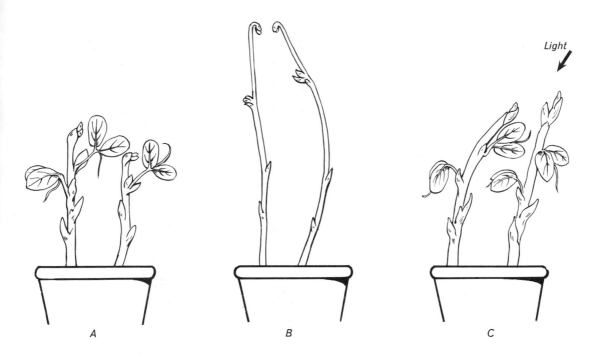

Light

FIG. 12.1 All the pea seedlings are the same age. A was in full light; B was in the dark, and C under a dark box with a slit on one side.

each in a polythene bag and leave in complete darkness for about six days. All subsequent examination of the seedlings must be made in red light to which the coleoptiles are insensitive. When several coleoptiles in each pot are just over 10 mm long remove the bags and carry out the treatments as shown in Fig. 12.3. Record your results carefully. In which region does the bending or uneven growth occur? Is this region covered up in pot 3? Which part of the coleoptile do you deduce to be sensitive to light?

OAT COLEOPTILES

FIG. 12.2 Photograph of seven-day-old oat coleoptiles.

12.2 Geotropism

When you sow seeds you do not usually bother to see that they are placed in the ground so that the radicle of each one is pointing downwards, yet the root always grows down into the soil and the shoot up into the light. This is because both the root and the shoot are sensitive to gravity: the roots are **positively geotropic** and grow downwards, but the shoots are **negatively geotropic** and grow upwards.

To show that it is gravity which makes the root and the shoot behave in this manner, a special instrument called a **Klinostat** is used. This consists of an electric motor geared to turn slowly a metal rod to which a cork disc and a celluloid cover are attached. The cork disc is covered with moist cotton-wool, and pea seedlings with radicles about half an inch long are fixed to it by pins pushed through the cotyledons. The rod is rotated slowly (about one revolution in fifteen minutes) and this Klinostat then acts as a control for a similar apparatus which is not rotated. After about twenty-four hours you will usually find that the seedlings on the stationary disc have grown and that the radicles have curved downward and the

SENSITIVITY TO LIGHT

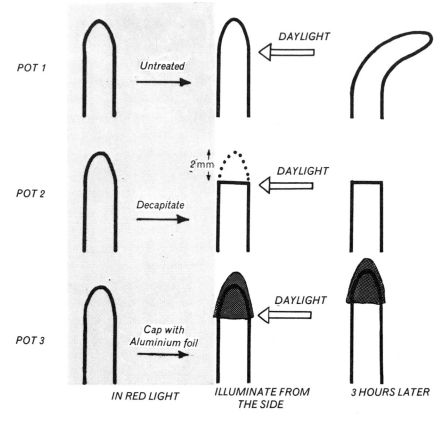

POT 1 — Untreated — DAYLIGHT

POT 2 — Decapitate — 2 mm — DAYLIGHT

POT 3 — Cap with Aluminium foil — DAYLIGHT

IN RED LIGHT — ILLUMINATE FROM THE SIDE — 3 HOURS LATER

Fig. 12.3 Summary of a series of experiments to study the sensitivity to light of the tip of the oat coleoptile.

plumules upwards. The rotating seedlings have also grown, but both the radicles and plumules have continued to grow in the original direction. The Klinostat, by its rotation, equalises the effect of gravity on opposite sides of the radicle and plumule.

It is the tip of the root which is sensitive to gravity, but it is the growing region just behind the tip which responds to the stimulus. This can be shown by cutting the tip off the radicle of a broad-bean seedling and marking the stump with indian ink before placing it in a horizontal position in a moist chamber. Remember to set up a control experiment. The decapitated radicle, though it does not curve at once, does respond after several days, since a new root tip is formed.

12.3 Chemotropism

The direction of growth of some parts of plants is also influenced by the presence of a concentration gradient of various substances. For example,

the hyphae of many fungi will grow towards nutritive substances but away from the waste products of their own metabolism. Similarly the growth of the pollen tube on the stigma of a flower, through the style towards the ovule which it fertilises, is directed by a chemical stimulus and on p. 183 there is a description of an experiment by which this **positively chemotropic** response of the pollen tube can be demonstrated.

12.4 Control of Growth—Tropic Mechanisms

In 1910 Boysen Jensen, a Danish botanist, discovered that the tip of an oat coleoptile (see Fig. 12.2) produces a substance which is soluble in water, diffuses readily, and controls its growth in length. Later, in 1926, Frits Went, a Dutchman, confirmed this by further experiments and even established a biological test (a bioassay) for measuring the amounts of the growth substance present in coleoptile tips which he had cut off. Following from this work the substance was isolated and identified as indolyl acetic acid

(I.A.A.). Many other growth substances, or **auxins**, as they were first called, are now known and they are all highly potent, producing responses in plant organs out of all proportion to their concentration. One-millionth of a milligram of I.A.A., for instance, will cause an obvious curvature in an oat coleoptile if it is placed in contact with one side about 10 mm below the tip.

Now examine Fig. 12.4 in which some of Went's experiments are described. Note that all treatment of the coleoptiles was carried out in red light, to which they are insensitive, and that at all other times they were kept in the dark. What conclusions can you draw about the properties of auxin from these experiments? To repeat these experiments is not quite as simple as Fig. 12.4 makes out and you may not carry out the various treatments satisfactorily first time. However, you can easily test the action of I.A.A. on oat coleop-

tiles as shown in Fig. 12.5. Be careful not to damage the delicate coleoptiles when you apply the lanolin and remember to carry out the treatments in red light only.

The question now remains whether the curvatures of stems and roots made in response to light and gravity are also controlled by auxins, for clearly, if uneven light and the stimulus of gravity can somehow produce uneven auxin distribution in these organs, this could be the basis of the response mechanism.

If you examine Fig. 12.6, which summarises the results of experiments to study the effect of one-sided illumination on the growth of an oat coleoptile, you will see that such an uneven distribution of auxin does occur. The higher concentration of auxin in the darker side of the shoot accelerates the growth in this region, hence the shoot bends towards the light—i.e. it is positively

AUXIN CONTROLS GROWTH

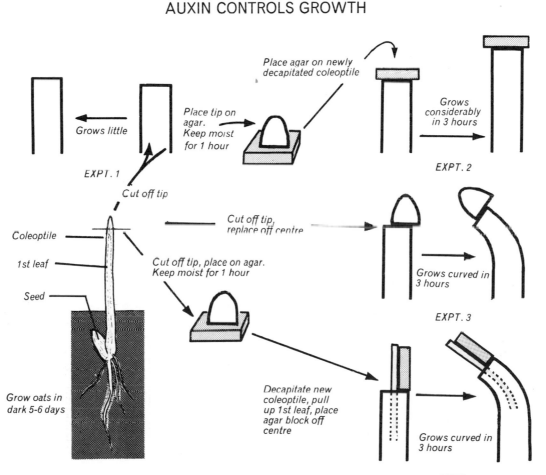

Place agar on newly decapitated coleoptile

Place tip on agar. Keep moist for 1 hour

Grows little

EXPT. 1

Cut off tip

Grows considerably in 3 hours

EXPT. 2

Coleoptile

1st leaf

Seed

Cut off tip, replace off centre

Cut off tip, place on agar. Keep moist for 1 hour

Grows curved in 3 hours

EXPT. 3

Grow oats in dark 5-6 days

Decapitate new coleoptile, pull up 1st leaf, place agar block off centre

Grows curved in 3 hours

EXPT. 4

Fɪɢ. 12.4.

IAA CONTROLS GROWTH

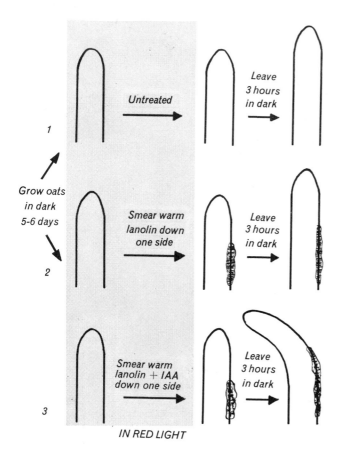

Fig. 12.5 Technique for testing the action of I.A.A. on the growth of oat coleoptiles. Why is it important to carry out treatment 2?

phototropic. Does this explain why the bent plumule of the broad bean seed straightens out as it reaches the surface of the soil (see sec. 11.1)?

Fig. 12.7 summarises the results of experiments to study the effect of gravity on the direction of growth of the shoot and root. Gravity will alter the distribution of the auxin when the seedling is placed in a horizontal position. Note that the response to the increased concentration in the shoot is an increased growth rate, but in the root the response is a slower growth rate, hence the response of these two organs to gravity is different.

The mechanism whereby the auxin causes these varied growth responses is unknown. Many different chemical compounds have been found to act as substitutes and some of them have proved useful horticultural and agricultural aids. The development of selective weed killers is based on our knowledge of auxins. It has been found that some of these chemical compounds have auxin-like properties if applied in very low concentrations, though they are toxic to many plants if used at higher concentrations. One such substance known as 2.4-D, which is readily absorbed by the leaves so that it can be applied as a spray or powder, is found to be less effective against narrow-leaved plants such as grasses and cereals. It can, therefore, be used as a selective weed killer for wheat or on lawns.

Horticulturists increase their stocks of plants by taking cuttings, and they have often found it difficult to induce such stems to produce adventitious roots. The production of these roots has been found to be stimulated by auxins, and the soaking of the cuttings in chemical substances which have

MECHANISM OF PHOTOTROPISM

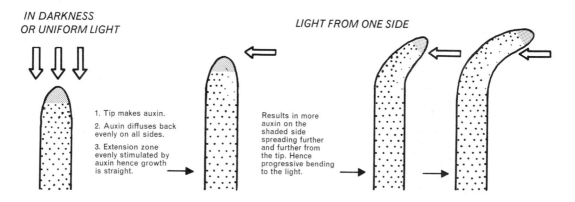

Fig. 12.6 Summary of the results of experiments to study the effect of one-sided illumination on the direction of growth of oat coleoptiles.

MECHANISM OF GEOTROPISM

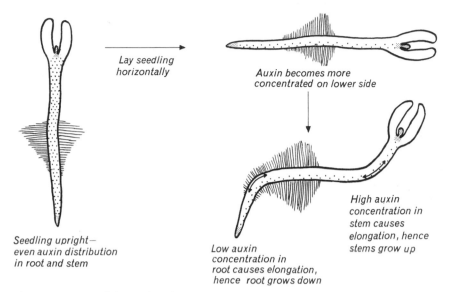

Lay seedling horizontally

Auxin becomes more concentrated on lower side

Seedling upright— even auxin distribution in root and stem

Low auxin concentration in root causes elongation, hence root grows down

High auxin concentration in stem causes elongation, hence stems grow up

Fig. 12.7 Summary of the results of experiments to study the effect of gravity on the direction of growth of the shoot and root.

he same effect has proved a great boon to the horticulturist.

OTHER MOVEMENTS

12.5 Nastic Movements

You have probably noticed that daisies close up in the evening and open again in the morning.

This is an example of a nastic movement made in response to a stimulus, but the direction of the movement is not controlled by the stimulus as in the case of a tropism. Similarly nastic or sleep movements are shown by the leaves of the clover, which fold downwards at night, and the evening primrose, which opens its flowers at night. In *Mimosa pudica* (sensitive plant) the leaflets are

NASTIC MOVEMENT

A

B

Fig. 12.8 Leaves of *Mimosa pudica* before and after shock. (*Photographs by W. J. Garnett, by courtesy of Longmans, Green & Co. Ltd. for the Nuffield Foundation.*)

folded together and the whole leaf hangs down-wards in the sleep position, but this same change can be brought about during the day if the plant is subjected to the stimulus of shock—e.g. shaking the whole plant or singeing a leaflet (see Fig. 12.8).

Most of these movements are the result of changes in the turgidity of special groups of cells, and the mechanism is similar to that found in the stoma (see sec. 6.9). Look at the photograph in the heading to this chapter: it shows leaves of oxalis. Do they behave like clover leaves?

12.6 Tactic Movements

A very few plants are able to move about from place to place by means of cilia or flagella. These movements are generally directed by a stimulus.

Phototactic movements are shown by many small green organisms (e.g. Euglena, see sec. 2.4) which are able to swim freely. This ensures that they remain in a position suitable for photosynthesis.

Chemotactic movements are exhibited by bacteria and motile gametes.

QUESTIONS 12

1. Describe in detail how you would demonstrate the effects of light on the rate and direction of growth of the shoot of a named seedling. How would you proceed to show that the effects you describe are not due to water or gravity?   [O & C]

2. To what external stimuli do young growing organs of plants react? Give a careful description of a controlled experiment which illustrates the response made by either a young root or a young shoot to one of these stimuli.   [L]

3. What do you understand by the term auxin? How could you demonstrate the part played by auxin in the response of an oat coleoptile to unilateral light? Briefly describe two horticultural uses of auxins or auxin-like substances.   [O & C]

13 Animal Movement

Animals generally have to move about in order to find food. This they can do only by pulling or pushing against something in the environment which resists displacement. The necessity for this resistance in the environment will be clear if you consider the difference between walking on a firm concrete surface and walking on loose gravel or sand. Whales, fish, water-fleas, worms, and many other animals move efficiently in water, even though this medium is more easily displaced than solid matter. Only a few groups of animals, e.g. birds, bats, and some insects, have mastered movement in the air, which is even more easily displaced than water.

So far we have considered movement only from an external aspect; now we must examine what happens within an animal during movement. We shall therefore study the bony, jointed skeleton of mammals and the muscular system whose function is to move the skeleton at the joints.

SKELETAL SYSTEM OF MAMMALS

13.1

The internal skeleton of mammals serves three main purposes:

1. To support the soft parts of the body.
2. To protect delicate organs.
3. To provide an anchorage for the muscles.

In an adult mammal the skeleton is composed mainly of **bone,** but **cartilage,** a flexible tissue with a smooth, glossy appearance, is present in the joints where two bones move against each other. In embryos, however, the only skeletal tissue present is cartilage, but it is largely replaced by bone during growth. Both these tissues contain a high proportion of non-living material or ground substance which has been secreted by the living cells of the tissue.

The ground substance of bone is hard and contains calcium salts (mainly phosphates) and fibres of a protein called collagen; that of cartilage is composed of chondrin, another protein. In bone, the cells which make the ground substance are often arranged concentrically round blood vessels so that the bony material is laid down in the form of concentric tubes following the course of the vessels. In cartilage, the cells are either solitary or in small groups (see Fig. 13.1).

Joints are the places where two or more bones meet each other. The structure of those in the limbs is such that great freedom of movement is permitted (see Fig. 13.2). The ends of the bones are covered with **articular cartilages** whose smooth surfaces are lubricated by a fluid, the **synovial fluid,** secreted by the synovial membrane. The joint is enclosed by a **capsular ligament** which is tough, fibrous, and inelastic, and limits the degree of movement possible. This type of joint can be classified according to the movement permitted.

BONE AND CARTILAGE

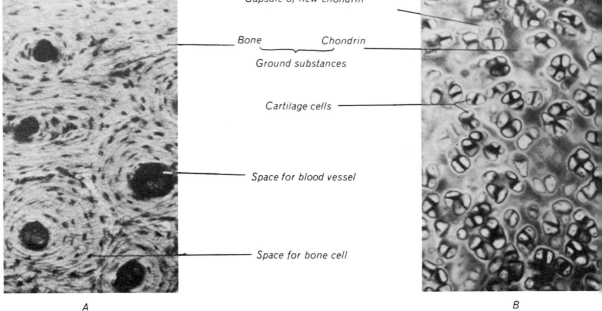

Fig. 13.1 A—Photomicrograph of dead bone cut transversely across the blood vessels. B—Photomicrograph of section of cartilage from a joint. The cells which have recently divided are still enclosed in a common capsule of new chondrin.

1. Ball and socket joint—universal movement, e.g. hip and shoulder joints.
2. Hinge joint—movement in one plane only, e.g. knee and elbow joints.
3. Gliding joint—two flat surfaces glide one over the other, e.g. most of the joints in the wrist and the ankle.

Other joints, such as occur in the vertebral column, involve two bones separated by a pad of cartilage and bound together by bands of ligaments so that very little movement is possible. Another type, where the bones are firmly bound together by fibrous material, allowing no movement at all, is found in the skull.

Before continuing with the detailed anatomy of the skeleton it may be advisable to revise section 4.1.

13.2 The Vertebral Column of the Rabbit

The vertebral column forms the central axis of a rabbit's skeleton. It is composed of about forty-six bones, or vertebrae, most of which are able to move a little relative to their neighbours, so making the column as a whole flexible. The vertebrae are specialised in various ways in different regions

of the column, but they usually possess the following parts (see Fig. 13.3):

1. A ventral more or less cylindrical rod of bone, the **centrum**. Adjacent centra are separated at

SYNOVIAL JOINT

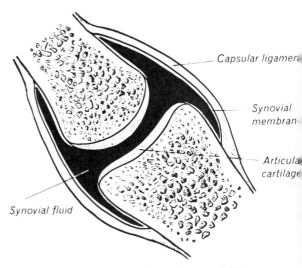

Fig. 13.2 Section through a synovial joint.

VERTEBRAE

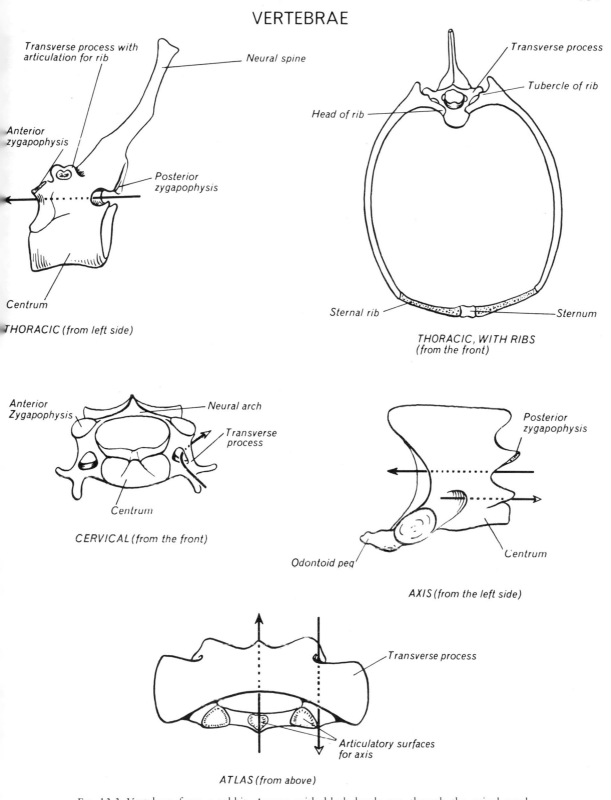

Fig. 13.3 Vertebrae from a rabbit. Arrows with black heads run through the spinal canal, arrows with open heads run through the vertebrarterial canals of cervical vertebrae.

RABBIT SKULL

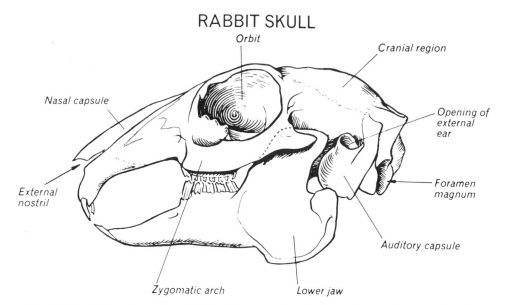

FIG. 13.4 Left side view of a rabbit's skull. Note the fine jagged suture lines marking the boundaries of some of the bones.

their flattened ends by a cartilage pad known as an intervertebral disc.

2. An arch of bone, the **neural arch**, arising from the centrum dorsally and enclosing the spinal canal in which lies the spinal cord.
3. A left and a right **transverse process** arising from the junctions of the neural arch and centrum.
4. A **neural spine** projecting dorsally from the centre of the neural arch.
5. Four short **zygapophyses** projecting from the neural arch. The anterior pair of zygapophyses has flattened surfaces facing upward which articulate with the downwardly directed faces of the posterior pair belonging to the vertebra in front.

Cervical Region, Seven Vertebrae

The first cervical vertebra, known as the **atlas**, is specialised to articulate with the skull. Its centrum and zygapophyses are absent, but it forms a complete ring of bone with strong broad transverse processes used for anchoring muscles which move the head. Only a nodding movement is possible at the atlas/skull joint.

The second cervical vertebra or **axis** is specialised to permit a rotational movement of the head. This is achieved by an odontoid process projecting from the anterior end of the centrum into the spinal canal of the atlas. When the head is turned, the skull and atlas pivot round this process. The axis has no anterior zygapophyses.

Both these cervical vertebrae share with the others the possession of a small canal on each side of the large spinal canal. Because these canals carry an artery they are known as **vertebrarterial canals**, and they are a feature by which cervical vertebrae can be easily recognised.

Thoracic Region, usually Twelve Vertebrae

Thoracic vertebrae are characterised by their long backwardly directed neural spines, and by the fact that each has a pair of ribs articulating with it in two places on each side. The head of a rib articulates in a shallow depression formed where the centra of two vertebrae are close together, but the tubercle, a short ill-defined projection of a rib near its head, articulates with the transverse process of the vertebra. The bony ribs are continued ventrally as cartilages, and in the case of the first seven, join a row of bones which collectively make up the sternum. The eighth and ninth ribs join the seventh but the remainder 'float' at their ventral ends. The whole skeleton of the thorax protects the heart and lungs, and the ribs are able to be moved in breathing as explained in section 9.6.

Lumbar Region, usually Seven Vertebrae

The lumbar vertebrae are large and strong, and have several extra projections which provide the necessary anchorage for the powerful back-muscles. Their neural spines and their long transverse processes point forward.

Sacral Region, Four Fused Vertebrae

The first sacral vertebra has stout transverse processes which are attached to the pelvic girdle and it therefore has to transmit the full force applied by the hind legs in jumping and running. The other three sacral vertebrae are fused to the first and to each other, but are without the well-developed transverse processes of the first. These fused vertebrae are sometimes called the sacrum.

Caudal Region, usually Sixteen Vertebrae

The caudal vertebrae are small and show a progressive loss of transverse processes, zygopophyses, and neural arch the nearer they are to the tip of the tail. The terminal ones are simply a tiny centrum.

13.3 The Skull of the Rabbit

The skull (see Fig. 13.4) is a complicated mass of bones, most of which are firmly joined together by fibrous tissue allowing no movement. The lines of these joints are called sutures and are easily seen in the rabbit, where complete fusion of the skull bones does not occur, as it does in many other mammals. The skull is formed of the following parts:

1. The **cranium**, which houses the brain and forms roughly the posterior half of the skull. At its posterior end is a large opening, the foramen magnum, through which the spinal cord issues from the brain. Also present are many smaller openings through which pass nerves and blood vessels. A flat bar of bone, the cheek bone or zygomatic arch, projects from each side of the skull just below the eye socket or orbit.

FORE-LIMB & PECTORAL GIRDLE

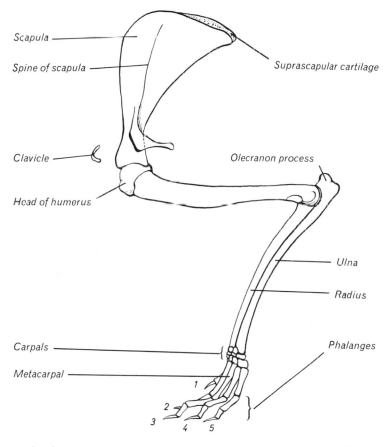

Fig. 13.5 Left side view of pectoral girdle and left fore-limb. Only the left half of the pectoral girdle is shown. The digits are numbered from the mid-line.

HIND LIMB & PELVIC GIRDLE

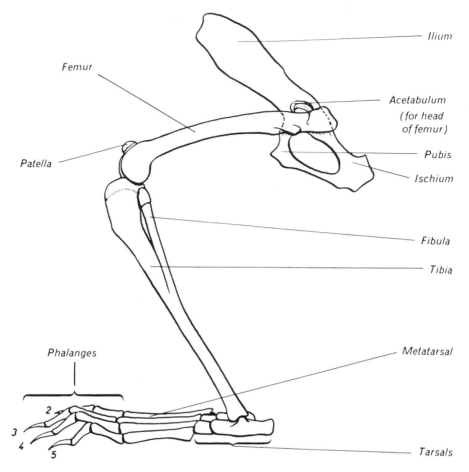

Fig. 13.6 Left side view of pelvic girdle and left hind-limb. Only the left half of the pelvic girdle is shown. The digits are numbered starting from the mid-line.

2. The paired **sense capsules**. The auditory capsules are bony cases which protect the middle and inner ears, and are fused into the cranium near its posterior end. The nasal capsules, which are fused to the front of the cranium, make up most of the facial region and contain the delicate sensory lining of the nasal passages. Both these pairs of sense capsules have openings (external nostrils, and the openings of the external ear) which put them in communication with the environment, so that the sense organs which they contain can do their job of detecting any changes in it. The eyes are without bony capsules, but are accommodated in the orbit, a pair of deep recesses between the nasal and cranial regions of the skull.

3. The **jaws**. The upper jaw is fused to the anterior half of the floor of the skull. The lower jaw articulates with the cranium on each side just in front of the auditory region.

13.4 The Limbs and Girdles (Appendicular Skeleton) of the Rabbit

The pectoral girdle (Fig. 13.5) is mainly composed of two flat triangular bones, the **scapulae,** one on each side of the thorax. The apex of each points down and forward, and has a concave surface, the glenoid cavity, which forms a ball and socket joint with the head of the first bone of the fore-limb, the humerus. On the outer surface of each scapula runs a ridge or spine which becomes free from the main part of the bone near its ventral end, and is directed backward. On each side, between this backwardly directed part of the spine and the sternum, a tiny **clavicle** is suspended in a ligament.

Apparently this bone is functionless in the rabbit. The scapulae, since they are anchored by muscles to the ribs and vertebrae, form a suitable base for the articulation of the fore-limbs. A third bone, the **coracoid** (see Fig. 13.14), is present on each side of the girdle in birds, reptiles, and amphibia.

The **humerus** is the long bone of the upper part of the fore-limb and has a rounded head, which is partly divided in front by a tendon of the biceps muscle of the arm, and a grooved distal end or trochlea (*L. pulley*) which articulates at the elbow joint with two parallel long bones, the **ulna** and the **radius**. The ulna is the outer of these two bones and has a projection beyond the elbow joint known as the olecranon process. When the limb is straight or extended, this process fits into a depression in the posterior surface of the humerus.

Eight small **carpals** in the wrist region articulate with the lower ends of the radius and ulna, and distally with five **metacarpals**. The fore-limb ends in five digits, each of which, except the first, contains three short cylindrical bones known as **phalanges** (sing. phalanx). Note that the digits are numbered starting from the mid-line of the animal. Horny claws, formed from the skin, cover the ends of the terminal phalanges of the digits.

The pelvic girdle (pelvis) is firmly attached to the sacral region of the vertebral column and, with it, forms a complete ring of bone. When the skeleton is boiled and cleaned, the girdle falls into two halves which were attached to each other in the mid-ventral line. Each half consists of three bones which fuse into one early in life:

1. The **ilium** (dorsal), which is attached to the first sacral vertebra.
2. The **pubis** (anteroventral), which is attached to the same bone of the other side.
3. The **ischium** (posteroventral), which, along most of its length, is separated from the pubis by a large opening.

Externally, where the three bones of each side meet, the girdle has a deep socket, the **acetabulum**, in which the hemispherical head of the femur articulates.

The **femur** is a long bone with a strong straight shaft. Near its head are three irregular projections to which muscles are attached; at the opposite (distal) end are two rounded condyles separated by a groove in which the **patella** or knee-cap moves. The **tibia** is the larger of the two bones of the lower leg and possesses two shallow concavities which articulate with the condyles of the femur. External to the tibia, and fused with it

distally, is a long thin bone, the **fibula**. Six small **tarsals** are found in the heel region and the two largest articulate with the fused tibia and fibula to form the ankle joint. There are only four **metatarsals** in the foot and each has three phalanges articulating distally. It is believed that the first digit of the hind limbs has been lost in the evolution of the rabbit.

Vertebrates, with the exception of fish, possess limbs which all conform to a general plan (see Fig. 13.7). Such limbs are described as **pentadactyl** (*Gk. pente, five; dactylos, finger*), but many have been so modified in evolution that they have fewer digits than five. The horse, for example, has one digit in each limb, deer and cattle have two.

13.5 The Human Skeleton compared

Many of the differences between a rabbit's skeleton and that of a man are connected with man's habit

PENTADACTYL LIMB

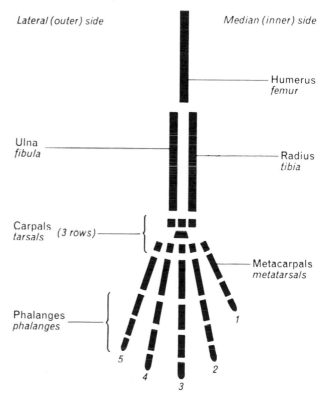

FIG. 13.7 Diagrammatic anterior view of right limb. Names of bones of the fore-limb are in normal type; those of the hind limb are given below in italics.

HUMAN SKELETON

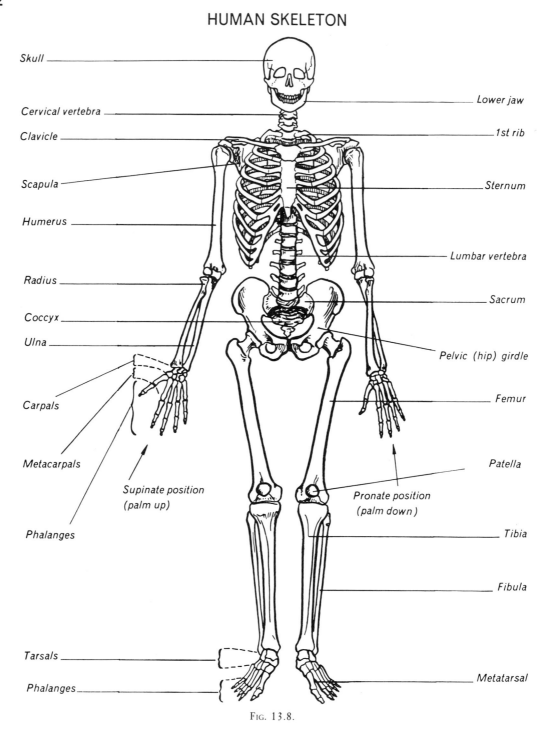

Skull

Cervical vertebra

Clavicle

Scapula

Humerus

Radius

Coccyx

Ulna

Carpals

Metacarpals

Supinate position
(palm up)

Phalanges

Tarsals

Phalanges

Lower jaw

1st rib

Sternum

Lumbar vertebra

Sacrum

Pelvic (hip) girdle

Femur

Patella

Pronate position
(palm down)

Tibia

Fibula

Metatarsal

FIG. 13.8.

of standing erect and moving about on his hind limbs, thus freeing his fore-limbs for the handling of objects. The most important differences are listed below (see Fig. 13.8):

1. Man's cranium is relatively much larger, he has a flattened face instead of a long snout, and his eyes face forward.

2. Man's vertebral column is relatively shorter and the coccyx is all that remains of the caudal vertebrae.

	Rabbit Vertebrae	Human Vertebrae
Cervical	7	7
Thoracic	12	12
Lumbar	7	5
Sacral	4 fused	5 fused
Caudal	16	4 fused

3. The clavicles are large and useful in climbing. They act as props bracing back the shoulder joints so that the arms can swing clear of the trunk.

4. The radius at its distal end can be rotated about the end of the ulna, so twisting the hand from the pronate (palm down) position to the supinate (palm up) position. This ability, coupled with the greater mobility of the humerus in its articulation with the scapula, and of the first metacarpal in its articulation at the wrist, allows the development of a high degree of manual dexterity in man.

5. The fibula is not fused with the tibia.

6. There are five digits in each hind limb.

SKELETAL MUSCLE

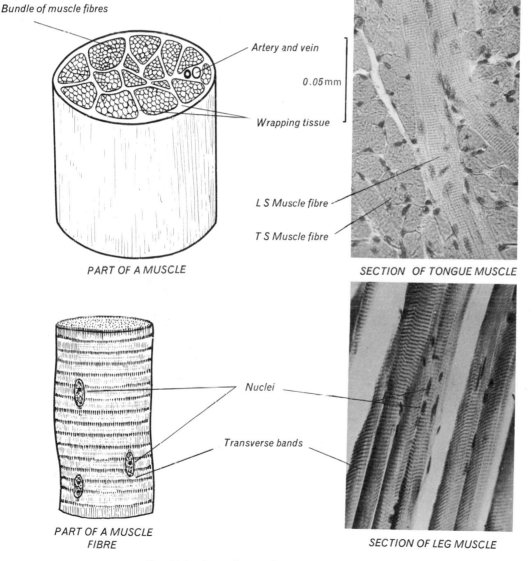

Bundle of muscle fibres

Artery and vein

0.05 mm

Wrapping tissue

L S Muscle fibre

T S Muscle fibre

PART OF A MUSCLE

SECTION OF TONGUE MUSCLE

Nuclei

Transverse bands

PART OF A MUSCLE FIBRE

SECTION OF LEG MUSCLE

FIG. 13.9 Photomicrographs of skeletal muscle.

MUSCULAR SYSTEM OF MAMMALS

Three types of muscle tissue can be distinguished in mammals:

1. Plain or involuntary muscle, found mainly in the gut.
2. Striped, skeletal, or voluntary muscle.
3. Heart muscle.

Skeletal muscle tissue, the type we are concerned with in this chapter, is commonly known as flesh and forms the bulk of meat purchased at the butcher's shop. Its functioning is under the control of the will, hence the name 'voluntary'. It is composed of parallel cylindrical fibres with an average length of 25 mm and an average diameter of 0.05 mm. The fibres are covered with a sheath, which is not easy to observe, and are not true cells since their cytoplasm contains many nuclei. When examined with a microscope, fresh or stained material exhibits prominent transverse bands (see Fig. 13.9). Many fibres are wrapped together in bundles with blood capillaries branching among them, and these bundles are again wrapped together in large numbers to make a single muscle. Muscles are not attached directly to bones, but end in strong **tendons** whose minute fibres, cemented to the muscle fibres at one end, pass into the ground substance of the bone at the other, so making a firm union of bone and muscle.

The three muscle tissues differ in their structure and function, but they are all contractile. When a muscle **contracts**, its fibres shorten and fatten so that the two parts of the body to which it is attached are pulled nearer together. During the opposite process, **relaxation**, a muscle does not push parts of the body farther apart, it merely stops pulling them together.

13.6 Muscles at Work

Example 1. The Human Elbow Joint (see Fig. 13.10)

Two large muscles, the **brachialis** in front of the upper arm and the **triceps** behind it, help to control the movement at the elbow joint. The brachialis is attached by a short tendon to the shaft of the

ELBOW JOINT

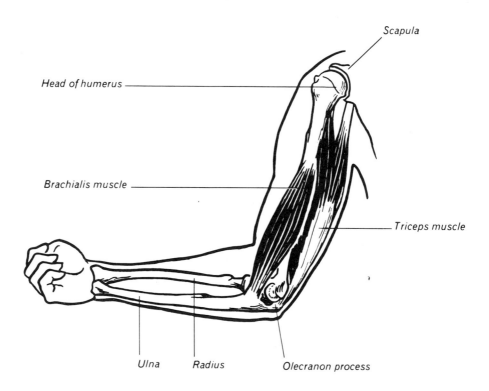

Head of humerus

Scapula

Brachialis muscle

Triceps muscle

Ulna   Radius   Olecranon process

Fig. 13.10 Diagram of the human elbow joint to show the brachialis muscle, which flexes the arm at the elbow, and the triceps muscle, which acts antagonistically and extends the arm at the elbow.

humerus, and at its other end by one tendon to the ulna. The relatively fixed end of the muscle (i.e. that attached to the humerus) is known as the origin, and the relatively movable end (i.e. that attached to the ulna) is known as the insertion. When the brachialis contracts it becomes shorter and fatter, so bending the arm at the elbow, i.e. flexing the elbow joint. The triceps, having a triple origin on the scapula and humerus and a single insertion on the olecranon process of the ulna, opposes the flexing of the joint by the brachialis muscle. When the triceps contracts it straightens the arm at the elbow, i.e. it extends the arm. The extensor muscle (triceps) and the flexor muscle (brachialis) are therefore **antagonistic**, and each is lengthened, provided it is relaxed, by the contraction of the other. However, both muscles can be relaxed at the same time, as for instance when the fore-arm is supported, and both can be contracted at the same time if the fore-arm is held tense in one position. Notice that the radius and ulna, being bound together, work as a lever in the lifting of weights held in the hand. The brachialis has its insertion near to the joint and, therefore, works at a mechanical disadvantage, but it is able to move the hand quickly through a wide arc.

There is another muscle, the **biceps** of the arm, which helps to flex the elbow joint. It has a double origin on the scapula and an insertion on the radius. On contraction it can be seen to bulge visibly, since the main mass of muscle lies in front of the humerus, and it both flexes the elbow joint and twists the fore-arm to bring the palm uppermost. If the arm is flexed with the palm held downward the biceps does not contract, a fact which can be checked by feeling the muscle during the movement.

Example 2. Walking in Man

When man walks he uses almost all the muscles of the legs (about sixty in each), and many which extend into the trunk. Simplified details of only the more important muscles are given in Fig. 13.11 and in the table. Study them carefully and feel in the places where they can be detected, at the same time making the movement listed under 'action'. You must get to know the muscles in this way before you can hope to understand how they are used in walking.

Now take a long stride with the left foot and from that position take a walking pace (first pace) with the right foot and another with the left foot. Repeat this slowly several times, trying to decide which of the leg muscles contract and in which order. Then check your findings with the account on page 147.

Muscle	Origin on	Insertion on	Action(s)	Where detectable
1. Quadriceps femoris group (includes four muscles)	Pelvic girdle and femur	Tibia, by patellar tendon	(a) Protracts leg (= flexes hip joint) (b) Extends leg at knee	On the front of the thigh
2. Biceps femoris	Pelvic girdle and femur	Fibula, by lateral hamstring tendon	Flexes leg at knee	Tendon is prominent behind the knee on the outer side
3. Gluteus maximus	Pelvic girdle	Femur	(a) Retracts leg (b) Straightens body by rotating it about the head of the femur	On the buttocks
4. Gastrocnemius (calf muscle)	By two tendons on both condyles of femur	By the Achilles tendon on a heel bone	Extends leg at ankle (= raises heel off ground)	Behind the tibia
5. Anterior tibial muscle	Tibia	First metatarsal	Flexes leg at ankle	Half-way down the tibia on the outer side

Note: Antagonistic pairs of muscles are 1 and 2, 1 and 3, 4 and 5.
Protractor muscles move a limb forward, retractor muscles move it backward.

SOME HUMAN LEG MUSCLES

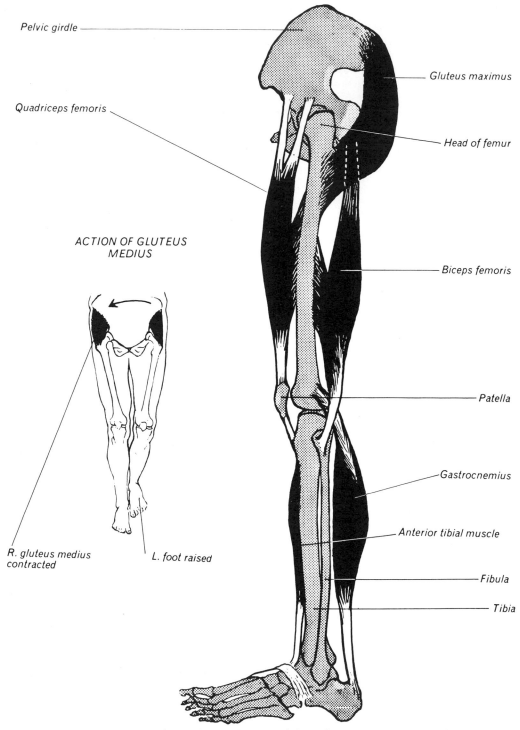

Pelvic girdle

Gluteus maximus

Quadriceps femoris

Head of femur

ACTION OF GLUTEUS MEDIUS

Biceps femoris

Patella

Gastrocnemius

Anterior tibial muscle

R. gluteus medius contracted

L. foot raised

Fibula

Tibia

FIG. 13.11 Diagram of the left human leg viewed from the outer side, to show the origins, insertions, and positions of the muscles listed in the table on p. 145. The median origin of the gastrocnemius is not shown in this diagram.

First Pace

First, the right heel is raised from the ground (r. gastrocnemius contracts) so pushing the body forward whilst supported on the left leg as a pivot. The right leg is now protracted (r. quadriceps femoris contracts), and the right toe lifted off the ground by slight flexing of the knee joint (r. biceps femoris contracts). As the right leg swings past the left, it is extended (r. biceps femoris relaxes, r. quadriceps femoris continues to contract), and the right toe raised (r. anterior tibial muscle contracts) so that the heel of the right foot touches the floor first. During this movement the left leg has been used as a strut, taking the weight of the body, and the extensor muscles (l. quadriceps femoris and l. gastrocnemius) have been contracting to prevent the collapse of the limb. Also the left gluteus maximus contracted to help pivot the body forward with the left foot on the ground (this latter action is the same as the retraction of the left leg except that the foot is a fixed point instead of the pelvic girdle).

Second Pace

As for first pace, but for right read left and vice versa.

Whilst the left or right foot is off the ground the body is pulled to the opposite side to bring the centre of gravity over the other foot. This is achieved by the gluteus medius (shown in the inset of Fig. 13.11) of the side in contact with the ground, contracting and rotating the body about the hip joint of that side.

13.7 Accommodation in Muscular Exercise

Complicated chemical changes resulting in the liberation of energy accompany the contraction of muscles. The energy is obtained by the muscle oxidising its glycogen to carbon dioxide and water (see sec. 9.10). Only a small quantity of glycogen (1 per cent by weight) is present in resting muscle, and this is soon consumed during exercise, after which the muscle has to rely on glucose supplied by the blood as the source of its energy. Oxygen is, of course, necessary for the oxidation. During exercise, increases in the following occur:

1. Pulse rate.
2. Volume of blood expelled at each stroke of the heart (stroke volume).
3. Respiration rate.
4. Depth of respiration.

These changes, as well as the redistribution of blood from the gut to the skeletal muscle, all help to increase the oxygen supply to the muscles. Sometimes, in spite of these changes, the oxygen supply to a muscle lags behind its requirement and it starts to carry out anaerobic respiration in which lactic acid is produced. If the exercise is severe and

$$C_6H_{12}O_6 \longrightarrow 2C_3H_6O_3 + \text{Energy}$$

Glucose Lactic acid Energy

prolonged, accumulation of lactic acid may prevent the muscle from contracting further, when it is said to be fatigued. During recovery the muscle continues to use oxygen at a fast rate until the lactic acid has disappeared. By training, an individual can increase the maximum rate at which the lungs and circulatory system can supply oxygen to the muscles.

OTHER METHODS OF LOCOMOTION

13.8 Birds

Nearly every organ of a bird can be regarded as adapted for flight. The wings are, perhaps, the most obvious specialisation, but there are many others such as the lightweight bones, the large air sacs connected with the lungs, and the enormous keel-shaped sternum which provides a suitable anchorage for the strong wing muscles. It is these muscles which one removes from the sternum when carving the breast of a bird.

The wing skeleton is made up of the usual humerus, radius, and ulna, but in the remainder, reduction of the number of digits and fusion of bones makes it difficult to recognise the pentadactyl plan (see Fig. 13.12). There is no recognisable hand region and the second digit is the only well-developed one. A carpo-metacarpus (formed from fused carpals and metacarpals) is an important bone, since it has several flight feathers attached to it.

Feathers, of which there are several types, are formed in the skin. A flight feather (see Fig. 13.13) is made up of a rigid stem, part of which (the quill) is embedded in the skin, and part of which (the rachis) has a series of long, thin, flexible plates or barbs attached on each side. The barbs are held together by barbules, some of which bear hooks and some of which are plain. This arrangement prevents the passage of air through the vane of the feather. The vanes of adjacent flight feathers overlap when the wing is extended, so that the whole under surface of it is impervious to air.

WING OF BIRD

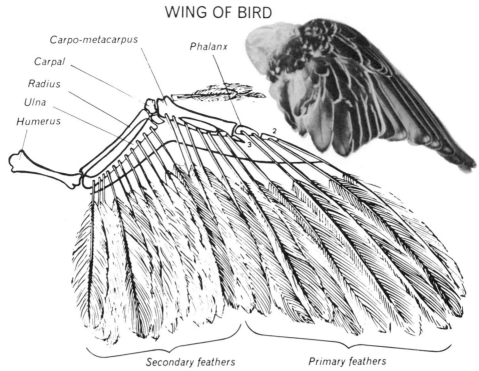

FIG. 13.12 Diagram of a bird wing, showing bones and flight feathers. The three digits are numbered, but only the second is well developed.

FLIGHT FEATHER

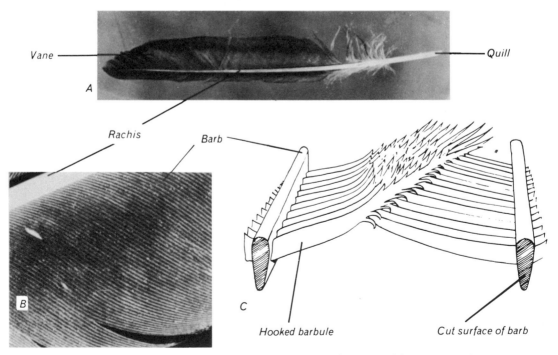

FIG. 13.13 A—Photograph of a flight feather. B—Photograph of part of the vane. C—Diagram of a small piece of two barbs showing their barbules.

FLIGHT MUSCLES

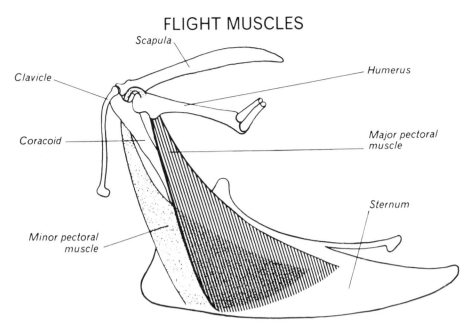

Clavicle

Scapula

Humerus

Coracoid

Major pectoral muscle

Sternum

Minor pectoral muscle

Fig. 13.14 Diagrammatic view of the left flight muscles of a bird as seen from its left side. Note that the two pectoral muscles are antagonistic in their actions although they have their origins and insertions on the same bones. The minor pectoral muscle elevates the wing, and the major pectoral muscle depresses the wing.

The two largest wing muscles (left and right pectoralis major) (see Fig. 13.14) run from the sides of the sternum (origins) to the humerus of each side (insertions), so that, on contraction, they pull the wings obliquely downward and backward giving the bird both vertical lift and forward thrust. Antagonistic to these is a smaller pair of wing muscles (pectoralis minor) which take origin from the sternum. They also have their insertions on the humerus of each side but, when they contract, they pull the wings up in preparation for the down stroke. Notice that their tendons pass through an opening (foramen triosseum) on each side of the pectoral girdle, which acts like a pulley altering the direction in which the muscles pull. On the upstroke, the bird does not neutralise all the lift and thrust it has gained because the flight feathers are moved so as to permit the passage of air between them. Tail feathers are also important in flight for steering and braking.

The breathing mechanism of birds is very different from that of a mammal. Inspiration, the active phase of mammalian breathing, occurs passively as the sternum falls and the air first enters a series of large air sacs. Expiration is caused by the contraction of muscles which draw up the sternum and force air from the air sacs through the lungs to the exterior. Such a mechanism will obviously be speeded up by the working of the wing muscles in flight, the time when the bird requires its greatest oxygen supply.

Bats have solved the problem of flight without feathers. They support a thin membranous wing with four extremely long fingers, the other one being left free and having a claw (see the photograph in the heading to chapter 4). It is interesting to note that there is also a keel on the sternum of bats.

13.9 Fish

Most fish have a smooth surface and a streamlined shape, which helps them to move easily through water. The main organs used by fish in swimming are a strong muscular tail usually occupying half the body length, and a series of fins also operated by muscles. The tail, with its fin at the end, is used like a scull to drive the fish through the water. This is achieved by alternate contractions and relaxations of the longitudinally arranged muscle fibres on opposite sides of the vertebral column, which causes S-shaped waves to pass down to the tail. The other unpaired fins help to prevent rolling and the paired fins are put to a variety of uses. In dogfish, for instance, they are used as elevators. In many bony fish, the whiting and roach for instance, they can be used as brakes, since a swimbladder is present. This organ contains a mixture

STICKLEBACK

Median spines and fin   *Lateral line canal*

Caudal fin

Operculum   *Pectoral fin*   *Pelvic fin (Spine)*   *Anal fin*

FIG. 13.15 Photograph of a stickleback.

of gases similar to air and helps the fish to keep its station at a certain depth in the water.

13.10 Earthworms

Earthworms have already been mentioned in section 5.6, but here we must consider their movements rather than their effect on the soil. They rarely move far on the soil surface except when disturbed by digging. They usually forage for vegetable matter at night leaving their posterior ends firmly anchored in the burrow. If, with the aid of a torch, you find them doing this, you will be surprised how quickly they can return to their burrows when touched. This type of movement and the slow movements within the burrows are natural ones, but are difficult to study in detail. However, observations on earthworms crawling on a piece of rough paper can be quite instructive and provide information about the mechanisms involved in more natural movements.

In each segment of the body, with the exception of the first and the last, are four pairs of **setae** (*L. seta, bristle*) as shown in Fig. 13.17. Muscles in each segment can extrude these, so anchoring the segment, or withdraw them. The setae can easily be seen and felt, and the noise which they produce can even be heard when the worm crawls on rough paper. In the body wall of each segment is a **circular** band of muscle which, when it contracts, makes the segment long and thin. Strips of muscle with the fibres arranged **longitudinally**

EARTHWORM MOVEMENT

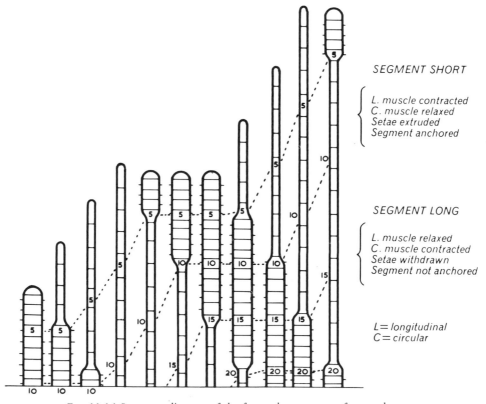

SEGMENT SHORT

L. muscle contracted
C. muscle relaxed
Setae extruded
Segment anchored

SEGMENT LONG

L. muscle relaxed
C. muscle contracted
Setae withdrawn
Segment not anchored

L = longitudinal
C = circular

FIG. 13.16 Sequence diagram of the forward movement of an earthworm.

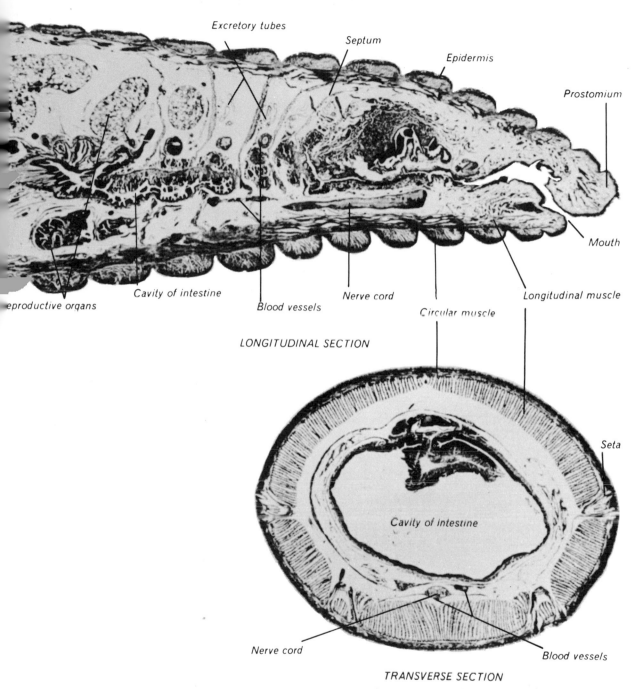

Excretory tubes

Septum

Epidermis

Prostomium

Mouth

eproductive organs

Cavity of intestine

Blood vessels

Nerve cord

Circular muscle

Longitudinal muscle

LONGITUDINAL SECTION

Seta

Cavity of intestine

Nerve cord

Blood vessels

TRANSVERSE SECTION

Fig. 13.17 Photomicrographs of sections of the earthworm. (*Photomicrographs by M. I. Walker.*)

run the whole length of the body and antagonistically to the circular muscle. As a rule, the setae in a segment are extruded to grip the ground or paper only when the longitudinal muscle in that segment is contracted. By watching one segment at a time in a large earthworm moving on paper, try to analyse the stages of a forward movement. (It is essential to keep the worm moist during your observations.) Compare your results with Fig. 13.16.

Movements of Amoeba (sec. 2.2) and Hydra (sec. 2.10) have already been described and would appear to be simpler than any of the types of animal movement discussed in this chapter, but it should be remembered that all animal movements, even those of spermatozoa (sec. 17.2), depend upon the ability of cytoplasm to change its shape, and we are still ignorant of much of the mechanism of this change.

With the experience of the animals already mentioned you should not find it difficult to make progress in the study of movement in other animals. Insects, snails, tadpoles, and frogs are suitable material to start with.

QUESTIONS 13

1. What is meant by (*a*) a voluntary muscle and (*b*) an involuntary muscle? Draw the skeleton of the pectoral girdle and fore-limb of a named mammal. Add two main muscles concerned in the movement of the limb, showing their origin and insertion. State how the limb functions and how its movements are controlled. [O]

2. Distinguish between voluntary and involuntary muscles. State where in the body the following muscles are found and give an account of their functions: (*a*) sphincter, (*b*) intercostal, (*c*) ciliary, (*d*) radial. [O]

3. Describe the structure of a movable joint which you have studied, and illustrate it by a labelled diagram. Explain the action of the muscles on the bones which causes movement at the joint. Every muscle has a blood supply and a nerve supply. Give reasons for this. [C]

4. Briefly describe the main parts of the skeleton of any mammal you have studied. Show how the skeleton appears to be suited to carry out its chief functions. (Do not describe the skeleton in detail.) [L]

5. Make a labelled diagram of a generalised mammalian vertebra. Describe how this generalised structure is modified in the different regions of the vertebral column of a mammal which you have studied. [L]

6. Describe concisely with the aid of labelled diagrams how movement is produced (*a*) in the limb of a mammal, and (*b*) in the shoot of a seedling which is making a tropic response. [L]

14 Sense Organs

Animals and plants can detect many of the changes which occur from time to time in their external environment. All animals, except the very simplest, possess specialised sensitive cells or **receptors** for this purpose. Sometimes a great many receptors are aggregated together to form an organ, as in the eye and the ear. Plants are without such obvious specialisations, but are nevertheless highly sensitive.

The detection of environmental changes or stimuli enables the organism to respond in a way advantageous to itself. For example, many aquatic animals, such as caddis-fly larvae which cling to the underside of stones, will sense the increased intensity of light falling on their eyes when their stones are turned over, and move about until they have once more found protection under a stone. Similarly, the stem tip of a pea plant, when illuminated from one side, can perceive the un-even illumination and respond by bending its stem toward the light, a response which is advantageous to the plant, since it brings the leaves where they will be able to photosynthesise. This stem tip does not possess any specialised sense organ comparable with the eyes of the caddis-flies mentioned above.

It is generally stated that man has five senses: sight, hearing, smell, taste, touch. (The so-called sixth sense is not definable biologically.) The organs in which these senses reside are the eyes, ears, nose, tongue, and skin respectively. When senses are carefully investigated, it has to be admitted that far more than five exist in man, and that the sense organs listed above are concerned with more than one type of sensation. In fingers, for instance, there are sense cells sensitive to heat, touch, and cold: the ear is sensitive to sound and to movement of the head. The eyes are sensitive to changes of light intensity, to colour, to distance, to shape, and to movement, but few would allow all these as separate senses. Taste and smell are both chemical senses, and what is commonly experienced as taste certainly involves the nasal sense organs as well as those on the tongue. Yet another sense is the muscle sense (see sec. 15.3), which is vital in co-ordinating movement.

Many mammals probably have senses very similar to those of man, since they possess the same sense organs: but, as one moves farther from this group, to fishes for instance, where there are sense organs not found in man, one's appreciation of their sensations must get less and less accurate. For example, in most fish there are sense organs in the lateral-line canals (see Fig. 14.9) which allow the detection of disturbances in the water such as may be produced by the movements of other fish or water currents.

The mammalian eye and ear have been studied more than any other sense organs and are dealt with at length in this chapter.

THE MAMMALIAN EYE

14.1 Structure

Every mammal has two eyes which are hollow, nearly spherical bodies, each protected in a bony

HUMAN EYE

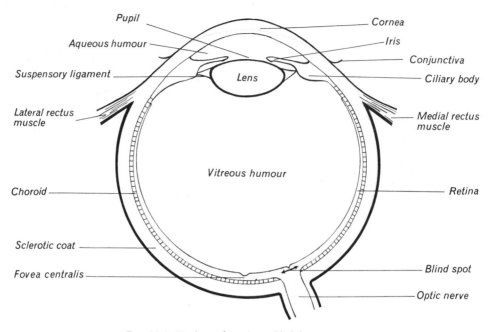

FIG. 14.1 Horizontal section of left human eye.

recess of the skull. Anteriorly, in the part exposed when the eyelids are open, the eye has a shorter radius of curvature and so bulges forward (see Fig. 14.1). The outer surface of the eye is lubricated by tears formed by the **lachrymal glands** whose ducts open inside the eyelids. The wall of the eye is composed of three layers:

The Fibrous Coat

The fibrous coat is the tough outer covering of the eye. Its anterior transparent part is known as the **cornea**, and the remainder, which is white and opaque, is the **sclerotic coat**. The **conjunctiva** is a thin transparent membrane which covers and protects the cornea externally and is continuous with the lining of the eyelids. Part of the sclerotic coat can be seen surrounding the coloured iris which is visible through the cornea. Where the optic nerve leaves the eye it perforates the sclerotic coat.

The Vascular Coat

Over the greater part of the eye the vascular coat is a very thin middle layer containing black pigment and blood vessels. This part of the vascular coat is the **choroid**. Anteriorly, the vascular coat is specialised to form the muscular ring-shaped **ciliary body** to which the suspensory ligament of the lens is attached. The disc-shaped **iris** with a

circular aperture, the **pupil**, in its centre lies in front of the lens and is also part of this layer. The anterior surface of the iris is variously pigmented, but the posterior surface which is in contact with the lens is black.

The **lens** itself is a transparent elastic body enveloped in a fibrous sheet, the **suspensory ligament**, which attaches it to the ciliary body all round its inner perimeter. The lens separates the anterior chamber of the eye, which contains a watery fluid known as the **aqueous humour**, from the posterior chamber, which is filled with a jelly-like transparent material called the **vitreous humour.**

The Retina

The retina is the innermost layer and forms a lining over almost the whole internal surface of the posterior chamber of the eye. It is everywhere thicker than the choroid with which it is in contact, and contains two types of light-sensitive cell as well as nerve cells and fibres (see sec. 15.1). These two cell types are called **rods** and **cones**. Rods are more numerous near the periphery of the retina and cones are more numerous near its centre. When stimulated by light, 'messages' from these cells are propagated along nerve fibres which all run toward the **blind spot** (see Fig. 14.2), where they leave the eyeball in the optic nerve on their

BLIND SPOT

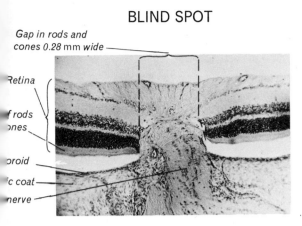

Gap in rods and cones 0.28 mm wide

Retina

f rods

ones

oroid

c coat

nerve

FIG. 14.2 Eye of rat sectioned through the blind spot.

way to the brain. The blind spot is devoid of light-sensitive cells and lies on the nasal side of the optical axis. It can be detected by looking at the cross below with the right eye only and moving the book slowly to and fro at a distance of approximately 0.2 m from the eye. Notice that the black spot disappears and then reappears.

+ ●

When it is not visible its image is falling on the blind spot.

On the optical axis of the eye is a small depression in the retina (**fovea centralis**) where only cones are found. This is the region of the retina which allows the most detailed and distinct vision as well as the most efficient rendering of colour. Normally, when concentrating one's gaze on a certain spot, the six eyeball muscles of each eye adjust its position so that the images of the spot fall on the foveas.

Cones are stimulated only by comparatively high light intensities and they enable us to distinguish different qualities of light, i.e. different colours. At low light intensities, such as are experienced at night, only the rods are able to work and they do not give rise to colour sensations; everything at night appears black or white or an intermediate shade of grey.

14.2 Focusing (Accommodation)

The eye forms an optical system focusing light rays on the retina, and it is at the air/cornea boundary that most refraction occurs. This refraction is constant in amount whether distant or near objects are being observed. Fine adjustment of the focusing system is provided by changing the shape of the lens. When it is least convex, converging power is at the minimum and the eye is accommodated for distant vision; when it is most convex, converging power is at the maximum and the eye is accommodated for near vision (see Fig. 14.3). Two factors affect the shape of the lens: firstly the natural elasticity of the lens, and secondly the tension in the suspensory ligament which is controlled by the muscles in the ciliary body.

When the eye is observing distant objects, circular muscles in the ciliary body relax and the ciliary body is extended to a large circle by the pressure of the fluid contents of the eyeball. This causes the suspensory ligament to become taut, which squeezes the elastic lens to a flatter shape, so reducing the converging power of the system and bringing light rays from distant objects to a focus on the retina. In this state the eye is at rest and is accommodated for distant vision.

When the eye is observing near objects the ciliary muscles contract, constricting the ciliary body to a slightly smaller circle. (The muscles are contracting against the pressure of the fluid contents of the eyeball.) This change in diameter of the ciliary body is sufficient to slacken the suspensory ligament, which ceases to flatten the lens and allows it to assume a more convex shape under its own elasticity. The converging power of the system is thus increased and rays of light from near objects are brought to a focus on the retina. In this state the eye is accommodated for near vision.

ACCOMMODATION

for

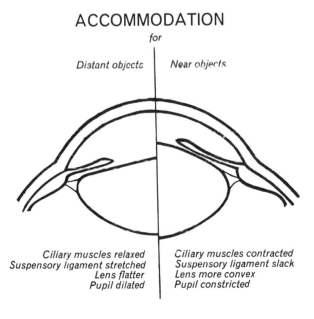

| Distant objects | Near objects |

Ciliary muscles relaxed
Suspensory ligament stretched
Lens flatter
Pupil dilated

Ciliary muscles contracted
Suspensory ligament slack
Lens more convex
Pupil constricted

FIG. 14.3.

156

When one is observing near objects, circular muscles in the iris also contract, reducing the pupil's diameter and admitting only a narrow cone of light. The effect of this is to allow many near objects at different distances from the eye to be in focus at the same time, i.e. to increase the depth of focus.

The iris also acts as a diaphragm reducing the pupil's diameter when the intensity of light falling on the retina is increased, as happens when a light is suddenly shone into the eye. This response is protective, since very intense light can damage the delicate retina.

The response of the iris to increased light intensity is easy to demonstrate, but a little practice may be necessary before you can demonstrate iris movements in accommodation for near vision. The latter demonstration involves focusing on a distant object with one eye, and then interposing a sheet of paper with a cross on it, about eight inches from the eye. Do not focus on the cross at first, but gaze right through the paper at infinity.

Then, when your partner taps you on the shoulder, look at the cross. If you do this correctly your partner will see the pupil diameter decrease when you look at the cross. It is, of course, important to keep the light intensity constant during the demonstration. You should, therefore, keep the paper perfectly still once it is in position.

The mammalian eye and a camera are compared in the table below.

14.3 Defects and their Correction

In such a complicated delicate structure as the human eye it is hardly surprising that abnormal vision occurs. The four commonest abnormalities are described below.

Long Sight (Hypermetropia)

Hypermetropia can be defined as a condition in which rays of light from distant objects are focused behind the retina in the unaccommodated eye. In such cases distant objects can be seen

Eye	*Camera*
1. Sclerotic coat	Camera case
2. Choroid	Black paint
3. Retina	Sensitive film
4. Iris	Iris diaphragm
5. Most refraction of light occurs at air/cornea boundary, a variable small amount at the lens	Refraction of light only at the lens
6. Focusing objects is automatic and is achieved by altering the focal length of the lens	Focusing different objects is achieved by moving a lens of fixed focus toward and away from the film
7. Depth of focus automatically increased when looking at near objects	Iris diaphragm has to be moved by hand

LONG SIGHT AND ITS CORRECTION

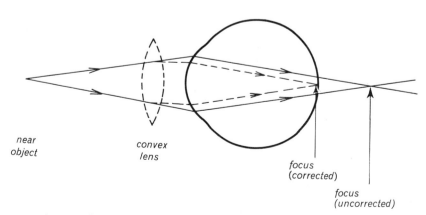

near object   convex lens   focus (corrected)   focus (uncorrected)

Fig. 14.4 Diagram of a long-sighted eye focusing rays of light from a near object behind the retina. For simplicity the lens has been omitted. The dotted lines represent the rays of light after correction by a convex lens. The eye, in this example, is fully accommodated.

SHORT SIGHT AND ITS CORRECTION

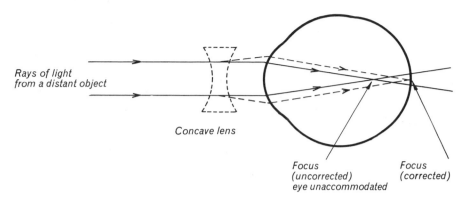

Fig. 14.5 Diagram of a short-sighted eye focusing rays of light from a distant object in front of the retina. For simplicity the lens has been omitted. The dotted lines show the rays after correction by a concave lens.

clearly by accommodating (i.e. by contracting the ciliary muscles and increasing the converging power), but further accommodation to focus the divergent light from near objects is not possible (see Fig. 14.4). Thus distant objects can be seen clearly, but near objects cannot.

All infants are long-sighted because the lens reaches the adult size long before the rest of the eyeball has grown. Thus light from near objects is focused behind the retina and a clear image is not obtained. This condition usually disappears as growth continues.

A frequent result of long sight is eye-strain owing to continual use of the ciliary muscles even when looking at distant objects. Long sight is corrected by the wearing of spectacles with suitable convex lenses which give the light rays an initial convergence before entering the eye.

Short Sight (Myopia)

Short sight, or inability to focus distant objects clearly, is defined as the condition in which rays of light from distant objects are focused in the vitreous humour by the unaccommodated eye (see Fig. 14.5). It can arise from several causes, but usually it is the result of the eyeball becoming too long. This condition often develops at school age when a lot of close work is being done with the eyes. The correction, in this case, is to wear spectacles with suitable concave lenses which diverge the light rays slightly before they enter the eye.

Old Sight (Presbyopia)

This condition is caused by a progressive hardening of the lens with age, which results in a gradual loss of accommodating power. It is usually first noticed when the near point has receded so far that a book or newspaper has to be held at arm's length in order to be read. It is corrected by using suitable convex lenses in spectacles for reading, which increases the converging power sufficiently to focus the divergent rays of light.

Astigmatism

Astigmatism is caused by a deformation of the cornea, which causes rays of light from a point source to form either a vertically or horizontally elongated image of the point source on the retina. One effect of this abnormality is that the vertical and horizontal bars of a window cannot both be seen clearly at the same time. It is corrected by the use of cylindrical lenses. This condition can occur with any of the previous defects.

THE MAMMALIAN EAR

14.4

The ear of a mammal is commonly associated with hearing, but it is not generally known that it is an important sense organ concerned with posture and balance.

Structurally it is divisible into three parts: the outer ear, the middle ear, and the inner ear. (In

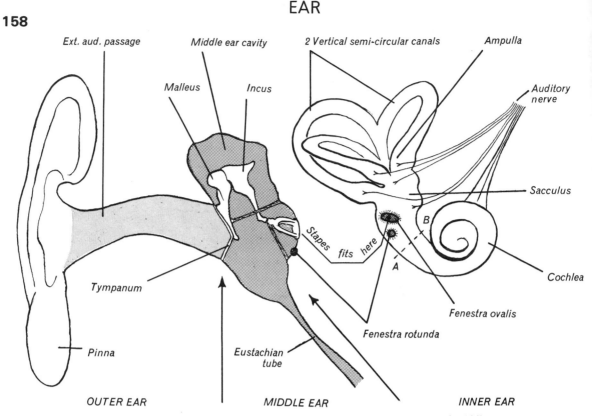

FIG. 14.6 Diagram of the right human ear seen from in front. The outer and middle ears are seen in section, but the inner ear is drawn entire without its bony protection and displaced inwards.

reading the following account constant reference should be made to Fig. 14.6.)

The Outer Ear

The outer ear, composed of the **pinna** and the **external auditory passage**, is responsible for collecting sound waves which cause the eardrum (**tympanum**) at the inner end of the passage to vibrate. The tympanum marks the boundary between the outer and middle ears.

The Middle Ear

The middle ear lies near the posterior end of the skull completely embedded in bone. In the cavity of the middle ear are three small bones or ossicles articulated together, which transmit vibrations from the tympanum to the oval window (**fenestra ovalis**), an opening into the inner ear. The **malleus** (*L. hammer*), the largest of the three ossicles, is attached to the tympanum at one end and articulates with the **incus** (*L. anvil*) at the other. The innermost of the three ossicles, the **stapes** (*L. stirrup*), is shaped remarkably like a stirrup and, at its inner end, fits into the fenestra ovalis across which a delicate membrane is stretched. The

T.S. COCHLEA

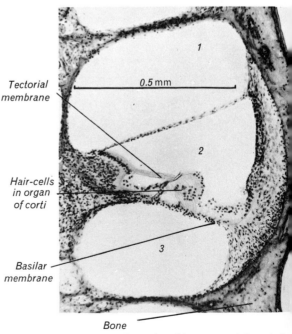

FIG. 14.7 Transverse section of cochlea at AB (Fig. 14.6). The numbered cavities are full of fluid.

eustachian tube is a narrow tube which leads down from the middle ear cavity to the pharynx, and has the function of equalising the air pressure on both sides of the tympanum. This tube is opened to the pharynx every time swallowing occurs.

The Inner Ear

The inner ear or labyrinth is, like the middle ear, encased in bone. It is, however, filled with a fluid and not air. Vibrations of the stapes are transmitted to the fluid through the membrane stretched across the fenestra ovalis. One part of the labyrinth, the **cochlea** (*L. snail*), is a spirally coiled tube, blind at its apex, and divided into three cavities by two membranes which run along its length (see Figs. 14.7 and 14.8). When a certain sound is heard, the thicker of these two, the **basilar membrane**, vibrates in a certain part in sympathy with the incoming sound, i.e. it is made to resonate. On the basilar membrane is the **organ of Corti**, which contains delicate hair cells, the hairs of which are embedded in the **tectorial membrane**. Movement of these hairs causes electrical 'messages' to be sent along fibres of the auditory nerve to the brain, with the result that the noise is heard. The part of the organ of Corti nearest the apex of the cochlea is sensitive to low notes and vice versa. There are several safety devices which prevent damage to the organ of Corti when a loud sound is heard.

In addition to the cochlea, the labyrinth also contains three **semi-circular canals**. These are arranged in three planes at right angles to each other and at one end of each is a swelling, the **ampulla**, where sensory hairs from a small disc of

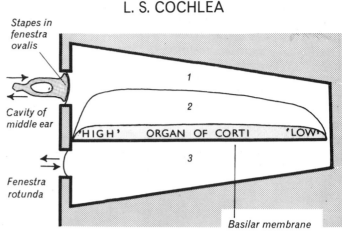

L. S. COCHLEA

Stapes in fenestra ovalis

Cavity of middle ear

'HIGH' ORGAN OF CORTI 'LOW'

Fenestra rotunda

Basilar membrane

FIG. 14.8 Diagrammatic longitudinal section of an imaginary uncoiled and shortened cochlea. The middle-ear cavity is on the left of the diagram.

cells project into the fluid inside. Attached to these hairs is a mucilaginous body which is moved when the head is suddenly rotated in the plane in which the canal lies. This happens because the wall of the canal moves with the head, but the fluid lags behind owing to its inertia, thus moving the mucilaginous body and bending the sensory hairs. When the hairs are bent electrical 'messages' are set up which pass from there to the brain in the **auditory nerve**. The brain thus gets 'information' from the semi-circular canals about changes in rotational movement of the head. It does not, however, receive any input of 'messages' which allow steady rotations of the head to be detected, as can easily be shown with a blindfold person

SEMI-CIRCULAR CANALS

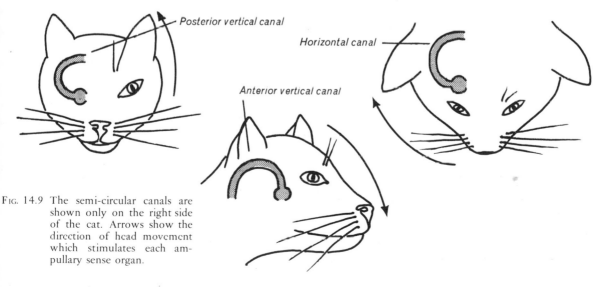

FIG. 14.9 The semi-circular canals are shown only on the right side of the cat. Arrows show the direction of head movement which stimulates each ampullary sense organ.

seated on a rotatable stool. There are also similar structures in the part of the labyrinth between the cochlea and the canals, but these are concerned with detecting position of the head; they are gravity receptors rather than rotary receptors. Sea-sickness originates from the irregular stimulation of these gravity receptors. From these messages which are continuously being received, the brain is able to control muscular movement necessary to keep the body upright or the posture adjusted. Can you explain why giddiness occurs when the body is stopped after rapid rotation? What happens to the eye movements immediately after rotation?

The sense organs in the **lateral-line canals** of fishes (see Fig. 14.10) are of the same type as those in the ampullae of mammals, and they work in a similar way. When the water near a fish is disturbed the eddies cause movement of water in the canals, which moves the hairs causing 'messages' to be sent to the brain. In most fish it is this mechanism which renders them difficult to catch in a net.

LATERAL LINE CANAL

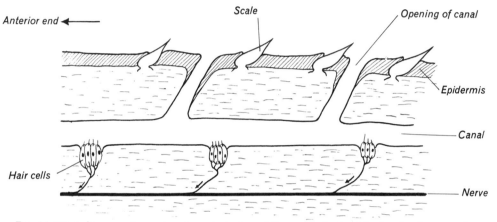

FIG. 14.10 A diagram of a section through the lateral-line canal of a dogfish, showing its external openings and nerve supply.

QUESTIONS 14

1. Make a labelled diagram to show the structure of a mammalian eye. How does the eye produce a record of a page of print in the brain? Show diagrammatically the way in which the defect of short sight is corrected. [C]

2. Draw a large labelled diagram of a section through a vertebrate eye. How does the eye control the amount of light entering it? Explain the functions of (a) the lens and (b) the retina. [L]

3. (a) Make a large labelled sectional diagram to show the structure of the mammalian eye.

(b) A person of normal vision is reading a book in the garden and he suddenly looks up to see a passing aircraft. Explain how his eyes are able to focus clearly on the printed page and an instant later on the passing aircraft. He then goes into a room, the curtains of which are drawn. At first he sees very little, but gradually he sees more of the room's contents. What is happening in the eyes during this change in light intensity? [N]

4. Explain how the following function: (a) ciliary muscles; (b) the iris; (c) semi-circular. canals; (d) the maculae (gravity perceptors) in the utriculus; (e) the cochlea [O]

5. Make a large labelled diagram of the ear of a mammal. Explain briefly the functions of (a) the bones of the middle ear, (b) the semicircular canals, and (c) the eustachian tube. [O & C]

15 Co-ordinating Responses

Mammals and most higher animals possess muscles and glands which can respond to changes in the environment. Such organs are known as **effectors** because they are capable of producing a response. The many various sensory cells of an animal are known as **receptors** because they are sensitive to (or receive) the stimuli. The effectors do not act independently; their responses are co-ordinated by a nervous system whose branches are connected to them and to the receptors. The working of the nervous system of vertebrates is extremely complicated and, particularly in the case of the brain, is very imperfectly understood. Nevertheless, it is true that the nervous system, when a stimulus is received by one or more sensory cells, brings into action a group of effectors which will produce a suitable response. For example, when a person touches a very hot object with a finger several muscles in the arm contract to close the hand, to flex the arm at the elbow, and to retract the whole arm from the shoulder. Each of these individual responses helps to remove the finger from contact with the hot object and the whole response is a useful one.

Co-ordination of the effectors is also carried out by chemical messengers or **hormones** (*Gk. hormao, I set in motion*) which are carried in the blood from the glands which secrete them to the effectors which give the response.

THE NERVOUS SYSTEM

15.1 Structure

The nervous system of a mammal (see Fig. 4.3) is divisible into the following parts:

1. **The central nervous system** (commonly abbreviated to C.N.S.) composed of the brain and spinal cord.
2. **The peripheral nervous system** composed of the paired cranial (from the brain) and spinal (from the spinal cord) nerves which connect receptors and effectors with the central nervous system.

The central nervous system lies in the mid-line of the body, protected inside the skull and vertebral column. It is extremely soft and delicate and receives further protection from three membranes, the meninges, the outermost of which is quite strong and serves to attach it to the inside of the cranium and spinal canal.

The **spinal cord** is roughly cylindrical, tapering toward its posterior end, and it has a narrow cavity running down its centre. It gives off paired spinal nerves which arise by both a dorsal and a ventral root (see Fig. 15.1) and pass out between adjacent vertebrae.

The **brain** (see Fig. 15.2) is continuous with the spinal cord through the posterior opening of the

SPINAL CORD

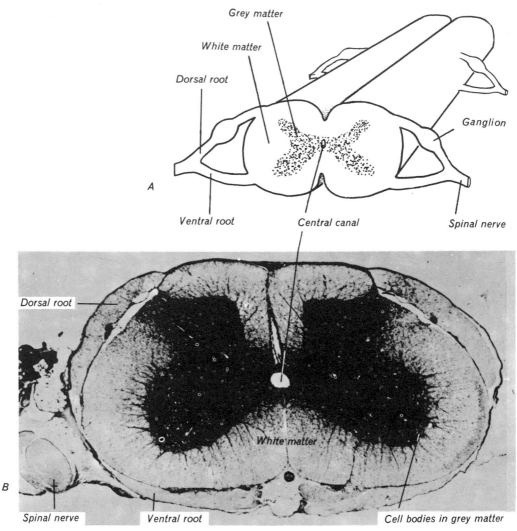

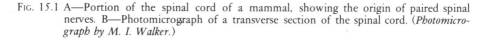

TRANSVERSE SECTION

Fig. 15.1 A—Portion of the spinal cord of a mammal, showing the origin of paired spinal nerves. B—Photomicrograph of a transverse section of the spinal cord. (*Photomicrograph by M. I. Walker.*)

cranium, the foramen magnum, and is a part of the hollow central nervous system which has become so much expanded and folded in parts that its original tubular structure is no longer obvious. Its central cavity is extensive and divided into four interconnected chambers or ventricles. The most prominent parts of the brain of a rabbit are the paired **cerebral hemispheres** (= cerebrum), which are large expansions of the anterior part of the **brain stem**, which they cover dorsally. The cerebral hemispheres of man (see Fig. 15.3) are corrugated externally with deep grooves and are

relatively larger than those of a rabbit, extending back far enough to overhang the cerebellum and hide it from above. In the rabbit the **cerebellum** is a five-lobed outgrowth from the posterior part of the brain stem, but in man it is more compact and composed of two hemispheres.

The characteristic cell of the nervous system is the long, thin **nerve cell** or **neuron** which is capable of transmitting a nervous (electrical) impulse in one direction from one end to the other. The neuron is typically composed of a **cell body** (5 to 120 μm in diameter) and one or more fine

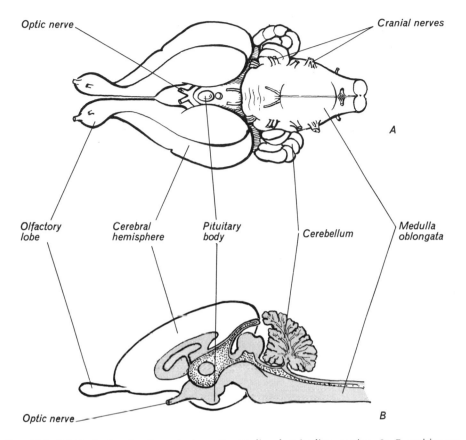

Optic nerve — Cranial nerves

A

Optic nerve — *B*

Olfactory lobe | Cerebral hemisphere | Pituitary body | Cerebellum | Medulla oblongata

Fig. 15.2 Rabbit brain. A—Ventral view. B—Median longitudinal section. In B cavities are stippled, cut surfaces are grey. Note that the brain stem is straight.

processes (from 5 μm to nearly 1 m long). Two types of neuron are shown in Fig. 15.4. The processes which conduct the nerve impulse toward the cell body are known as **dendrites** (*Gk. dendron, a tree*) and those which conduct away from the cell body are called **axons**. Note the differences between the motor neuron which conducts from the C.N.S. to effector organs, and the sensory neuron which conducts from the sense organs to the C.N.S. It is important to realise the great length of the processes in some neurons. For instance, a motor neuron may have its cell body in the lumbar region of the spinal cord and its terminal branches in a muscle of the foot.

The long processes are usually covered by a thick layer of **myelin** which is fatty (see Figs. 15.5 and 15.6) and is thought to be important in increasing the rate at which the nerve impulse travels. The myelin sheath is formed from cells about

1 mm long which grow spirally around the long nerve cell processes and is thus interrupted where the cells meet at the nodes of Ranvier. The **neurilemma** (*Gk. neuron, nerve; eilema, sheath*) is the outer membrane of these cells and their nuclei lie close underneath. A long nerve cell process, its myelin sheath and neurilemma are collectively known as a **nerve fibre**.

The arrangement of cell bodies and nerve fibres is not the same in the brain as in the spinal cord. In the former, most cell bodies are near the outside and fibres coming from them run internally to other parts of the central nervous system. The fatty nature of these fibres makes the central part look **white** in contrast to the **grey** tissue where the cell bodies are accumulated. The position of the grey and white tissues is reversed in the spinal cord (see Fig. 15.1).

The dendrites and terminal branches of adjacent

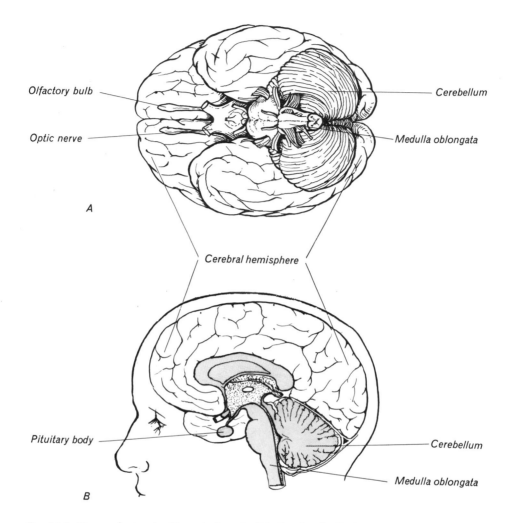

Olfactory bulb

Optic nerve

Cerebellum

Medulla oblongata

A

Cerebral hemisphere

Pituitary body

Cerebellum

Medulla oblongata

B

Fig. 15.3 Human brain. A—Ventral view. B—Median longitudinal section. In B cavities are stippled and cut surfaces grey. Note that there is almost a right-angle bend in the brain stem. Compare with Fig. 15.2.

neurons end in very close contact with one another, but it is believed that there is no protoplasmic continuity between them. Nevertheless, a nerve impulse passes across from one neuron to another at these places which are known as **synapses** (singular, synapse).

15.2 The Nerve Impulse

When a nerve impulse passes along a fibre an electrical change which propagates itself rapidly can be detected. It is not surprising, therefore, that artificial electrical stimulation of a nerve fibre can set off a nerve impulse. The impulses travel at

a maximum velocity of thirty metres per second in a frog and one hundred and twenty metres per second in man.

The passage of nerve impulses along nerve fibres can be demonstrated in the following way. Kill a frog and quickly dissect out one of its gastrocnemius muscles and the long nerve which supplies it. Keep the preparation moist during the whole experiment with Ringer's solution, which is similar in composition to the plasma of the frog's blood. Anchor the knee joint, which has been cut out with the muscle, by pinning it to a sheet of cork, thus fixing the origin of the muscle. Tie the

NERVE CELLS - NEURONS

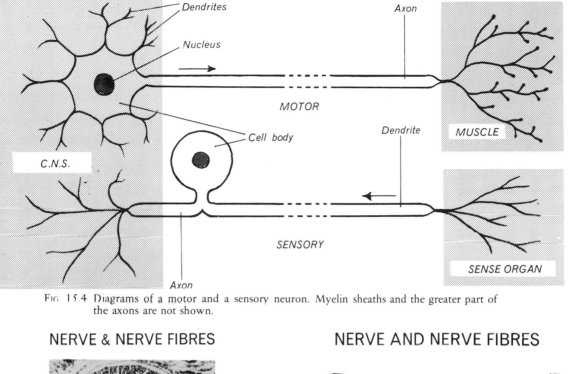

FIG. 15.4 Diagrams of a motor and a sensory neuron. Myelin sheaths and the greater part of the axons are not shown.

NERVE & NERVE FIBRES

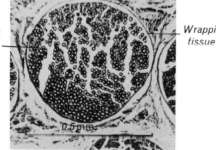

T. S. PART OF BRACHIAL NERVE OF GUINEA PIG

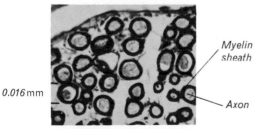

PORTION OF THE ABOVE SECTION HIGHLY MAGNIFIED

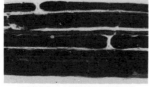

TEASED NERVE FIBRES

FIG. 15.5 Photomicrographs of nerve and nerve fibres. Compare with Fig. 15.6.

NERVE AND NERVE FIBRES

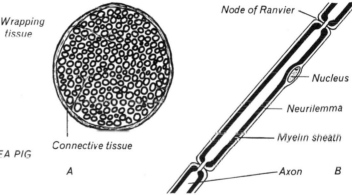

FIG. 15.6 A—Diagrammatic transverse section of a small nerve. B—Diagrammatic longitudinal section of part of a single nerve-fibre.

tendon of insertion of the muscle to a small lever with thread and then stimulate the nerve electrically as far from the muscle as possible. This stimulation can be carried out either with an induction coil or with a mains-powered stimulator (see Fig. 15.7). Notice that the muscle contracts when the nerve is stimulated electrically and also when the free end of the nerve is pinched with forceps. Thus, nerve impulses set up in the nerve fibres by the electrical and mechanical disturbances have passed from the point of stimulation to the muscle. Records of the muscle's contraction can be made by a lever which writes on a rotating smoked drum. There is ample evidence that this conduction takes place in the nerve in a living frog.

15.3 Reflex Action

Now we must examine the working of the nervous system in a living animal. It is easiest to consider first how the nervous system co-ordinates simple responses known as reflex actions. **Reflex actions** are quick, automatic responses given on receipt of a definite external stimulus and they do not involve any conscious effort.

The knee-jerk reflex can be demonstrated in the sitting position by crossing the right leg over the left and allowing the former to hang freely from the knee, thus stretching its quadriceps femoris muscle (see sec. 13.6) slightly. Then, ask

someone to give a smart tap with the edge of a thin book just below the knee-cap on the patellar tendon. The quadriceps femoris immediately contracts, jerking the lower leg upwards. If the response is not convincing, it can usually be increased by clenching the fists and setting the teeth.

Some of the nerve cells involved in this reflex action are shown in Fig. 15.8. You will see that the **sensory** or **afferent neuron** has its origin in the quadriceps muscle. Its fine branches are actually wrapped around the muscle fibres, forming a **stretch receptor** which, when stretched slightly as by a tap below the knee-cap, sends off a nerve impulse to the terminal branches of the cell in the lumbar region of the spinal cord. From here the impulse crosses a synapse to a **motor** or **efferent neuron**, down the nerve fibre of which it passes to the terminal branches. These endings are in contact with several muscle fibres which contract when the impulse reaches them. Notice that the

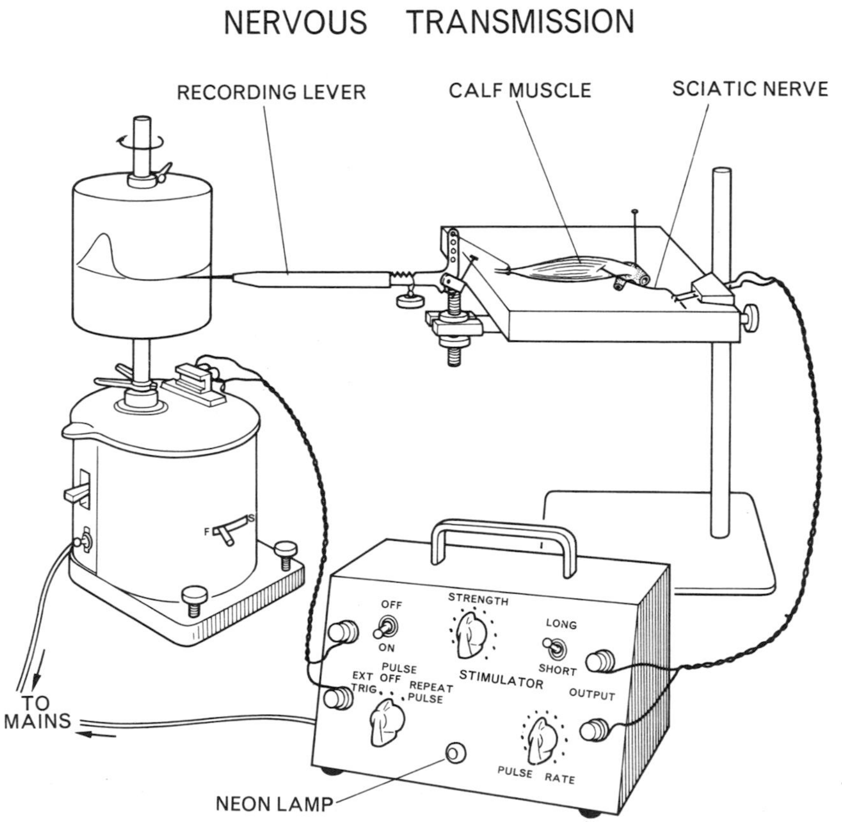

NERVOUS TRANSMISSION

FIG. 15.7 Diagram of the apparatus used to demonstrate nervous transmission in the sciatic nerve of a frog. On the body of the recording drum is a switch which triggers the stimulator to send an electrical pulse to the nerve once every revolution of the drum. The tendon of the muscle is tied with cotton to the lever whose point makes a tracing on the carboned paper attached to the drum. The neon lamp provides a visual check of pulses sent out by the stimulator.

KNEE-JERK REFLEX

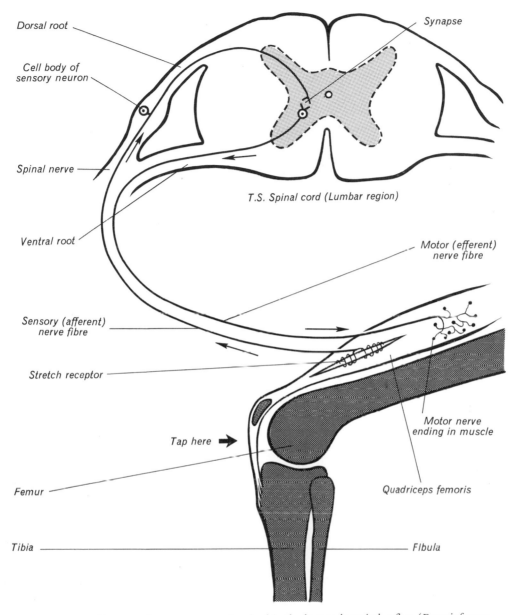

Fig. 15.8 A diagram of some structures involved in the human knee-jerk reflex. (*From information supplied by Dr. D. Taverner.*)

sensory fibre carries the impulse into the spinal cord by the **dorsal root** of the nerve and that the motor fibre carries the impulse out by the **ventral root**. The sense organ, the effector, and the two neurons are often referred to as a '**reflex arc**'.

No reflex action involves quite such a simple arrangement as that just described, for more than one stretch receptor is involved, and each sensory neuron 'connects' with several motor neurons.

It can be shown that the nervous pathways mentioned above are involved in this reflex action, since when the dorsal or ventral roots supplying the hind legs of a mammal are severed, the response is no longer given. A decapitated mammal can still give the response, however, so we conclude that the brain is not essential for the response to be given.

This stretch reflex helps one to stand upright

REFLEX ACTIONS

Name	Stimulus	Response
Coughing reflex	Foreign particles in the respiratory tract	Coughing, i.e. sudden contraction of abdominal muscles and expiratory intercostal muscles, and relaxation of diaphragm muscles
Swallowing reflex	Food, saliva, or tongue touching the posterior wall of the pharynx	Soft palate and larynx raised, epiglottis closed, peristaltic wave passes down oesophagus
Blinking reflex	Solid objects touching the conjunctiva	Contraction of sphincter muscle which closes the eyelids together
Withdrawal reflex	Painful stimulation of hand or foot, e.g. pricking or burning a finger	Contraction of flexor muscles, which bends limb at elbow or knee so withdrawing hand or foot
Pupil reflex	Increased light intensity falling on the retina of the eye	Circular muscle fibres of iris contract and reduce the diameter of the pupil

without a voluntary effort since the slightest bending of the knee joint as the legs begin to collapse will stretch the receptors and so cause extension of the leg at the knee.

One further complication remains to be explained—voluntary control over the extension of the leg at the knee is provided by nerve cell bodies in the cerebral hemispheres, whose fibres run down through the brain stem and the white matter of the spinal cord to 'connect' at synapses with the motor neurons already described. Also the sensory neurons involved in this reflex 'connect' with nerve cells whose fibres run up the spinal cord to the brain, and carry impulses which make us conscious of the response having taken place. Normally, the brain does not play an effective part in the knee-jerk reflex.

Stretch receptors are present in muscles as well as in tendons, and they are responsible for man's **muscle sense**. They are constantly sending nerve impulses to the brain, which helps us to know exactly how our limbs are positioned without looking to see.

A great many reflexes occur in all mammals, in fact these constitute a large part of the total responses or behaviour of which we are normally unaware. Such reflexes are operating when we blink, when we breathe, when we secrete saliva, when we swallow, and when we reduce the size of our pupils in strong light (see sec. 14.2).

15.4 Conditioned Reflexes

Those reflexes we have considered in the previous section are simple instinctive ones which are inherited. All mammals which have been investigated exhibit another type of reflex, which is not inherited but is acquired as a result of certain conditions in the environment they have met. An example will make this clear.

A dog, one of whose salivary ducts has been made to open on the cheek so that salivary flow can easily be observed, is found to salivate when food is placed in the mouth. On several successive occasions a tuning fork is struck when food is given, and each time the dog salivates. Then the tuning fork is struck without giving food and the dog still salivates. Thus, the sound, which previous to the experiment had no association for the dog with food, has acquired new stimulative properties. In short, the dog has become conditioned, or has established a conditioned reflex. Striking the tuning fork without feeding would eventually lead to the disappearance of the response. You can probably think of several other examples of domesticated animals establishing conditioned reflexes connected with feeding.

There is no doubt that the establishment of conditioned reflexes is continuously playing a part in the life of humans and other mammals, adjusting their behaviour to the environment rapidly and automatically. It is also an important factor in the learning exhibited by young children.

15.5 More Complex Behaviour

The way in which man maintains his equilibrium and controls fine degrees of muscular movement is an instructive example of co-ordination. The brain receives impulses through sensory nerves from four sources:

1. The semicircular canals.
2. The eyes.
3. The pressure receptors in the skin.
4. The stretch receptors in the muscles and tendons.

MOTOR AND SENSORY AREAS

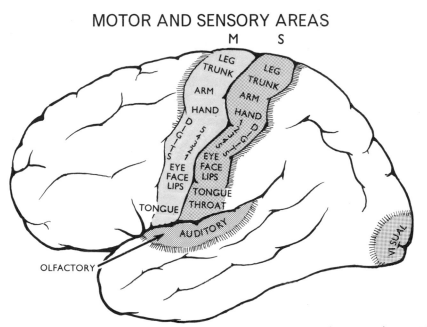

FIG. 15.9 Left side view of cerebrum of man to show motor (M) and sensory (S) areas. The olfactory area is not visible externally but lies close to the mid-line at the end of the arrow. The remaining areas, known as silent areas, contain association neurons which interconnect sensory and motor areas.

This sensory input, as we may call the impulses referred to above, is relayed in various tracts of fibres to the cerebellum, a part of the brain containing **nerve centres** (i.e. aggregations of nerve-cell bodies) important in the control of equilibrium and posture. From the cerebellum a tract of motor fibres runs down the spinal cord making 'connections' with final motor neurons leading to muscles. Now consider the following experimental facts:

1. When the cerebellum is damaged, staggering, inability to perform fine degrees of muscular movement (e.g. such as in threading a needle), and tremor of the limbs occur.
2. A blind or blindfolded person can control his posture and maintain his equilibrium with ease.
3. A person, one or both of whose auditory nerves have been chilled out of action by pouring ice-cold water into the external ear, is quite unable to walk along a white line, is very unsteady on his feet, and even falls over.
4. A blindfolded person who tumbles into deep water is unsure which way up he is for a very short time or until some part of his body touches the bottom. (Note that water stimulates all pressure receptors equally, since it presses all round the body.)

Apparently the part of the sensory input we can dispense with least is that from the semicircular canals. As a result of the sensory input it receives, the cerebellum is constantly sending out impulses in motor nerves adjusting the degree of contraction of muscles and hence the posture. However, it should be stated here that there are higher motor centres in the cerebral hemispheres which 'have the final word' in what happens—one can lie down or fall down voluntarily!

In complex activities such as walking, running, skipping, or cycling, activities which involve almost the whole muscular system of the body, a great many local reflexes are involved as well as the controlling action of the cerebellum superimposed upon them. It is essential that each reflex action should be brought into play at exactly the right moment if the normal rhythm of such movements is not to be upset. One has to learn such movements laboriously, but it is not known what is happening within the nervous system when this learning is taking place. Once learnt, however, the movements can be performed without any conscious effort.

Part of the surface of the cerebral hemispheres can be divided into well-defined **motor areas** which, when stimulated electrically, cause movements of localised parts of the body on the opposite side to that stimulated (see Fig. 15.9). This is explained by the fact that there are nerve-cell bodies

MUSCULAR RESPONSE TO A VISUAL STIMULUS

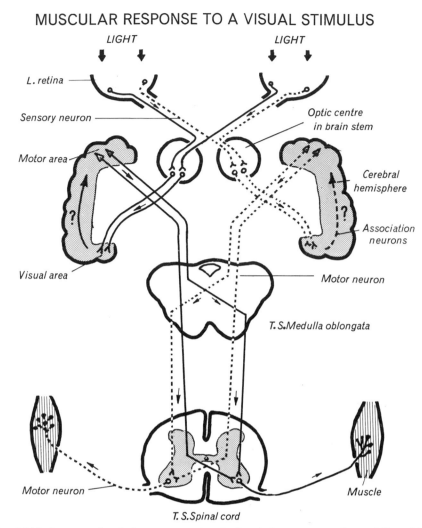

FIG. 15.10 An example of the structures involved and the nervous pathway followed by an impulse starting from a visual stimulus and resulting in a muscular response. Use this diagram to work out what happens when you try to catch a ball which has been thrown into the air.

in the motor areas whose fibres run down the brain stem and spinal cord in a great tract 'connecting' with the final motor neurons all the way down. It is believed that impulses are initiated in these cell bodies whenever a voluntary movement takes place. Posterior to the motor areas is a series of **sensory areas** whose electrical stimulation gives rise to sensations of touch, warmth, or numbness in a part of the body opposite to the side stimulated. Other areas concerned with vision, hearing, smell, and eye movements have been located. The rest of the cerebral hemispheres is composed of association neurons which interconnect the motor and sensory areas. If you study Fig. 15.10 you will appreciate the parts played by the sensory and motor areas in a common example of human behaviour.

Man exhibits a high degree of intelligence which could be defined as the ability to associate facts, to reason or think, and to solve problems mentally. He is also possessed of consciousness which allows him to be aware of his environment and perhaps it is this awareness which leads him to spend so much time and energy in trying to control his environment for his own purposes. How far other animals are aware of their surroundings it is impossible to tell. The seat of all these higher activities is the cerebrum, whose structural development in man far surpasses anything found in other mammals, even in chimpanzees.

THE ENDOCRINE SYSTEM

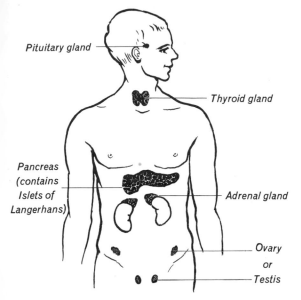

FIG. 15.11 Some of the glands which produce hormones in man. The pancreas, which normally lies in front of the kidneys and adrenal glands, is shown higher than its correct position.

THYROID DEFICIENCY

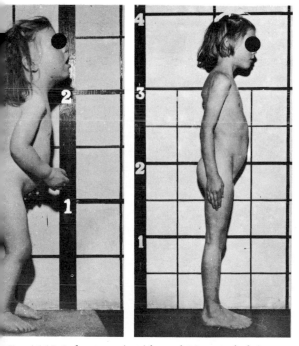

FIG. 15.12 Left—A cretin girl, aged 2½ years, before treatment. Right—The same girl five years later after treatment with thyroid-gland extract. (*Photograph supplied by Dr. Alex Russell.*)

So far, little has been said about the functions of the brain stem. Its most posterior part, the medulla oblongata, contains nerve centres which control many vital activities such as the pulsating of the heart, breathing movements, and the production and loss of heat. It also contains the tracts of nerve fibres leading to and from the cerebellum and cerebrum.

HORMONES—THE ENDOCRINE SYSTEM
15.6

You will remember several examples of glands which have ducts (pancreas, salivary glands, sweat glands, lachrymal glands) from previous chapters. There is also in the mammalian body an important series of **ductless glands**, collectively known as the endocrine system, which pass their secretions into the blood in their capillaries. The secretions contain **hormones** which circulate in the blood and, on reaching an effector organ, which may be far removed from the gland, they cause a response. This is yet another way in which mammals co-ordinate their activities.

Secretin was the first hormone to be discovered (1902). It was shown to be secreted into the blood by the lining of the duodenum when food enters this organ from the stomach. On reaching the pancreas it caused this organ to pour out its juice, thus ensuring the presence of the juice at the right time. Since 1902 a great many hormones have been discovered and their actions described. They occur not only in mammals, but in all vertebrates and many invertebrates, notably insects. The more important glands of the human endocrine system are shown in Fig. 15.11 and some details of them are given below.

The **thyroid gland** in the neck secretes a hormone, **thyroxin**, which contains iodine (see sec. 7.4). Such secretion is vital to life, hence the need for small quantities of iodine in the diet. It does not cause any one organ to respond but has the following widespread effects:

1. Control of growth in man and metamorphosis in frogs.
2. Stimulation of the rate at which glucose is oxidised to produce energy.

If, as sometimes happens, the gland is underactive in children, stunted growth and a failure of mental development result. Children so affected are known as **cretins**, but nowadays they can usually be cured by feeding with tablets prepared from the thyroid gland of sheep (see Fig. 15.12). Overactivity sometimes occurs in adults and may be treated successfully by the surgical removal of part of the gland or by the injection of thiouracil, a drug which prevents the formation of the hormone, thus reducing the amount of thyroxin which is passed into the blood.

Near the kidneys is a pair of small **adrenal glands**. Their central part or medulla secretes a hormone known as **adrenalin** but, unlike thyroglobulin, which is produced continuously, adrenalin is secreted only in moments of stress. The effect of adrenalin is a widespread one and its combined effects can be described as 'preparation for flight or fight'. Its more important effects are:

1. Raising the pulse rate and stroke volume of the heart.
2. Increasing the rate and depth of respiration (restricted sense).
3. Constricting blood capillaries in the alimentary canal and skin, and dilating those in the muscles.
4. Stimulating the rate at which glucose is oxidised to produce energy.

If you have ever experienced a weird 'sinking feeling' in the abdomen just before an important event, such as a race on Sports Day or a visit to a person in authority about a disciplinary matter, it was probably due to your adrenal medulla pouring out adrenalin into the blood.

Sugar diabetes, like cretinism, is a disorder caused by the underactivity of a ductless gland. The gland, in this case, is not compact but is composed of small globular masses of tissue embedded in the pancreas. They are known as the **Islets of Langerhans** (see Fig. 15.13) and secrete the hormone **insulin** (see sec. 7.2), which has the effect of keeping the glucose content of the blood down to its normal value of 0.1 g per 100 cm³ of blood. It achieves this by:

1. Stimulating the conversion of glucose to glycogen, which is stored in the liver.
2. Stimulating the oxidation of glucose to produce energy.
3. Inhibiting the production of glucose from protein in the liver.

If the gland is underactive, the amount of glucose in the blood rises because of the absence of the effects listed above. The kidney tubules cannot reabsorb all this glucose from the filtrate which passes down them and so glucose appears in the urine and is lost. Finally the body will be drained of its glucose and death will occur unless insulin, which can be isolated from the pancreas of cattle, is administered. All diabetics had to inject insulin

PANCREAS

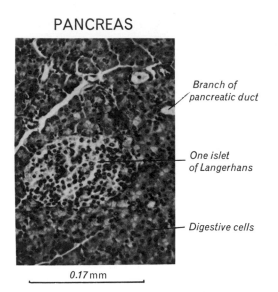

Branch of pancreatic duct

One islet of Langerhans

Digestive cells

0.17 mm

FIG. 15.13 Photomicrograph of a stained section of pancreas.

regularly until recently, when a new preparation which can be given to some by mouth became available.

These examples make it clear that the hormonal or chemical method of co-ordinating the functions of the parts of the body is quite different from the nervous method. Nevertheless, they work together intimately in integrating the activities of the body. Many endocrine glands are under nervous control (e.g. the adrenal medulla); others (e.g. the thyroid gland) are controlled by hormones from the **pituitary gland**, which is itself under nervous control from the brain. Because it controls other glands by its secretions, and because it produces a large number of hormones affecting activities such as reproduction, water regulation, growth, and the metabolism of fat and carbohydrate, it is sometimes called the 'master gland'. The pituitary gland is located on the ventral side of the brain stem (see Fig. 15.3).

Transmission of the nerve impulse across a synapse is known to be the result of the nerve cell secreting either adrenalin or acetyl choline from its terminal branches. These hormones are capable of starting off a new nerve impulse in the dendrites of an adjoining neuron.

Throughout this chapter we have been dealing with mammals which divide their labour amongst many specialised organs. It is not surprising, therefore, that they possess a complex co-ordinating system to integrate the functioning of these organs. In plants, where there is relatively little division of labour, and no rapid movements to execute, there is nothing like a nervous system and only a system of chemical co-ordination (see sec. 12.4, auxins). It seems apparent that the greater the division of labour, the more highly developed the co-ordinating mechanism has to be in order to ensure the harmonious working of the whole organism.

QUESTIONS 15

1. Describe clearly with the aid of labelled diagrams how the nervous and muscular systems enable a mammal to move a limb. [L]

2. Describe the principal features of the nervous system of a mammal. State the functions carried out by the various main parts you have described. [L]

3. Give an account of one ductless gland and the effects of its secretion on the body. Co-ordination is one of the main functions of both hormones and nerves. What important differences are there between their modes of action? [O & C]

4. What are the functions of a nervous system? Make a large labelled diagram to show the structure of the mammalian ear. State the two functions of the ear, and indicate clearly which parts are concerned with each function. [W]

5. What is meant by a reflex action? Describe, with the help of a large, fully-labelled diagram, how a simple spinal reflex action takes place. [W]

16 Plant Reproduction

It was stated in chapter 1 that perhaps the most remarkable activity of living organisms is their ability to create new ones like themselves. Plants and animals are not immortal and must, therefore, reproduce before death overtakes them in order to ensure the perpetuation of the species. This process of reproduction is characteristic of all living things; it is achieved either as the result of two special nuclei from different sources meeting and fusing, or by the separation of some part from the parent. The first type is termed **sexual reproduction** and the second **asexual reproduction**. In sexual reproduction the cells containing the special nuclei are termed **gametes** and their fusion **fertilisation.**

Most flowering plants rely on the flower for reproduction. The flower contains stamens which form pollen (see sec. 3.3); within this pollen the male gametes are produced. The female gamete is in the ovule. After fertilisation the seed develops and this may germinate to form a new plant.

In asexual reproduction there is no fusion of gametes and the new plant does not grow from a seed produced by a flower. You probably know that gardeners increase their stock of many plants by dividing them, or by taking cuttings, rather than by sowing seeds. These are simple examples of asexual reproduction but they are similar to the natural methods employed by many flowering plants as an additional means of reproduction.

BUTTERCUP FLOWER

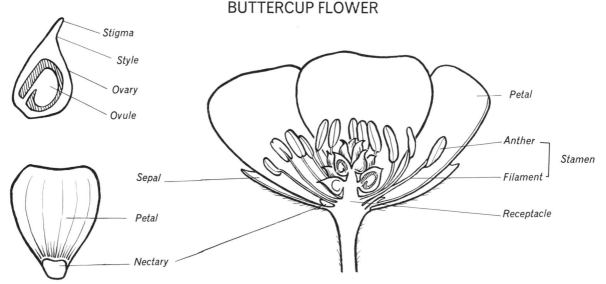

FIG. 16.1 Meadow buttercup (*Ranunculus acris*). The flower is shown cut in half. The two small drawings show the upper surface of a petal and a half-section of a carpel.

Other plants which do not form flowers, such as ferns, mosses, fungi, and algae, also reproduce by similar methods. Many of them form gametes which fuse together and eventually grow into a new plant, though the gametes which fuse are often alike. Some of these plants reproduce asexually by division but many form special **spores** which grow into new plants. These spores are quite different from gametes, for each one can grow into a new plant and does not have to fuse with another one first.

SEXUAL REPRODUCTION IN FLOWERING PLANTS

16.1 Flower Structure

The flower of the wallflower was described in section 3.3, which should now be re-read in order to revise the terms used to describe the various parts. Although these same parts are present in most flowers they are arranged in a great variety of ways. The following brief descriptions of the flowers of the buttercup, white dead-nettle, sweet pea, and ox-eye daisy will help you to appreciate some of these variations.

1. *Meadow buttercup* (*Ranunculus acris*). This has a cup-shaped flower in which the floral leaves are arranged in the same four whorls as in the wall-flower (see Fig. 16.1). There are two other common species of buttercup but all have a similar flower. The calyx consists of five boat-shaped sepals which are hairy on the outside and papery at the edges. (They become reflexed in the bulbous buttercup as the flower opens.) The corolla has five petals which alternate with the sepals and these form the most conspicuous part of the flower. Each one has a small pouch at the base in which nectar is formed. They are bright shiny yellow on the upper surface but much duller on the under surface. Within the cup formed by the corolla are numerous stamens each with a thin yellow filament and a darker yellow anther. In a newly opened flower the whole central region seems to be filled with stamens, but in an older flower you will see a green knob in the centre; this is the gynaecium. All the flower parts are attached to a cone-shaped receptacle and the tip of this is covered by a number of green pear-shaped carpels. This type of gynaecium, made up of several separate carpels, is quite different from that found in the wallflower in which the carpels were joined together. Each of these separate carpels contains one ovule and, therefore, after pollination and fertilisation the flower forms not a single fruit containing many seeds as in the wallflower, but a head of several fruits each containing one seed.

2. *White dead-nettle* (*Lamium album*). This has a flower with a long corolla tube and the floral parts are not arranged as symmetrically as in the previous examples (see Fig. 16.2). Both the calyx and corolla are really made up of five parts, but they are joined together to form tubes. The calyx tube is funnel-shaped and remains after fertilisation to protect the developing fruit. The corolla tube is much longer, and at the mouth the petals form two lips—the upper one forms a hood over the stamens and style, the lower one serves as an alighting platform for bees. There is a constriction toward the base of

WHITE DEADNETTLE FLOWER

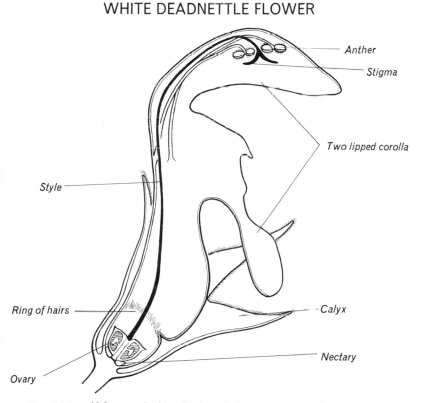

- Anther
- Stigma
- Two lipped corolla
- Style
- Ring of hairs
- Ovary
- Calyx
- Nectary

FIG. 16.2 Half-flower of white dead-nettle (*Laminium album*). An irregular flower pollinated by humble bees.

SWEET PEA

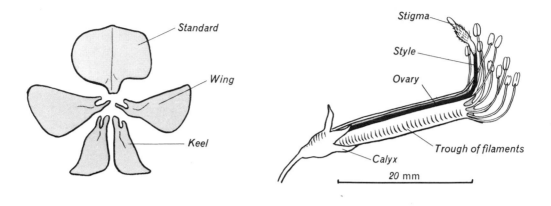

Fig. 16.3 Sweet pea (*Lathyrus odoratus*). The petals, which have been removed from the flower on the right, are drawn separately on the left. The keel is shown split in two.

the corolla tube, and here the tube is lined with short hairs. The four stamens are joined to the corolla tube and not to the receptacle, and their filaments are curved so that all four anthers lie below the hood of the corolla. The gynaecium apparently consists of four small fused carpels. There is a long style ending in a two-lobed stigma. It arises from the base of the ovary and comes up between the four carpels of the ovary. Surrounding the base of the ovary is a nectary.

3. *Sweet pea (Lathyrus odoratus)*. This flower appears to be very different from those previously described, but it is made up of the same four whorls, though, as in the white dead-nettle, the arrangement of the parts is irregular (see Fig. 16.3). The calyx is green, hairy, and forms a short tube with five pointed tips which curl backwards as the flower opens. The corolla consists of five petals, but the two anterior ones are fused along their lower edges to form a '*keel*' which encloses the androecium and gynaecium. At each side of the keel is a '*wing*' petal and at the back of the flower there is a larger petal termed the '*standard*'. Pull off the petals and in doing so notice how they are locked together at the base. The wings are really quite free, but if they are pressed downwards the keel moves also. The androecium consists of ten stamens, nine of which are joined together for the basal two-thirds of their filaments to form a trough. This trough is open along its upper edge where the tenth free stamen lies.

Nectar is secreted at the base of the filaments and collects in the trough. The gynaecium is composed of one carpel. The ovary, which lies in the trough, is hairy, flattened from side to side, and contains a single row of ovules along the upper side. The style is curved upwards and is hairy just below the stigma which is at its tip. After fertilisation the ovary enlarges to form a fruit, known as a pod, which splits to scatter its seeds.

4. *Ox-eye daisy (Chrysanthemum leucanthemum)*. The so-called 'flowers' of the ox-eye daisy and dandelion are not really single flowers but collections of many small flowers or **florets** all packed close together on a large flat receptacle. This collection of flowers is enclosed in the bud by tiny green scales called **bracts** and the whole collection of them is an **involucre**. In the ox-eye daisy the florets toward the centre of the receptacle have a tubular corolla but those toward the edges have a strap-shaped one (see Fig. 16.4). In both these disc and ray florets the calyx is much reduced and often absent. In the dandelion, the calyx consists of a ring of fine white hairs; such a calyx is termed a **pappus**. The ray florets have four lines on the corolla and the disc florets have five points at the mouth of the corolla tube, showing that in both types of florets there are five joined petals. The ray florets have no stamens, but in the disc florets there are five stamens which are joined to the base of the corolla tube. These stamens have their anthers joined to form a tube round the style.

A COMPOSITE FLOWER

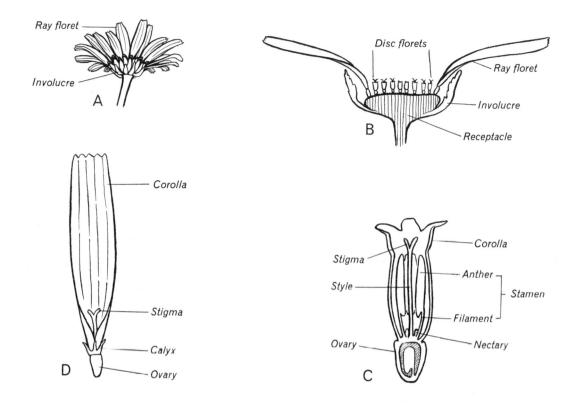

Fig. 16.4 Ox-eye daisy (*Chrysanthemum leucanthemum*). A shows the whole inflorescence; B is a diagrammatic representation of a section through it. C shows a single disc floret cut in half; D shows a single ray floret.

The gynaecium is the same in both types of florets, but in this case the ovary is not within the corolla but below it and is, therefore, described as **inferior**. (The previous examples all had a **superior** gynaecium.) The ovary is made up of two joined carpels but it contains only a single ovule. The style is long and forks above the anther tube. The actual stigmatic surface is on the upper sides of the fork, and when the flower first opens these two surfaces are pressed together.

16.2 Anthers and Ovules

The anthers of the buttercup and wallflower are both bilobed, and between the two lobes is the **connective** which contains a vascular bundle carrying food and water to the developing pollen. The pollen develops in tubular cavities, called **pollen sacs**, within the anthers; each lobe contains two such cavities. Fig. 16.5 shows the appearance of a section across an anther. Eventually a longi-

tudinal split appears between each pair of pollen sacs and this gradually enlarges as the anther wall curls back. This splitting and curling is the result of the drying and uneven contraction of the cells surrounding the pollen sacs. The pollen, formed from cells within each sac whose nuclei divide in a special way, consists of hundreds of minute grains, each grain being a single cell. The cell wall is two-layered and in it are several thin places. Within the wall is a mass of cytoplasm and, at first, a single nucleus which divides before the grains are shed from the anther. One of the nuclei so formed is the pollen tube nucleus while the other divides to form two **male gametes** (see Fig. 16.6).

In the wallflower and lily, the ovules develop in two rows on ridges of the ovary wall called the **placentae**. In the buttercup, a single ovule is attached to the placenta of each ovary. As the ovule grows it becomes enclosed in a covering of

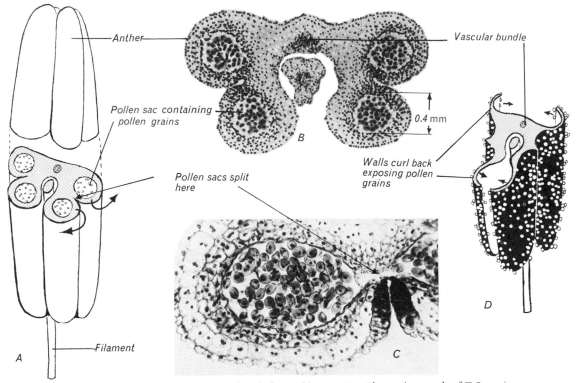

Fig. 16.5 A—Stereogram of an anther before dehiscence. B—Photomicrograph of T.S. unripe anther. (*By M. I. Walker.*) C—Photomicrograph of part of T.S. ripe anther. D—Stereogram of an anther dehiscing.

cells except for a small hole at the apex which persists as the **micropyle** (see sec. 11.1). At the base a stalk or **funicle** forms containing a vascular bundle which carries food and water to the developing ovule. The centre of the ovule consists of a large cell, the **embryo sac**, the nucleus of which is produced in a special way and which divides to form several nuclei, one of which is the **female gamete** (see Fig. 16.12).

The normal division of the nucleus was described in section 11.7. If you read this section again you will realise that if the nucleus always divided in this manner during the formation of the male and female gametes, then one of the results of fertilisation would be a doubling of the normal chromosome number. This does not occur, as the nucleus divides once in a special way (meiosis) during the formation of the pollen grains and embryo sac. As a result each daughter nucleus contains only half the original number of chromosomes. Thus the gametes, unlike the other cells of the parent, contain only half the normal number of chromosomes, and hence fertilisation results in a return to the normal chromosome number and

not in a doubling of it. This type of nuclear division will be referred to again in chapter 28.

16.3 Pollination Mechanisms

A flower may be pollinated by its own pollen (**self-pollination**) or by pollen from another flower of the same plant or another plant of the same species (**cross-pollination**). The pollen grains are unable to move, so they have to be transferred from the anthers to the stigmas by some agent, such as insects and wind.

Insect Pollination

Insects, some of which are entirely dependent on flowers for their food, visit them to collect pollen and nectar, and unconsciously transfer the pollen during this search for food. Small insects such as flies and beetles crawl over the flowers and their whole body becomes covered with pollen; larger insects such as hive bees, humble bees, hover flies, butterflies, and moths have a long proboscis for sucking nectar (see sec. 18.3). Flowers which rely on insects for pollination generally have bright petals, a strong scent, and form nectar or abundant pollen for the insects' food.

POLLEN & POLLEN TUBES

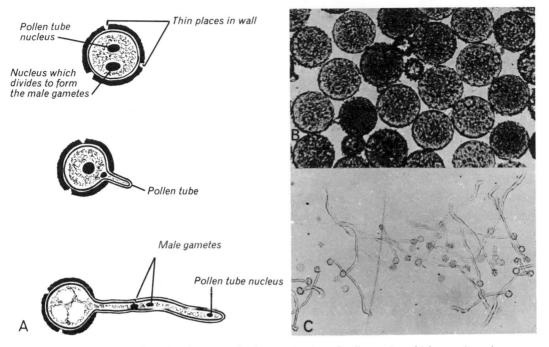

Fig. 16.6 A—Drawings showing stages in the germination of pollen grains which contain male gametes. B—Photomicrograph of pollen of Canterbury bell. C—Photomicrograph of pollen of *Primula wanda* which has been germinated in sucrose solution.

1. *Meadow buttercup.* The bright yellow of the petals of the buttercup attracts insects, such as small beetles, which crawl over the flower in search of the nectar and carry away pollen on their bodies. This pollen is brushed on to the stigmas of the next flower they visit. The stamens usually ripen before the stigmas, and as they do so the filaments, which are at first bent inwards toward the middle of the flower, straighten so that there is less chance of self-pollination, a possibility which is still further reduced by the fact that the anthers shed their pollen outwards.

2. *White dead-nettle.* Some flowers are dependent on the visits of special insects to carry out cross-pollination. The lower lip of the corolla of the white dead-nettle provides an alighting platform, and as a humble bee pushes its head and proboscis into the corolla in search of nectar the back of its head first meets the stigma, which thus receives pollen from another flower. The bee's back next comes in contact with the stamens so that their pollen may be carried to the next flower the bee visits. There is thus a good chance of cross-pollination, but self-pollination may take place, as the stigmas and anthers sometimes ripen at the same time. It is only the humble bees with a proboscis

longer than 10 mm. that can reach the nectar at the base of the corollar tube. Some humble bees with a proboscis too short to reach the nectar in the usual way pierce the corolla tube and take the nectar in this way. Honey bees may use these holes too. Cross-pollination is not effected by these insects. The ring of hairs in the corolla tube forms a barrier which prevents small insects from creeping down and 'stealing' the nectar.

3. *Sweet pea.* The sweet-pea flower is also adapted for pollination by special insects as it is visited for its nectar, which is well hidden. The large standard petal makes it conspicuous, and as the bee, which must have a proboscis of at least 15 mm, alights on the upper sides of the wings, the keel is depressed. This, besides exposing the stamens and style, makes it possible for the bee to insert its proboscis under the free stamen into the trough to reach the nectar. In doing so its under surface is struck first by the stigma and cross-pollination may be effected, and later by the style from which pollen is brushed. The pollen is swept on to the hairs of the style when the stamens are forced out of the keel. Many of the larger cultivated sweet peas are not visited by bees, probably because of the difficulty of depressing the heavy keel, but

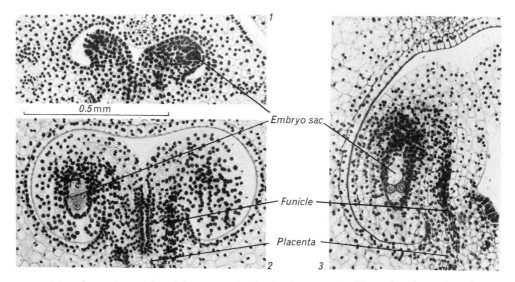

FIG. 16.7 Photomicrographs of three stages in the development of a lily ovule. The sections do not pass through the micropyle.

they still set seed so that self-pollination is obviously possible.

4. *Ox-eye daisy.* The composite flowers, like the ox-eye daisy, are really adapted for cross-pollination by insects, but if this fails they may be self-pollinated. The stamens ripen before the stigmas, and the anthers split, so that the pollen is shed into the tube made by the joined anthers. At this stage the style is short and the stigmas press close together. As the style grows the tip pushes the pollen out of the anther tube so that it is exposed

SELF POLLINATION

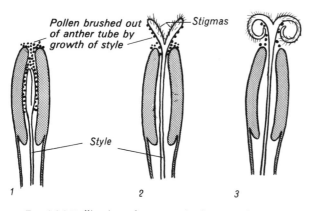

FIG. 16.8 Pollination of a composite flower such as the ox-eye daisy. In all the drawings the anther tube has been cut in half to expose the style. The stigmas are shown in the position for cross-pollination in 2 and for self-pollination in 3.

and eventually the tip opens to reveal the two stigmas. Insects crawling over the inflorescences in search of nectar may effect cross-pollination, as the outer florets develop before the inner ones. If the floret is not cross-pollinated the stigmas may curl back and pick up any pollen which remains on the style (see Fig. 16.8). One of the commonest composite flowers, the dandelion, has this arrangement of its flower parts, but is peculiar in that it can form seeds which are able to germinate without either pollination or fertilisation.

Wind Pollination

Wind-pollinated flowers (e.g. hazel, plantain, and grasses) do not form nectar but produce great quantities of pollen, since much will inevitably be wasted. This pollen is light and usually formed in anthers on long filaments which sway in the wind. These flowers are often inconspicuous and have no brightly coloured corolla. The distribution of pollen is not hindered as the flower buds either open before the leaf bud or they are borne on long stalks which grow well above the leaves.

1. *Hazel (Corylus avellena).* If you examine a hazel twig toward the end of February you will see that it has three different kinds of buds. Most of the buds are small and show no signs of opening. These are the leaf buds and, like the one of the horse-chestnut described in section 11.8, will eventually grow into leafy shoots. Some of the buds have already opened and formed long

pendulous catkins, and others resemble the leaf buds but have a tuft of red threads at their tips. These open buds are the flowers of the hazel. Unlike the other plants which you have examined, the hazel forms two different kinds of flowers, one containing the androecium and the other the gynaecium. Use a hand lens to examine the long catkin and then pull off one of the small male or staminate flowers and compare it with Fig. 16.9. It is quite easy to see the anthers joined to the lower side of the yellowish-green bract, but you will need a good hand lens to see that there are two smaller bracts below this and that each one of these has four anthers joined to it. If you remove the large anthers you may be able to see the filaments which are branched. The outside of the female or pistillate catkin consists of a covering of brown scales and a few leaves. In the centre are four or five bracts each bearing two small flowers. The ovary is round and green with two long red stigmas. Joined to the top of the ovary is a minute perianth and at the base is a cupule formed from bracts. The male catkins sway in the wind so that the pollen is wafted on to the stigmas and the flower is pollinated.

2. *Greater plantain (Plantago major).* This common weed of lawns has flowers which are modified to aid the transfer of pollen by wind. It has a number of small flowers growing close together so that they form a compact inflorescence (see Fig. 16.10). Each flower develops from an axillary bud which has a tiny bract below it. There are four small sepals and petals, four stamens with very long filaments, and a long stigma covered with short hairs growing from an ovary. Examine an inflorescence; notice that it grows on a long stalk, that the lower flowers open first, and that the stigma, which is mature before the anthers, often protrudes from the flower while it is still in bud. By the time the flower opens completely the stigma has received pollen, probably from other plants, and is beginning to wither.

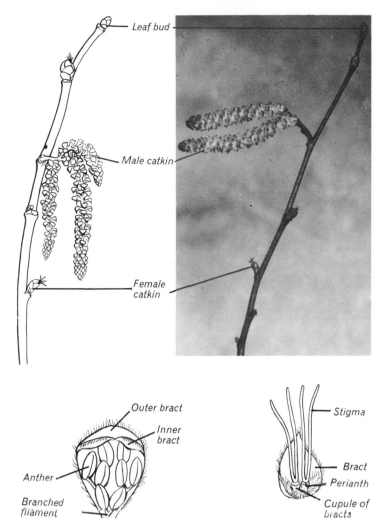

Fig. 16.9 Hazel (*Corylus avellena*). The twigs were drawn and photographed in February. (*Photograph by W. J. Garnett.*) The other drawings show separate male and female flowers.

16.4 Fertilisation

After being deposited on the stigma, the pollen grains absorb water and sugar from the sticky surface so that they swell and burst at one of the thin places in the wall (see sec. 16.2). A tube emerges from each pollen grain and grows through the style toward the ovary (see Fig. 16.11). The pollen tube nucleus passes into it and the two male gametes soon follow. The tip of the pollen tube secretes enzymes which break down the tissues of the style and help the pollen tube to absorb them. As the pollen tube feeds, it grows and the two nuclei (male gametes) can be seen

WIND POLLINATION

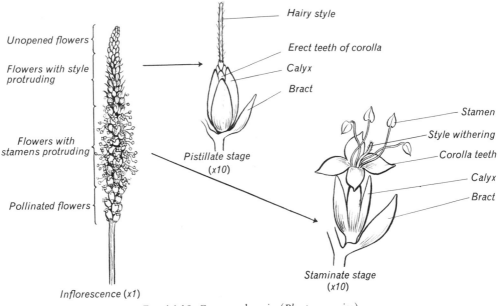

Unopened flowers

Flowers with style protruding

Flowers with stamens protruding

Pollinated flowers

Inflorescence (x1)

Hairy style

Erect teeth of corolla

Calyx

Bract

Pistillate stage (x10)

Stamen

Style withering

Corolla teeth

Calyx

Bract

Staminate stage (x10)

FIG. 16.10 Greater plantain (*Plantago major*).

near the tip. Eventually the pollen tube reaches the ovary and makes its way toward the micropyle of the ovule (see Fig. 16.12). The direction of its growth is controlled by substances secreted by the ovule, an example of chemotropism (see sec. 12.3). As the tip of the pollen tube passes through the micropyle and reaches the embryo sac it bursts, and the two nuclei enter the embryo sac. One of them fuses with the female gamete: this fusion of the two nuclei is **fertilisation.** As a result of fertilisation an embryo is formed within the ovule, and the latter eventually develops into the seed while the ovary which contained the ovule develops into a fruit. The other male nucleus in the pollen tube also enters the embryo sac and fuses with the secondary nucleus (see Fig. 16.12). The resulting nucleus divides to form a mass of food material termed the **endosperm.** In many plants this food store is used up before the seed is fully grown, but in some, such as the maize, it remains to provide food for later growth (see sec. 11.4).

Experiments to demonstrate the Growth and Function of Pollen

It is quite easy to show that pollination is an essential prelude to fertilisation and fruit formation by 'bagging' experiments. Suitable flowers for these are the daffodil, tulip, and sweet pea. The experiment should start just before the flowers are ready to open and they should be divided into three groups. All the flowers in groups A and B should have their stamens removed before the pollen sacs burst. The flowers of these two groups should be enclosed in muslin or polythene bags tied round the flower stalk with wool so that insects are prevented from visiting the flowers. The flowers of the third group C are left uncovered, and when these stamens start to shed their pollen the bags from group B should be removed and the stigmas of these flowers pollinated by rubbing mature stamens over them, after which the bags should be replaced. The flowers of all three groups should be examined as they begin to wither. In group A the ovaries, which received no pollen, are withered, but in groups B and C they are large, firm, and contain seeds.

POLLEN TUBES IN STIGMA

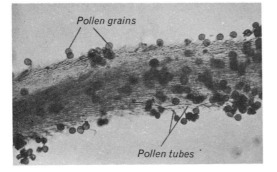

Pollen grains

Pollen tubes

FIG. 16.11 Photomicrograph of pollen grains growing into the stigma. (*Eschscholtzia.*)

The pollen of many plants germinates easily if it is put in sucrose solution of a suitable concentration (15 per cent for sweet pea, nasturtium, and Echeveria, 10 per cent for bluebell, 7 per cent for Narcissus, and 3 per cent for tulip). The pollen should be scraped off a dehisced anther, placed in a drop of solution on a clean cover-slip which is then inverted over a cavity slide or over an ordinary slide on which a ring of wire, string, or plasticine has been placed. In this way the pollen then inverted over a cavity slide or over an ordinary slide on which a ring of wire, string, or Plasticine has been placed. In this way the pollen take several hours. If the slides are to be left for some time they should be placed in a petri dish or covered with a saucer to maintain a moist atmosphere and kept in a warm place.

The chemotropic response of pollen tubes (see sec. 12.3) can be demonstrated by germinating pollen grains in this way in the presence of stigmatic tissue. A small piece of stigma and one or two ovules from the same flower as the pollen should be placed on the cover-slip with the pollen. As the pollen germinates it will be seen that the ends of the tubes are directed toward the piece of stigma or ovules.

FERTILISATION

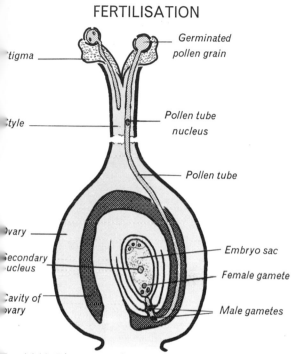

Fig. 16.12 Diagram to show fertilisation. One pollen tube is shown at the micropyle of the ovule.

Labels: Stigma, Style, Ovary, Secondary nucleus, Cavity of ovary, Germinated pollen grain, Pollen tube nucleus, Pollen tube, Embryo sac, Female gamete, Male gametes

16.5 Seeds, Fruits, and their Dispersal

After fertilisation the wall of the ovary or **pericarp** alters; when the fruit is ripe the pericarp may be either tough and leathery, woody and hard, or succulent. Fruits, like the flowers from which they develop, are very varied in their form, but it is possible to classify them according to the type of pericarp. The table shows one way in which this can be done. This table and Figs. 16.13 and 16.14 will give you some idea of the structure of fruits, but you should collect as many specimens as possible yourself so that you can examine them carefully.

TYPES OF FRUITS

OVARY	INDEHISCENT One seeded	Achene	Leathery pericarp	Buttercup, Geum, Clematis, Hip, Strawberry
		Samara	Winged pericarp	Ash
		Nut	Hard, brittle pericarp	Hazel, Acorn
		Cypsela	Inferior fruit. Plumed calyx	Dandelion, Burdock
	DEHISCENT Many seeded	Follicle	One carpel. Splits along one edge	Larkspur
		Legume	One carpel. Splits along two edges	Lupin
		Siliqua	Two joined carpels	Wallflower
		Capsule	Many joined carpels	Campion, Poppy, Willow herb
	SCHIZOCARPIC		Splitting into one-seeded parts	Geranium, White dead-nettle, Sycamore, Cleavers
SUCCULENT		Berry	Two-layered pericarp. Many seeds	Grape, Tomato
		Drupe	Three-layered pericarp. One seed in a 'stone'	Cherry, Plum, Blackberry, Coconut
		Pome	Fruit surrounded by juicy receptacle. Many seeds in 'core'	Apple

DRY FRUITS
Indehiscent types

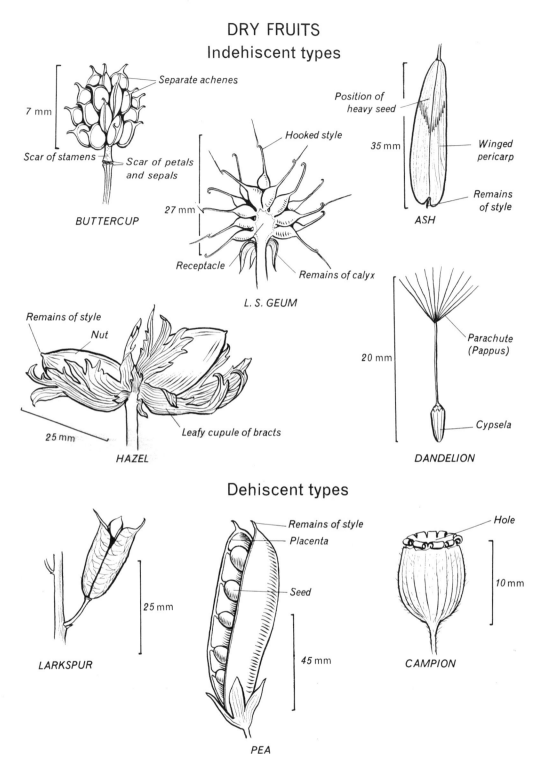

Separate achenes

7 mm

Scar of stamens

Scar of petals and sepals

BUTTERCUP

Hooked style

27 mm

Receptacle

Remains of calyx

L. S. GEUM

Position of heavy seed

35 mm

Winged pericarp

Remains of style

ASH

Remains of style

Nut

25 mm

Leafy cupule of bracts

HAZEL

Parachute (Pappus)

20 mm

Cypsela

DANDELION

Dehiscent types

25 mm

LARKSPUR

Remains of style

Placenta

Seed

45 mm

PEA

Hole

10 mm

CAMPION

Fig. 16.13 Fruits with a dry pericarp surrounding a single seed (indehiscent types) and fruits with a dry pericarp surrounding many seeds (dehiscent type).

SUCCULENT FRUITS

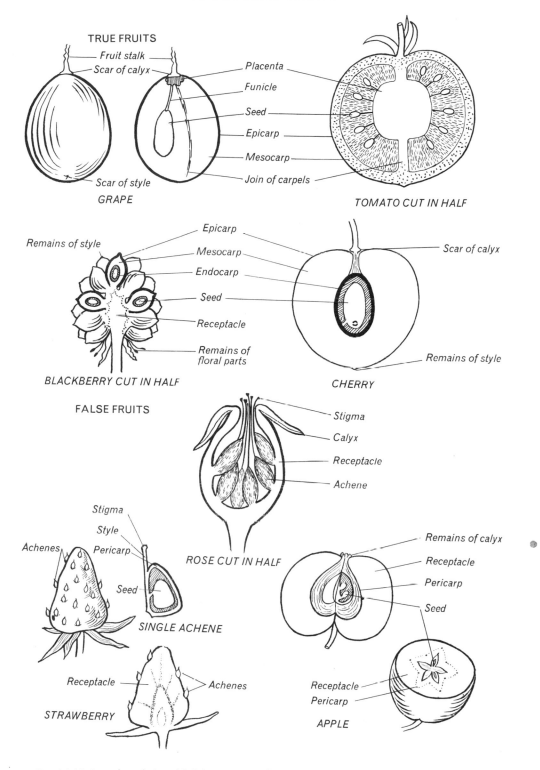

FIG. 16.14 Succulent fruits with juicy pericarps (berries and drupes) and juicy receptacles (false fruits).

WIND DISPERSAL

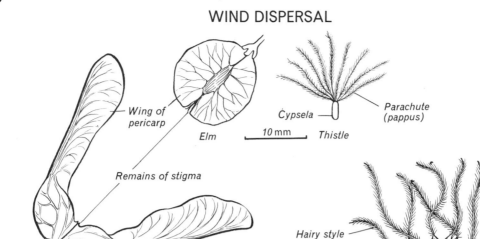

Wing of pericarp

Elm

Cypsela

Parachute (pappus)

10 mm

Thistle

Remains of stigma

Hairy style

Separate achenes

20mm

Sycamore

20 mm

Clematis

Fig. 16.15 Winged and plumed fruits.

CENSER MECHANISMS

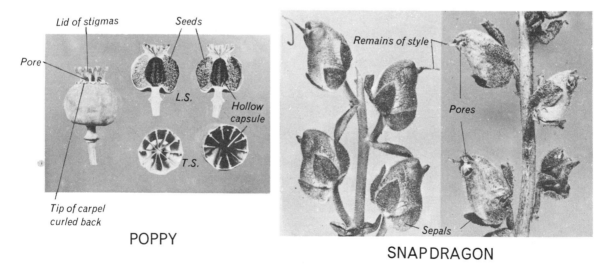

Lid of stigmas

Seeds

Pore

L.S.

Hollow capsule

T.S.

Remains of style

Pores

Tip of carpel curled back

POPPY

Sepals

SNAPDRAGON

Fig. 16.16 Photographs of two capsular fruits which disperse their own dust-like seeds when shaken by the wind.

Usually it is only the ovary which forms the fruit, but in some plants other parts besides the ovary enlarge after fertilisation. Such fruits are termed false fruits. Often it is the receptacle which enlarges and Fig. 16.14 shows some familiar examples of false fruits.

After the fruit containing the seeds has ripened, it is essential that the seeds should be scattered, as otherwise they will fall in a heap below the parent plant. If they do this they may germinate after a period of rest, but the seedlings will be crowded together and overshadowed by the

ANIMAL DISPERSAL

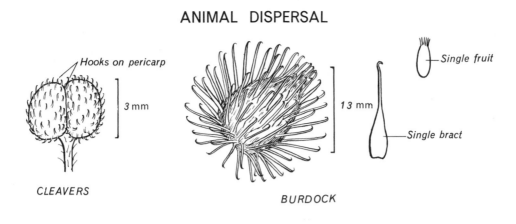

FIG. 16.17 Hooked fruits dispersed by furry animals.

parent plant. They would compete with one another for light, air, water, etc., and few plants would survive. Many plants have some special structure which aids seed dispersal by an external agent, such as the wind or animals, though there are some plants which are able to disperse their own seeds. Study all the specimens you have collected and try to decide how the seeds are dispersed and whether the plant has any special device to assist in the process. Refer to the table which summarises the different methods of dispersal.

DISPERSAL OF FRUITS AND SEEDS

Agent	Modification	Example
Wind	(a) Small and light	*Orchid seeds*
	(b) Special device to aid dispersal	*Willow herb*—capsule bursts open to release seeds each with a tuft of hairs
	(i) Parachute of hairs (usually low-growing plants; parachute helps wind to 'lift' fruits)	*Dandelion*—calyx enlarges after fertilisation to form a parachute (Fig. 16.13)
		Clematis—style lengthens after fertilisation and becomes hairy (Fig. 16.15)
	(ii) Wing-like structure (usually tree fruits; wing increases the surface area so slowing down the rate of fall)	*Ash*—pericarp enlarges, becomes flattened and twisted after fertilisation. Fruit spins as it falls (Fig. 16.13)
	(iii) Pepper-pot or censer mechanism (cup-like fruit on stiff stalk; dust-like seeds shaken out through pores when the wind blows)	*Poppy, Snapdragon* (Fig. 16.16)
Animals	(a) Hooked dry fruits (become hooked to fur of animals; fall off as fruit dries and hooks shrivel or are brushed off; usually grow at base of hedges)	*Cleavers*—hooked pericarp *Geum*—hooked style (Fig. 16.13) *Burdock*—hooked bracts (Fig. 16.17)
	(b) Succulent fruits (pericarp or some other part becomes thicker after fertilisation and develops colouring which makes it stand out from the foliage; eaten by birds and 'stones' or seeds excreted in faeces).	*Blackberry* (Fig. 16.14) *Hip* (Fig. 16.14) *Strawberry* (Fig. 16.14)
Plant Itself (Self-dispersal)	Parts of the fruit shrink unevenly as they dry, so that tensions are set up which make the fruit burst open explosively	*Lupin*—twisting pericarp *Geranium*—catapult action of style (Fig. 16.18)

SELF - DISPERSAL

Empty carpel

Style

Seed not ejected

Placenta

Calyx

Fig. 16.18 Photographs of the schizocarpic fruit of geranium.

ASEXUAL (VEGETATIVE) REPRODUCTION IN FLOWERING PLANTS

16.6 Natural Methods

Many flowering plants which reproduce by seeds are also able to reproduce by the separation of part of their vegetative structure. This is a form of asexual reproduction, since there is no fusion of gametes. A variety of organs is used in vegetative reproduction.

1. *Runners.* The strawberry plant has a short, thick, upright, underground stem and a rosette of leaves. Toward the end of the flowering season the buds in the axils of the leaves grow outward, forming long, thin, horizontal stems with a few tiny scale leaves (see Fig. 16.19). These runners may have internodes about 0.1 m long. The terminal bud of the runner turns upward and adventitious roots form, thus making a new plant, which may in turn form more runners. In time the runner connecting the young plant to the parent rots, so that the new plant becomes independent. This type of reproduction is probably a surer method than reproduction by seed formation, as it does not rely on the help of any external agent and the young plant is supplied with food by the parent plant until it is well established. The new plants are, however, close to the parent plant so that some overcrowding is bound to result, though the production of long internodes helps to reduce this. The creeping buttercup, ground ivy, and cinquefoil also reproduce by runners.

2. *Rhizomes.* Many herbaceous plants have part of their shoot system underground; if it is horizontal it is termed a rhizome. This rhizome provides the plant with a means of vegetative reproduction. Couch grass (twitch), which is often a troublesome weed in some gardens, spreads very rapidly because it has a rhizome which sends up aerial shoots at frequent intervals. These shoots grow from buds in the axils of small scale leaves and are anchored in the ground by adventitious roots which grow from the nodes. Mint spreads in the same way (see Fig. 16.19). Most rhizomes, however, are also storage organs and the horizontal stem is much swollen. The iris has such a typical rhizome (see Fig. 16.19). This is a stout horizontal stem with an erect shoot at the growing tip. The whole stem is marked with rings which are the scars left by the foliage leaves and the dots along these rings are the scars of the veins of the leaves. Close to each leaf scar is an axillary bud. Growing from the lower surface of the rhizome are numerous adventitious roots. The tip of the stem, which forms the shoot and bears flowers, marks the end of a year's growth. The growth of the rhizome is continued in the following year by an axillary bud; if more than one axillary bud grows several new branches will be formed. As the rhizome grows at the tip the old part rots away and so the whole plant gradually moves forward and divides into pieces. You may have noticed that the iris plants are in a slightly different position in your garden each year.

3. *Tubers.* The potato plant can be raised from seeds, though the fruits rarely ripen completely in this country (you may have seen the small green berries which look rather like small tomatoes on the fully grown plants). They are more frequently propagated by planting 'tubers', which is the correct term for the ordinary 'seed' potato (see Fig. 16.20). When the tuber is planted the 'eyes', which are really axillary buds, start to grow and form leafy shoots. Adventitious roots grow from the bases of these shoots and one tuber may give rise to many plants; in fact gardeners often cut their tubers into several parts before planting. The potato plant bears two types of leaves. Those near ground level are small and simple; the others are much larger and compound. The buds in the axils of the compound leaves develop into ordinary aerial shoots, but those in the axils of the simple leaves develop into long, slender rhizomes which grow downwards into the soil and bear small triangular scale leaves. The leafy shoots manufacture food, some of which is passed into these rhizomes for storage. The tips of these rhizomes

VEGETATIVE REPRODUCTION

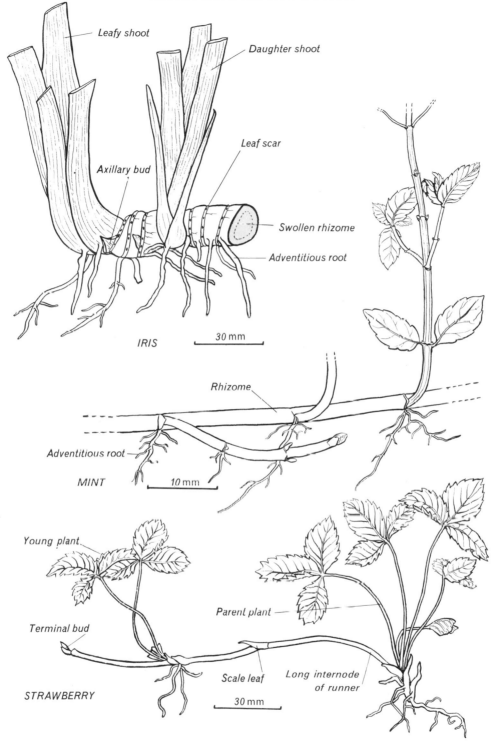

Leafy shoot

Daughter shoot

Axillary bud

Leaf scar

Swollen rhizome

Adventitious root

IRIS

30 mm

Rhizome

Adventitious root

MINT

10 mm

Young plant

Parent plant

Terminal bud

Scale leaf

Long internode of runner

STRAWBERRY

30 mm

Fig. 16.19 Stems modified for vegetative reproduction. The iris and mint have underground stems (rhizomes) but the runner of the strawberry is on the surface of the soil.

POTATO TUBER

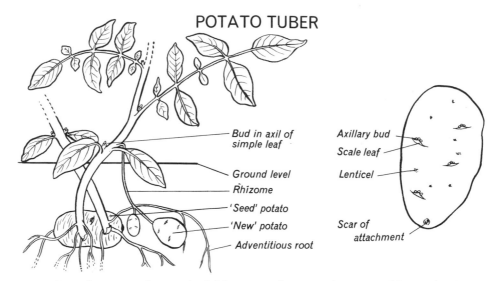

Bud in axil of
simple leaf

Ground level
Rhizome
'Seed' potato
'New' potato
Adventitious root

Axillary bud
Scale leaf
Lenticel

Scar of
attachment

FIG. 16.20 The potato plant on the left has grown from a 'seed' potato and is now forming a
crop of 'new' potatoes, one of which is shown enlarged on the right of the page.

swell as the food accumulates in them and the
swellings become the tubers, or 'new' potatoes.
You should now examine a tuber carefully with
a hand lens and note the scar which marks where
it became detached from the rhizome and the
scars left by the triangular scale leaves with their
axillary buds which form the 'eyes'.

4. *Corms.* You have probably all grown crocuses,
hyacinths, and daffodils in bowls during the
winter: these plants all exhibit vegetative reproduction. Though we often call all the structures we
plant in bowls 'bulbs', this is incorrect, for some
such as the crocus are solid, while others such as
the snowdrop and daffodil are arranged in layers.

CROCUS CORM

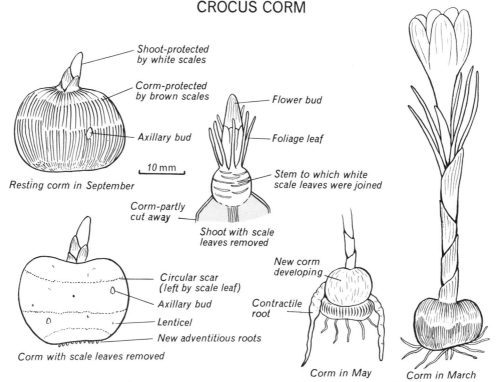

Shoot-protected
by white scales

Corm-protected
by brown scales

Axillary bud

10 mm

Resting corm in September

Flower bud

Foliage leaf

Stem to which white
scale leaves were joined

Corm-partly
cut away

Shoot with scale
leaves removed

Circular scar
(left by scale leaf)
Axillary bud
Lenticel
New adventitious roots

Corm with scale leaves removed

New corm
developing

Contractile
root

Corm in May

Corm in March

FIG. 16.21 Drawings to show the structure of a resting corm and stages in its growth.

SNOWDROP BULB

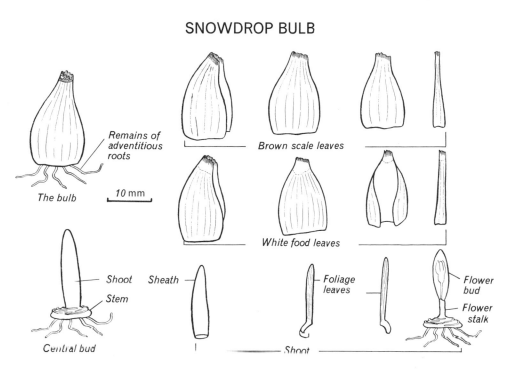

Fig. 16.22 Dissection of the snowdrop bulb.

Examine a crocus corm before you plant it and compare it with Fig. 16.21. Remove the circular fibrous scales and note their different lengths. Each scale leaf leaves a circular scar on the corm and a little above each one is an axillary bud. These leaves and buds show that the corm is really a stem (an erect underground one with short internodes and containing a food-store). At the tip of the corm you will see one or more larger buds. Cut away the corm on four sides so that the largest of these buds rests on a square piece of stem and then remove its surrounding white bud scales and compare the shape and number of them with the brown fibrous scale leaves you removed earlier. In the middle of the bud you will find about eight small, pale yellow foliage leaves, each with a wide sheathing base and surrounding one or more flower buds. Slices of the corm can be tested for starch and glucose (see sec. 7.1) to determine whether these materials are present in the food-store.

After the corm has been planted a number of adventitious roots grow out from the base of the corm and the buds grow above ground, forming flowers and foliage leaves. All this growth is accomplished at the expense of the food stored in the corm, which consequently shrinks. After flowering, the leaves manufacture food, most of which is passed down to the base of the stem to

which they are attached so that this swells and forms a new corm. If more than one bud has grown, more than one corm will be formed on top of the old one and the plant will have reproduced vegetatively. In June, the leaves wither and the scale leaves surrounding the new corm dry and become brown. From the base of this new corm thick roots are formed which grow deeply into the soil. They are contractile and pull the new corm down into the soil. In this state the new corm remains dormant until the following autumn.

5. *Bulbs.* The snowdrop bulb is very regular in its growth and in spite of its small size is a suitable one to study. Remove the outer brown scale leaves and arrange them in a row. There are really four, but often the outer ones come off in pieces or they may have been rubbed off during packing, etc. Note their different sizes and compare them with Fig. 16.22. The thick, white leaves below these scale leaves are also four in number and are similar in shape. They, too, should be arranged in a row. Notice that the tips of these leaves are shrivelled, indicating that something has withered. You will see that these leaves were attached to a very short stem which still has a bud joined to it. This bud is surrounded by a circular bud scale and contains two foliage leaves and a flower bud. Arrange these four parts in a row below the two rows of

scale leaves and compare the three sets. The first circular white scale leaf was the circular bud scale which protected the bud in the previous year. The next two are the bases of the foliage leaves, and the very narrow one is last season's flower stalk. The brown scales are similar structures but a year older.

When the bulb is in the soil, adventitious roots grow from the base of the stem and these absorb water which makes the inner bud swell and push upwards. The bud scale splits, the leaves lengthen and turn green, and the flower bud expands. Most of the food synthesised by the green leaves and green flower stalk is passed down to their bases which swell, and eventually the green part above ground level withers so that only the bases of the leaves remain in the ground. The flower bud develops from an axillary bud. Sometimes one or more buds in the axils of the white scale leaves develop into new bulbs which become separated and thus the plant reproduces vegetatively.

16.7 Artificial Methods

The gardener uses a variety of methods to increase his stock of plants, many of which are vegetative methods in that the new plant is formed by division of the parent plant.

1. *Cuttings.* Cuttings (short lengths of shoot) when placed in a warm, moist soil readily develop adventitious roots, so that they soon produce new plants. Pansies, fuchsias, and geraniums are easily propagated in this way. The layering of carnations and pinks is accomplished by pegging down a shoot and cutting at the node where it touches the soil. Adventitious roots are formed at this point so that a separate plant is eventually established.

2. *Budding and grafting.* Budding and grafting are more specialised methods used particularly in the cultivation of roses and fruit trees. In both cases the variety to be propagated (the scion) is made to grow on a well-established root system (the stock) (see Fig. 16.23). Cultivated roses are budded on to the root system of the wild rose. When the scion has finished flowering a dormant bud is cut from it so that it remains attached to a flat-pointed piece of stem which has the cambium exposed. A T-shaped slit is made in the bark of the stock and the edges turned back to expose the cambium. The bud is inserted so that its cambium rests on the cambium of the stock and it is then bound firmly but not tightly in position. If the two cambium surfaces are in very close contact, the bud will grow in the following spring.

Grafting is a similar technique but the scion is a twig. The way in which the cut surfaces of the stock and scion are arranged varies, but the aim in all methods is to bring the two cambium surfaces into close contact.

PERENNATION

16.8

Many flowering plants persist from year to year and may flower each season. These are termed **perennials** and their ability to live throughout the adverse winter season is termed perennation. They have to store sufficient food to start growth in the spring and somehow protect their buds against frost.

Most **herbaceous perennials** (plants with no permanent woody shoot system) have shoots too tender to survive the winter and they develop a special **perennating or hibernating organ.** The

BUDDING & GRAFTING

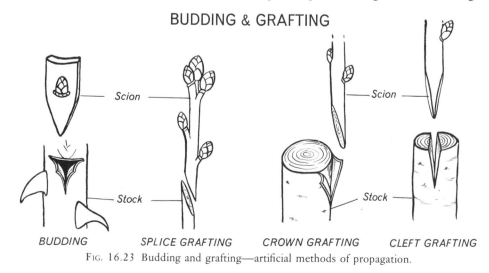

BUDDING   SPLICE GRAFTING   CROWN GRAFTING   CLEFT GRAFTING

FIG. 16.23 Budding and grafting—artificial methods of propagation.

LEAF FALL

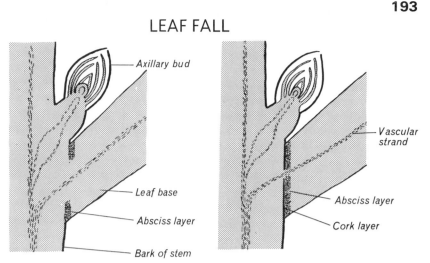

Fig. 16.24 Parts of longitudinal sections of a stem, passing through a leaf base and showing stages in leaf fall.

aerial shoots of the plant wither but the plant persists in the form of an underground food-store. Thus the rhizome, tuber, corm, and bulb described in the previous section all provide the plants which form them with a means of winter survival, as well as enabling them to reproduce vegetatively.

Woody perennials have shoots which are able to survive the winter, but many reduce their water loss by shedding their leaves and forming dormant buds (see sec. 11.8). Such plants are said to be **deciduous**. When the tree has finished growing in length the very numerous leaves synthesise food rapidly and much of this is stored in the woody shoot. As autumn approaches the leaves are shed. At the base of the leaf certain cells separate from one another and become spherical, forming an **absciss layer**. Below this a **layer of cork** is formed and the vascular bundles are blocked so that the leaf, which is now cut off from its water supply, withers and breaks off the stem in the region of the absciss layer (see Fig. 16.24). Some woody perennials retain their leaves throughout the year (**evergreens**). Their leaves are usually small, tough, and leathery, as they are protected by a waterproof epidermis. This allows them to remain unharmed by the winter conditions.

This problem of winter survival is common to all plants. Some survive only in the form of seeds: they are **annuals** and complete their growth, including seed formation, in one year and then die. Some plants, however, require two seasons to complete their growth; they are **biennials**. During the first season these plants (e.g. carrot) form leafy shoots which synthesise food that is stored in large roots. The shoots die at the approach of winter and the plant hibernates in the form of this food-store. If you cut a carrot verti-cally you will see the typical root structure and you should test a piece of it as explained in section 7.1 to show that it contains a large store of glucose. In the following spring this food-store is used to form a new aerial shoot which flowers, forms seeds, and then dies. Normally, the roots are harvested at the end of the first year, so you may not be familiar with this second year's growth.

REPRODUCTION IN OTHER PLANTS

16.9 Spirogyra and Mucor

The ways in which Spirogyra reproduces were described in section 2.8; this should now be read again. You will then realise that **fragmentation** is a method of asexual reproduction, since it involves no fusion of cells but is merely a division of the parent plant. **Conjugation**, on the other hand, does involve the fusion of two nuclei, or fertilisation, so that it is a method of sexual reproduction. The two gametes which unite are alike, though one moves toward the other and may be termed male. After fertilisation a zygospore is formed and after a period of rest this grows into a new filament. Spirogyra is able to withstand a prolonged period of drought in the form of the zygospore so that sexual reproduction also provides a method of survival during adverse conditions.

Mucor, too, is able to reproduce both sexually and asexually and the methods were described in section 2.15. During asexual reproduction a special reproductive body or spore is formed. These spores germinate quickly if they fall on suitable material. As a result of conjugation—a form of sexual reproduction—a zygospore is formed which, like that of Spirogyra, is able to withstand drought.

QUESTIONS 16

1. What are the characteristics of: (*a*) insect pollinated flowers, (*b*) wind pollinated flowers? Describe the changes which occur in a named flower after fertilisation has occurred. Confine your answer to those changes which can be observed with the aid of a hand lens. [L]

2. Describe, with the aid of labelled illustrations, a rhizome, a tuber and a bulb. What part does each play in the life of the plant? [L]

3. State the importance of pollination. By reference to two named flowers describe, with suitable illustrations, those features associated with (*a*) wind pollination and (*b*) insect pollination. [L]

4. Make two labelled drawings of a vertical section (or a half flower) of a named flower, one when fully opened and the other when the seeds have developed. State the functions of the parts you have labelled. How are the seeds dispersed in the example you have selected? [L]

5. Make two labelled diagrams of a named flower, one to show its external appearance and another to show its internal structure. Describe exactly how pollination is effected in this flower. [L]

6. (*a*) Explain fully the differences between annual, biennial and perennial plants and give one named example of each.

(*b*) For each example that you name describe (i) the condition of the plant as you would expect to find it in the autumn of the same year in which the seed germinated; (ii) the form in which the plant may survive the winter. [N]

7. (*a*) Make a large labelled diagram of a vertical section of the floret of a named composite.

(*b*) Give four differences between the flower you have drawn and a buttercup flower.

(*c*) Describe how (i) pollination and (ii) fruit dispersal is achieved in a named composite. [N]

8. Make large carefully-labelled drawings of a vertical section through a named bulb and a named corm at the time of flowering (omitting the structure of the flowers). From what structures will the succeeding bulb and corm be formed? Why is reproduction by vegetative means preferable to reproduction by seeds in many kinds of cultivated plant? [O]

9. Draw fully-labelled vertical sections of a named flower and of its fruit. Describe the events which occur in the development of the flower into the fruit. [O]

10. Using named examples, give an account of how the feeding habits of animals may aid the reproduction and dispersal of flowering plants. [O & C]

11. What are the differences between a fruit and a seed? Give an illustrated account of four fruits or seeds which have different methods of dispersal. [S]

12. Name four plants whose fruits or seeds are dispersed by different agencies. Make a large labelled drawing of the fruit or seed of each of the plants you have named. Describe briefly the method of dispersal in any two of the plants named, stating clearly the feature of the fruit or seed which assists in its dispersal. [W]

13. Explain how pollination is brought about in ONE named cross-pollinated flower. Make labelled diagrams to show the shape of this flower and the arrangement of its parts. [A]

14. (i) Draw a diagram to show the internal structure of a named bulb. Label your diagram. (ii) Describe an experiment which you could do to find out whether an onion bulb contains sugar. (iii) What is the main difference between a bulb and a corm? [S]

17 Animal Reproduction

17.1 Structure of Reproductive Organs of a Rabbit

Mammals can reproduce only by a sexual method and have special organs in which the male and female gametes are formed. The general arrangement of these organs is similar in all mammals, but the following descriptions are of those found in the rabbit.

The **male gametes or sperms** are formed in tubules in the two **testes** of the male rabbit. These develop in the abdomen close to the kidneys, but they soon migrate down the abdomen and come to lie in a pair of sacs which project between the hind limbs (see Fig. 17.1). These **scrotal sacs** are a continuation of the body cavity. Joined to the anterior end of each ovoid testis is a **spermatic cord** which leads back into the main abdominal cavity and contains an artery and a vein. Leading from the posterior end of each testis is a narrow curved tube—the **vas deferens**—which also leads back into the abdominal cavity where it loops over the ureter and opens into a small median sac—the **uterus masculinus**. This lies below the neck of the bladder and the two join together to form the **urethra**, which opens to the exterior at the end of a projecting organ—the **penis**. In the region of the urethra are the prostate and Cowper's glands, which secrete the fluid in which the sperms are discharged from the body.

The **female gametes or ova** are formed in the two small **ovaries** of the female rabbit. These develop in the abdomen close to the kidneys and remain attached to the dorsal wall. The ovaries, which are ovoid, have a blistered appearance in the adult owing to the presence of sacs or follicles within which the ova develop. The ripe ova, which are small and contain little yolk, are shed from the follicles at certain times. Close to each ovary is a funnel-shaped opening of the **oviduct** which is divided into a narrow, twisted, anterior part—the **Fallopian tube** (named after Fallopius, an Italian biologist)—and a wider posterior part—the **uterus**. The two uteri unite in the middle line of the body and form the **vagina**, which is joined to the neck of the bladder, so forming the **vestibule** which opens to the exterior at the **vulva**.

17.2 Mating and Fertilisation

Rabbits start to breed at the age of six months and do so at regular intervals during the spring and summer, producing four or five families a year. The female rabbit forms a nest at the end of a burrow and lines this with her own fur.

The union of the sperms and ova takes place inside the body of the female, and the sperms are introduced into the body during mating. The animals come very close to each other and the penis of the male is inserted into the vulva of the female and up into the vestibule. The sperms are moved from the testis along the vas deferens into the uterus masculinus and are discharged by muscular contractions through the penis in the fluid formed by the prostate and Cowper's glands. Each

REPRODUCTIVE ORGANS OF A RABBIT

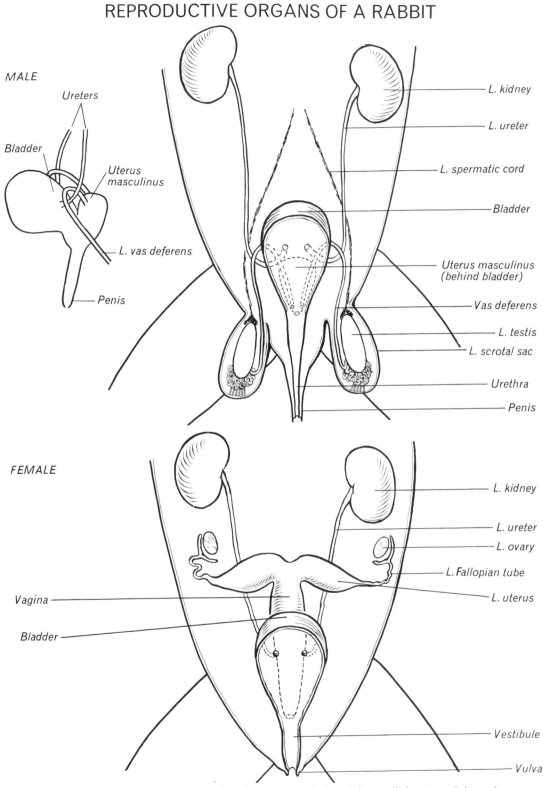

Fig. 17.1 Urino-genital organs of the rabbit in ventral view. The small drawing of the male organs shows their position when viewed from the left side.

testis (see Fig. 17.2) can form many millions of sperms, which are minute (less than 50 μ in length) and are easily damaged by heat. They are kept a little cooler than the rest of the body because the testes inside the scrotal sacs are outside the body cavity. Each sperm (see chapter heading for boar spermatozoa) consists of a head containing a nucleus, and a long whip-like tail which enables it to swim to an ovum, once it has reached a Fallopian tube. The sperms which are discharged in the vestibule are believed to be sucked up to the Fallopian tubes by muscular contractions of the vaginal and uterine walls. Many sperms fail to survive the journey to the penis and many perish during the movement up the Fallopian tubes.

In the rabbit, mating also results in the release of ova from the ovaries (see Fig. 17.3). These pass into the Fallopian tubes where they attract the sperms. Fertilisation takes place when one of the sperms penetrates the ovum and fuses with it (see Fig. 17.4). If an ovum is not fertilised it passes out of the body. Many ova may be fertilised at the same time, but usually only about five to eight of the eggs complete the development to a fully grown embryo—a process which takes about thirty days.

After the release of an ovum from the ovary the follicle which contained it forms a yellow sphere, the **corpus luteum**. This degenerates rapidly if pregnancy does not follow, but if the ovum is fertilised and starts to grow then the corpus luteum is retained throughout the period of development and secretes hormones which are essential for successful pregnancy and inhibit the release of further ova from the follicles.

The nucleus divides in a special way (meiosis) during the formation of the male and female gametes, so that fertilisation results in a return to the normal number of chromosomes (see sec. 16.2).

17.3 Development of Embryo, Birth, and Parental Care

The egg starts to divide into many cells immediately after fertilisation and passes from the Fallopian tube into the uterus, where it continues to develop, forming a ball of cells which grows into an embryo. The ball of cells sinks into the wall of the uterus, which in this region becomes swollen and develops a rich blood supply. This thickened area, the **placenta**, which adheres closely to the wall of the uterus, is formed partly from the tissue of the parent and partly from that of the embryo, which forms finger-like processes through which

TESTIS

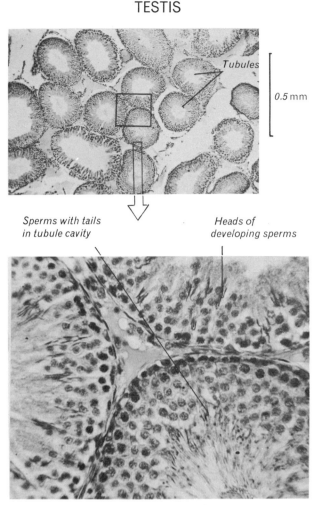

FIG. 17.2 Photomicrograph of stained sections of testis.

it absorbs some nourishment. Each embryo is contained in a fluid-filled membrane, the **amnion**, which is joined to the placenta and serves to protect the embryo from damage by pressure as well as enabling it to move freely.

The cells in the ball continue to divide and from them develop all the organs of the body. The blood system, which is formed at an early stage, is joined to the placenta by the **umbilical cord**, which contains branches of the embryonic blood vessels. These blood vessels are very close to a similar network of the parental system (see Fig. 17.5). The blood systems of the embryo and parent are quite separate, but in the placenta they are separated by very thin membranes so that dissolved substances can diffuse from one to the other. The placenta is the organ by which the embryo carries out the vital processes of nutrition,

OVARY & EGGS

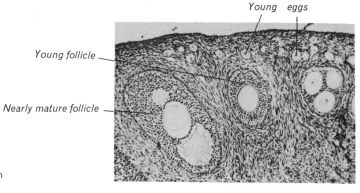

Young eggs

Young follicle

Nearly mature follicle

SECTION THROUGH THE OVARY OF A CAT

Follicle fluid   Follicle wall

0.5 mm

A MATURE FOLLICLE

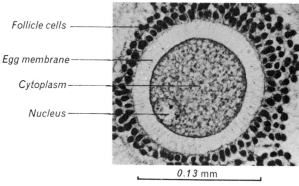

Follicle cells

Egg membrane

Cytoplasm

Nucleus

0.13 mm

THE EGG OF A CAT

FERTILISATION

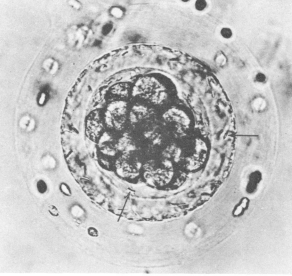

Fig. 17.4 A rabbit's egg (0.1 mm in diameter) recovered from the oviduct fifty hours after mating. The embryo has just passed the 16 cell-stage, and several sperms, two of which are arrowed, can be seen both inside and in the membranes around the egg. (*Photomicrograph by Dr. C. Adams.*)

MAMMALIAN EMBRYO

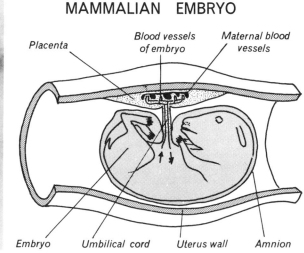

Placenta

Blood vessels of embryo

Maternal blood vessels

Embryo   Umbilical cord   Uterus wall   Amnion

Fig. 17.5 A rabbit's uterus cut open to show a well-developed embryo inside its transparent, protective membrane.

respiration and excretion. Digested food and oxygen diffuse into the blood of the embryo from that of the parent because of the concentration gradient (see sec. 9.8); waste substances, such as carbon dioxide and urea, diffuse from the blood of the embryo into that of the parent. The placenta also produces hormones (see sec. 15.6) which keep the uterus adapted to pregnancy.

Development of the embryo continues inside the uterus, which gradually enlarges, and the embryo is fully grown in about a month. At the end of this time the muscular walls of the uteri and vagina start to contract and relax alternately. Birth involves the detachment of the embryos from their placentae. The muscular movements of the uteri and vagina then squeeze them through the vagina and vulva, which dilate to allow this passage. The umbilical cord is bitten through by

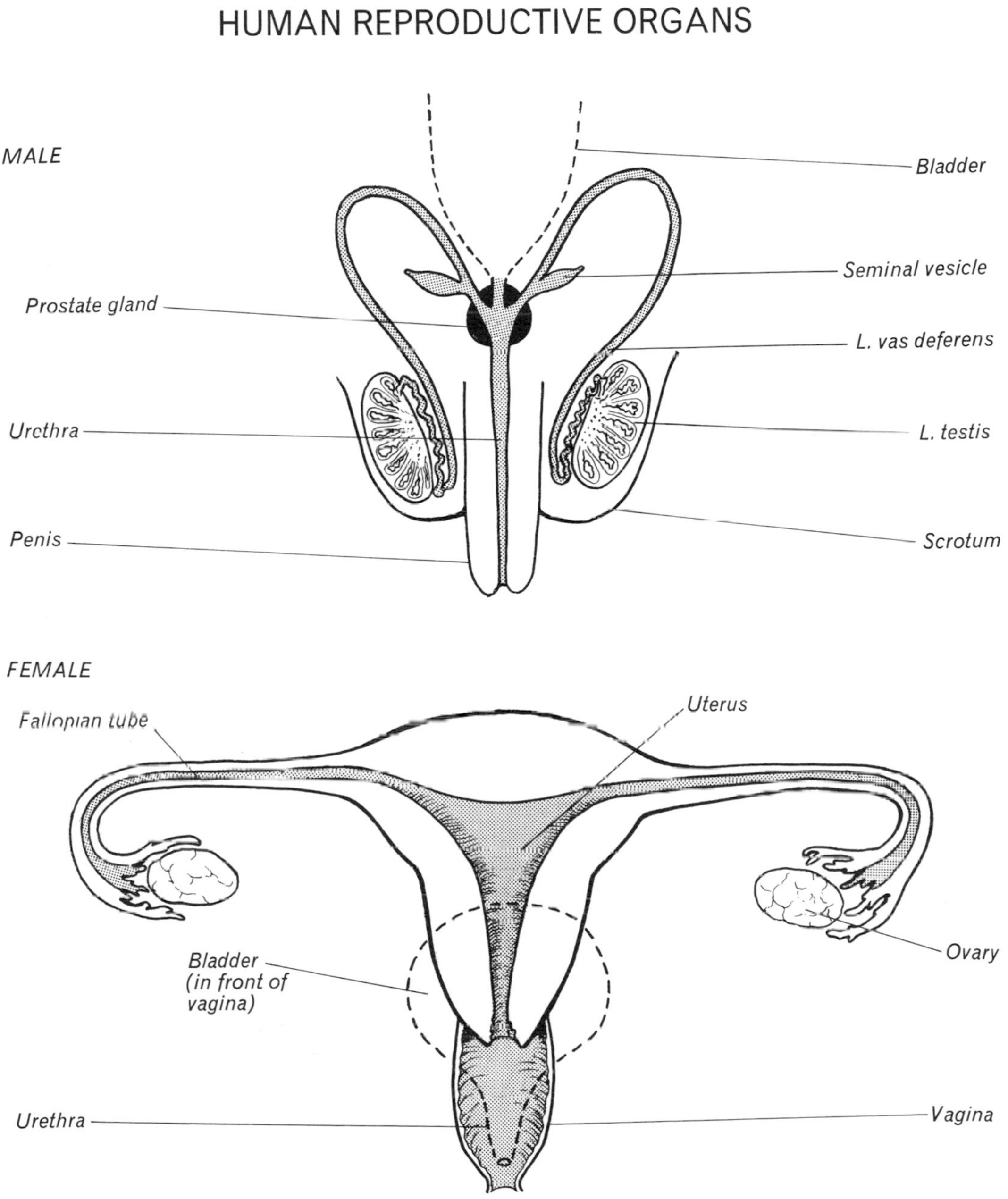

HUMAN REPRODUCTIVE ORGANS

FIG. 17.6 Sectional drawings showing in simplified form the human reproductive organs.

the female; the scar persists on the young mammal as the **umbilicus** or **navel**. The placentae are shed later, and are eaten by the female rabbit.

The newly born rabbits are naked and blind. Immediately after birth they must start to use their lungs, since they no longer receive oxygen from their mother's blood. For the first few weeks of their life they are fed on milk from the mammary glands on the thorax and abdomen of the mother. These glands start to develop after mating and the young rabbit sucks milk from them. This method of feeding the young is peculiar to mammals. Gradually the young rabbits develop fur, their eyes open, and they begin to feed themselves on grass.

17.4 Human reproduction

The human reproductive organs are similar to those of the rabbit (see Fig. 17.6). In the male, however. there is only one sac—the scrotum—behind the penis, and this contains both testes, and a pair of seminal vesicles is present; in the female there is one central uterus which is continuous with the vagina, and this opens to the exterior without joining the urethra.

Though the testes and ovaries are completely formed at birth they do not start to produce sperms and ova until much later. The stage at which this happens is termed puberty and is usually between the ages of eleven and fourteen. After this age sperms may be shed periodically, although many are either absorbed in the seminal vesicles or passed out with the urine. One or other of the ovaries sheds an ovum into the Fallopian tube at regular monthly intervals, until the female reaches the age of about forty-five. The majority of these ova are never fertilised. Also at monthly intervals, the thick lining of the uterus gradually breaks down and is discharged from the body with a little blood through the vulva. This discharge is called the menstrual flow and lasts about five days. Gradually the uterus develops a new lining, which becomes quite thick by the time the next ovum is shed. If fertilisation takes place the embryo becomes embedded in this thick lining and takes about nine months to develop, during which time the hormones secreted by the corpus luteum stop the menstrual cycle so that no further ova are shed from the ovary and the lining of the uterus does not break down. The cyclic changes in reproduction are co-ordinated by hormones formed by the pituitary gland and ovary.

The mating process is essentially similar to that already described, but as the purpose is to produce a fertilised egg which will grow to form a baby it is preceded, in human society, by marriage and the establishment of a home so as to provide the best environment for the development of the child. For social as well as physical reasons, therefore, it is not desirable that it should take place too early in life.

THE LIFE-HISTORY OF THE ROBIN
17.5

The reproductive organs of a bird are similar to those of a mammal, but only the left ovary and oviduct are functional in the female and there is no penis in the male. During mating the male opening is pressed close to that of the female and fertilisation is internal, occurring in the upper part of the oviduct. The ova are large, since they contain an abundant food-store (chiefly protein and fat)—the **yolk**. As the fertilised egg passes along the oviduct, the white, which is a solution of the protein albumen, the two shell membranes, and finally the shell, are formed round it. The shell is porous and allows gases to pass through it, and at the broader end of the egg there is an air space between the two membranes (see Fig. 17.7). By the time the egg is laid the embryo has already started to develop and appears as a small disc on the upper surface of the yolk. The yolk is heavy and floats on the albumen so that the embryo is kept in this position even when the egg is turned. As the egg is incubated by the parent bird, the embryo grows, forming blood vessels which spread over the yolk and absorb food from it. The blood vessels also absorb oxygen, which has diffused through the shell and the membranes. A hen's egg is fully developed after twenty-one days

BIRD'S EGG

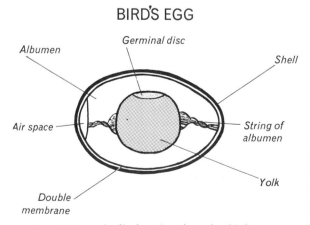

FIG. 17.7 Longitudinal section through a bird's egg.

and a robin's after thirteen to fifteen days. The young bird then pierces the inner-shell membrane with its beak, and as the air in the air space rushes into its lungs they expand and start to function. The chick pierces the shell with repeated blows of its beak, which has a sharp-pointed egg-tooth that is discarded after hatching.

Adult robins moult in July and August, and after this the cocks hold individual territory and begin to sing. Each cock takes possession of an area of land, about a hundred square yards, and defends it against other robins. This defence is carried out by a threatening display, by singing, and sometimes by actual fighting. Some hens also hold individual territories in the autumn, but many migrate over an area of a few square miles, though others leave England for the winter and return in the spring. Toward the beginning of January there is a revival of the cock's song and this attracts the hens. Cock and hen birds form pairs between January and March, and from then onwards the two birds share the same territory. About the middle of March the hen builds a nest, usually in a bank close to the ground. The nest, which is always well hidden, is cup-shaped, and built largely of moss with a foundation of dead leaves. The cup is lined with hair. The cock does not assist in building the nest and the hen is careful not to draw attention to herself during this period. She builds for only a few hours each day and the nest is completed in about four days (see Fig. 17.8).

Mating occurs after the hen has built the nest and the hen starts to lay. A clutch usually consists of about five eggs and is completed in about five days. The eggs are normally white, speckled with small reddish-brown spots, though variations occur. The hen robin starts to sit as soon as the clutch is complete, and incubation lasts from thirteen to fifteen days. The cock does not usually help to incubate, but during this period he frequently calls the hen off the eggs with a short song. Off the eggs, the hen repeatedly calls to the cock to feed her, which he does, though he does not often feed her while she is actually incubating. When the young robins hatch, the parents remove the egg shells from the nest and the hen spends much time brooding the naked young for the first few days after hatching. Both parents feed the young, chiefly on caterpillars. After a fortnight, the feathered but tailless fledgling leaves the nest, though it is another week before it begins to pick up food for itself. It does not become completely independent for another twenty-one days.

The female robin often leaves the young of the

Fig. 17.8 Photograph of a robin at its nest. (*By Eric Hosking.*)

first brood about the time they are fledged and begins to lay a second clutch. The cock continues to feed the fledglings. The pair usually rear two broods before moulting starts in July.

THE LIFE-HISTORY OF THE FROG

17.6 Mating, Egg-laying, and Fertilisation

The frog hibernates during the winter in the mud of ponds and ditches, emerging just before the breeding season, which is in March or April. Then the adult frogs return to the ponds, where they collect in large numbers and may often be heard croaking. At this season it is easy to distinguish the female frog from the male, as her body, which is brownish with yellow spots, is large and swollen owing to the mass of ova which it contains. The male, which is much thinner, has a darker, olive-brown skin and also a swollen pad under the first digit of the fore-limb.

The male frog mounts on to the back of the female and clings to the loose skin on her ventral surface with the pads on his fore-limbs. The two frogs may swim about the pond in this position for several days until they are ready to deposit their sperms and ova. The sperms are similar to those of a mammal and are formed in two testes which lie within the body cavity. They pass in the sperm fluid through tiny ducts to the kidney. This fluid flows with the urine from the kidney and collects in a store sac until it is required (see Fig.

FROG REPRODUCTIVE ORGANS

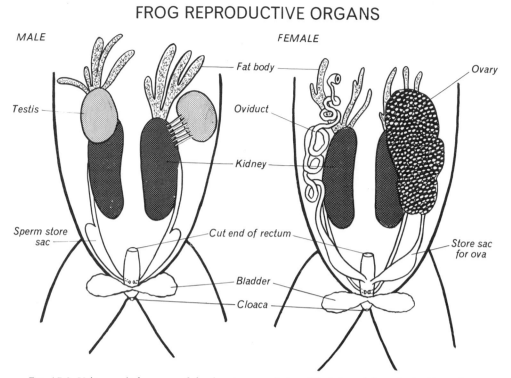

MALE   FEMALE

Fat body
Testis
Oviduct
Kidney
Ovary
Sperm store sac
Cut end of rectum
Store sac for ova
Bladder
Cloaca

FIG. 17.9 Urino-genital organs of the frog in ventral view. The left testis is pushed aside to show the ducts leading to the kidney; the right ovary is not shown.

17.9). The two ovaries are hollow sacs filled with ova, which, during the breeding season, pass from them to the exterior through the coiled oviducts which open near the base of the lungs. The ova, however, are much larger than those of a mammal, for they contain some yolk and are surrounded first by a delicate membrane and then by a thin layer of jelly which swells when it comes into contact with water. The sperms swim in the water toward the ova and fertilisation takes place in the water outside the body of the frog. The nearness of the male and female frogs serves to ensure the depositing of the sperms and ova in close proximity to one another, but in spite of this some ova are not fertilised. The pair of frogs form many hundreds of fertilised eggs, but of these only about two will complete their development to adult frogs under natural conditions. After egg-laying, the frogs separate and return to the marshy areas, though some frogs fail to survive the process.

17.7 Development of Egg and Growth of Tadpole

After fertilisation, the egg immediately starts to divide into a ball of cells which grows to form the tadpole, but no such change takes place in the unfertilised egg. At first, the egg (see Fig. 17.10)

is spherical and the upper part, which will grow into the tadpole, is black; the lower part, which is the food-store or yolk, is white. The egg is surrounded by jelly which swells as soon as it reaches

FROG SPAWN

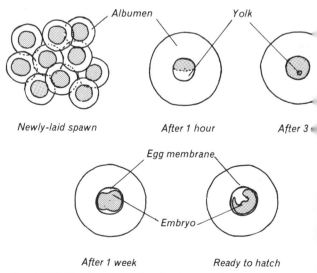

Albumen   Yolk

Newly-laid spawn   After 1 hour   After 3

Egg membrane
Embryo

After 1 week   Ready to hatch

FIG. 17.10 Development of a frog's egg. The drawings are all three times life-size.

TADPOLE STAGES

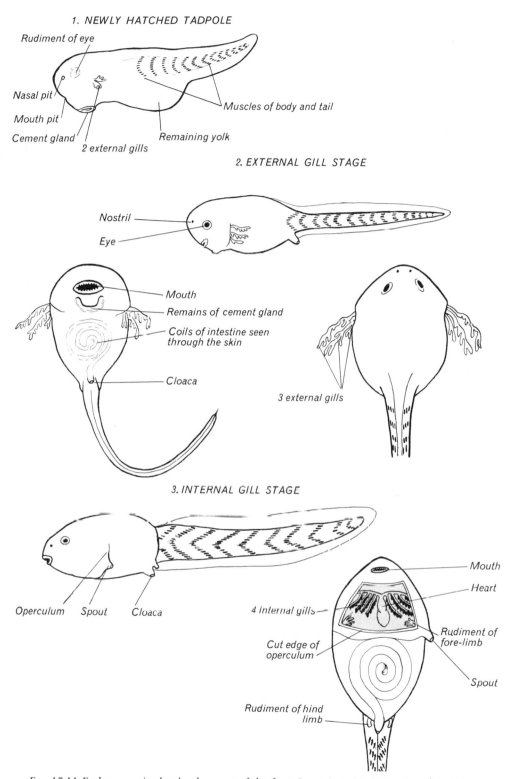

1. NEWLY HATCHED TADPOLE

Rudiment of eye

Nasal pit

Mouth pit

Cement gland

2 external gills

Remaining yolk

Muscles of body and tail

2. EXTERNAL GILL STAGE

Nostril

Eye

Mouth

Remains of cement gland

Coils of intestine seen through the skin

Cloaca

3 external gills

3. INTERNAL GILL STAGE

Operculum   Spout   Cloaca

4 internal gills

Cut edge of operculum

Rudiment of hind limb

Mouth

Heart

Rudiment of fore-limb

Spout

FIG. 17.11 Early stages in the development of the frog Stage 1. ×10. Stages 2. and 3. ×4.

the water and thus serves to spread the eggs out so that they are not overcrowded, and each one is able to obtain an adequate supply of oxygen. This jelly also serves to prevent damage by friction and its slippery nature makes it difficult for other pond animals to eat the eggs. If you examine the egg with a hand lens during the first week after it has been laid, you will see faint ridges on its surface, which indicate that it is developing to form a ball of cells. The developing embryo grows round the yolk so that toward the end of the first week the whole egg appears black; during the second week the egg changes shape, becoming ovoid and gradually lengthening until the head, trunk, and tail of the tadpole can be recognised. The tadpole, which is curved because of the limited space within its egg membrane, begins to wriggle inside the jelly, which by now has become less firm. Hatching (i.e. breaking out of the egg membrane and jelly) takes place about two weeks after fertilisation, and the young tadpole, which as yet has no mouth, is still dependent on the supply of food in the yolk. The tadpole (see Fig. 17.11), which is black, has a short, blunt tail by which it can wriggle, though this is too short to be used for swimming. On the under side of the head is a U-shaped cement gland which secretes a sticky mucus that enables the tadpole to cling to the jelly immediately after hatching, and later to pondweeds. Above this gland is a shallow pit which marks the position of the future mouth, and there are other places where the developing eyes, ears, and nostrils are visible. On each side of the head are two blunt projections which very soon develop into branching tufts of skin containing blood vessels; shortly afterwards a third develops. These are the **external gills** by which the tadpole absorbs dissolved oxygen from the water for a few days until the **internal gills** develop.

Soon after hatching, the mouth and anus develop at each end of the alimentary canal, which is at first straight, but as it lengthens it is thrown into coils which are clearly visible through the transparent skin. The mouth has a pair of horny jaws and a series of horny teeth arranged in rows above and below the jaws (see Fig. 17.12). These horny teeth are used for rasping the surface of pondweeds on which the tadpole feeds at this stage. Four slits are formed at about the same time as the mouth on each side of the tadpole's head, and the internal gills, similar to those of a fish, develop on these. Meanwhile, the external gills and the cement gland shrivel and disappear.

FROG METAMORPHOSIS

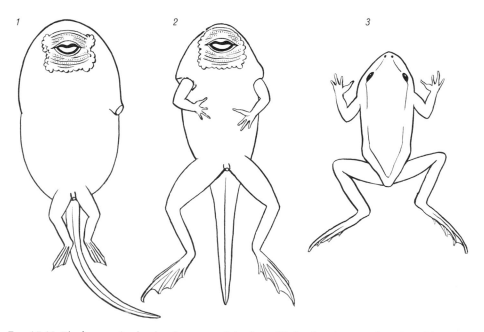

FIG. 17.12 Final stages in the development of the frog. All the drawings are four times life-size. 1 and 2 show the final tadpole stages in ventral view; 3 shows the dorsal view of the young frog.

The gill-slits do not remain visible for long, as a fold of skin—the **operculum**—forms in front of them at each side of the head. The operculum grows backwards, covering the gill-slits, and on the right side it fuses with the body, but on the left side the fusion is not complete, so that a small opening or **spout** is formed for the escape of the water that has passed over the gills.

By the time these changes are complete the tadpole is about a month old and is completely fish-like in its method of respiration and also has a fin-like tail by which it is able to swim. There is little change in form during the next three weeks, though the tadpole grows rapidly as it feeds, gradually changing to a carnivorous diet. It now uses its jaws rather than the teeth which it used for plants. If you are keeping tadpoles in an aquarium you must now begin to feed them on raw meat or fish food, as otherwise you will find that they begin to eat one another.

The hind limbs appear as tiny bumps at the base of the tail soon after the formation of the operculum and have become jointed by the end of the next week, and after a fortnight the five webbed digits are completely formed. The fore-limbs start to develop very soon after the hind limbs but they are hidden by the operculum. At this stage the lungs develop and the tadpole makes visits to the surface to gulp in air. As these visits become more frequent the internal gills gradually disappear.

17.8 Metamorphosis

When the tadpole is about three months old a fairly sudden change or metamorphosis takes place. It stops feeding and casts its skin. The mouth becomes much larger, the head more triangular, and the eyes very prominent (see Fig. 17.12). The right fore-limb appears through a weak spot in the operculum, and the left one pushes its way through the spout. During this time the body has become much thinner and the tail, which is gradually being absorbed by the white corpuscles, is shrinking. The small tailless frog is now ready to leave the water, though it is not fully grown until its third year.

Metamorphosis is a period of rapid transformation from the larval or tadpole stage to the adult one. The tadpole has a different structure, mode of life and method of feeding from the adult and does not therefore compete with it for survival.

THE LIFE-HISTORY OF THE STICKLEBACK 17.9

The three-spined stickleback is found in shoals in fresh-water streams and ponds. The torpedo-shaped body is about 0.06 m long (see Fig. 13.15). The pectoral fins are the principal organs of locomotion, the tail being used only for short spurts. Sticklebacks are carnivorous, feeding mainly on crustaceans, molluscs, worms, and insect larvae. Their main enemies are perch and dytiscus beetles. They can easily be kept in an aerated aquarium containing Elodea, and be induced to breed.

In March or April the males develop a reddish colour on the ventral surface and a green-blue eye tinge, but the females remain silvery though they become plumper. The male now occupies a definite territory and begins to build a nest under a canopy of weeds, by scooping out with biting movements

SUMMARY OF CHANGES IN THE LIFE-CYCLE OF FROG

(Approx.) Time	Change	Respiration	Feeding
Few hours after laying	Jelly swells	Dissolved oxygen diffuses in	Yolk
First week	Embryo grows round yolk	Dissolved oxygen diffuses in	Yolk
Second week	Hatching	Dissolved oxygen diffuses in	Yolk
2–3 days after hatching	Tail lengthens Mouth and external gills start to function	Dissolved oxygen through external gills and skin	Pondweeds
2 weeks after hatching	Operculum begins to grow	Dissolved oxygen through internal gills and skin	Gradual change to a carnivorous diet
1 month after hatching	Growth of operculum complete Formation of limbs	Dissolved oxygen through internal gills and skin	
2 months after hatching	Frequent visits to surface as lungs develop	Dissolved oxygen through internal gills and skin	Pond animals
3 months after hatching	Absorbs tail, casts skin, and leaves the water	Atmospheric oxygen through skin, mucous membrane of mouth and lungs	Worms, snails and flies

a depression about 5 cm. across. This is lined and roofed in with small pieces of weed which are pressed together by the snout and then glued together by an adhesive secretion from the anal region; the finished nest is like a tunnel with one opening. The male attracts a female to it by a series of swoops in which he displays his red ventral surface. The female enters the nest and is stimulated to lay eggs by the male pushing against the sides of her body. She leaves the nest after spawning and the male may attract several females to the nest before he enters to fertilise the eggs by shedding sperm over them. After this the male closes the nest and defends it from predators. He opens it from time to time to aerate the eggs, driving water through the nest by vibrating the pectoral fins. The eggs are surrounded by a membrane and contain a small amount of yolk. Incubation takes about thirty days, and just before the eggs hatch the male opens the top of the nest, making it bowl-shaped. For about a week after hatching the male keeps the young together in the nest; they are about 3 mm long, and the yolk sac remains attached to them for about three days. In the second week after hatching they begin to make longer and more frequent excursions from the nest. The fish are fully grown in three to five years.

This life-history is far from typical of all fish, but the stickleback is one which can be kept in an aquarium and its breeding habits observed. Most fish lay eggs which are fertilised in the water, and few show any parental care.

INSECTS, EARTHWORMS, HYDRA, AND AMOEBA

17.10

The insects, like the vertebrate types already described, have separate male and female reproductive organs, but the fertilised egg does not always develop into a form like the adult but may pass through a series of different stages before development is complete. A description of this metamorphosis in some insects will be given in the next chapter.

The earthworm is **hermaphrodite** (containing both male and female organs) but the ova are fertilised by sperms from another earthworm. Pairing usually takes place in warm moist weather and on the surface of the soil. The two earthworms lie with their ventral surfaces together but pointing in opposite directions. The **clitellum** secretes slime which binds the two bodies together. Sperms pass from the male opening on segment fifteen of one earthworm into special store sacs (which lie between segments nine and eleven) of the other earthworm. After this exchange of sperms the two animals separate and return to their burrows. There they form a **cocoon** in which the eggs are laid. This is formed from a thick band secreted by the clitellum. The earthworm wriggles out of the cocoon backwards and, as it passes along the body, ova from the opening on segment fourteen, and sperms from the store sacs, are shed into it. The cocoon is elastic and closes up when it is free from the body. It later becomes darker and from it hatch the young earthworms, which are similar to the parent but smaller and much paler in colour. In some species only one earthworm hatches from the cocoon.

The method by which Hydra reproduces was described in section 2.12, and if you read this again you will realise that its method of reproduction may well be controlled by food supplies. Rapid growth when food is plentiful during the spring and summer enables it to reproduce asexually by budding, but the onset of adverse conditions in the autumn stimulates it to reproduce sexually and form an embryo which can remain dormant for several months.

The life-cycle of Amoeba, which was described in section 2.3, also includes a spore stage which may hatch directly or remain dormant during a period of adverse conditions. This animal, however, reproduces only asexually.

ADVANTAGES AND DISADVANTAGES OF THE DIFFERENT METHODS

17.11

Sexual reproduction is the typical method in all multicellular animals, and though the details of the process differ in different groups, fundamentally they are all alike. The essence of sexual reproduction is the fusion of two special cells, and especially of their two nuclei so that the offspring receives chromosomes from both parents. This mingling of two sets of chromosomes provides an opportunity for variation (see sec. 20.5 and sec. 20.6). This method of reproduction is also an economical one, for large numbers of offspring can be started, often simultaneously (e.g. fish, frog), with comparatively little drain on the food reserves of the parent. Some groups (e.g. fish, frog) make no further provision for the new generation than the formation of a food-store or yolk in the ovum. This makes the ovum large and passive so that the sperms have to be active to reach it. Birds make greater preparation for the care of their offspring, for they build a nest in which the eggs are incubated and later feed the young until they become independent. In mammals, this parental care is extended even further and includes not only the feeding but also the training of the offspring.

The number of eggs formed is closely related to the degree of parental care and to the method by which fertilisation is ensured. Fish, which give no parental care and in which fertilisation is external and very haphazard, form a very large number of ova and sperms. Amphibians, which also give no parental care but which attempt to ensure external fertilisation by the close proximity of the male and female, also form a large number of ova though fewer than fish. The jelly gives some measure of protection to the egg. In both fish and amphibians only a very small proportion of the fertilised ova complete their development, since the offspring fall an easy prey to larger animals during their growth. Birds, which protect both the eggs and young and in which fertilisation is internal, lay fewer eggs and a greater proportion of these complete their development. Mammals, in which both fertilisation and development is internal, are still more successful in rearing their offspring; the placenta is of extreme importance in their success. An interesting feature in the development of the embryos of all these groups of animals is that they all require an aquatic environment, even though the adult is terrestrial.

QUESTIONS 17

1. How in a named mammal, a named fish, and a named bird is: (*a*) the egg fertilised? (*b*) the embryo nourished? (*c*) the young cared for both before and after birth?
[S]

2. Describe, with the aid of drawings, the external appearance of an adult frog, and explain how it is adapted to its way of life.
Describe the development of the animal from a fertilised egg. (Drawings *not* required.)
[C]

3. Describe the development from fertilisation to the stage just before the emergence of the fore-limbs in either a frog or a toad.
[C]

4. By means of illustrated diagrams give an account of the life-history of the frog. State how (*a*) the tadpole, and (*b*) the frog obtains its oxygen.
[L]

5. Many animals, such as frogs, produce numerous eggs each year and yet the numbers of adults remain fairly constant from year to year. How do you account for this state of affairs?
[L]

6. Compare and contrast sexual and asexual methods of reproduction. Give a concise description of the nutrition and respiration of the embryo of a mammal before birth.
[L]

7. Give a diagram showing the external structure of one of the following fish: dogfish, eel, herring, salmon, stickleback, trout. State the name of the fish selected and describe its life-history and habits.
[O]

8. Give an account of how a mammal protects and nourishes her young from the zygote stage till the time of birth. Briefly outline the parental care exercised by any one named animal.
[O & C]

9. What are the essential features of sexual reproduction? Describe with the aid of a large labelled diagram how protection and nourishment are provided for an embryo mammal.
[O & C]

10. Make a large fully labelled diagram of a fish in side view. By considering the external features only, compare and contrast the structure of a fish with that of a frog tadpole which has the hind limbs but not the fore-limbs developed.
[S]

11. Compare and contrast the frog, bird, and mammal, under the following headings: (*a*) mode of fertilisation; (*b*) protection of embryo; (*c*) the feeding of the embryo during development. (Details of development of embryo are *not* required.)
[W]

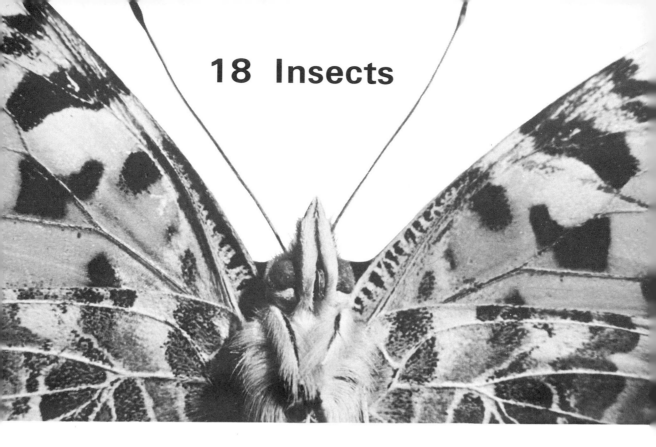

18 Insects

Insects belong to the largest phyla of the invertebrates called the Arthropoda. These animals all have an external skeleton, not of bone or cartilage, but made largely of horny material called **chitin**. The body is divided into segments and there are pairs of jointed legs. The Arthropoda include four chief classes: Crustacea (crabs, water-fleas, shrimps), Myriapoda (centipedes, millepedes), Arachnida (spiders, scorpions, mites), and Insects.

The insects form the largest division of the Arthropoda and are to be found almost everywhere except in the sea. They show a great variety of form, though all are built on the same general plan which can easily be seen in a large insect such as the cockroach. The other insects described in this chapter are all of great importance to man, either because they pollinate flowers and so help in the formation of fruit (*butterfly and hive-bee*), or because they destroy crops (*cabbage white butterfly*) or spread disease (*house-fly and mosquito*).

COCKROACH

18.1 General Structure and Life-history

Cockroaches, which are often called black beetles (though they are brown and not beetles), are a common pest found in kitchens, boiler-houses, and other warm places where food is plentiful. They remain hidden in crevices during the daytime and come out to feed at night. The fact that they can feed on almost any kind of material, including paper and whitewash on the wall, and can survive for a long period without nitrogenous food, probably accounts for their rapid spread. In the British Isles there are two kinds of cockroaches which are frequently found in houses. These are the common cockroach (*Blatta orientalis*), which is supposed to have been introduced through commerce in the sixteenth century, and the German cockroach (*Blatella germanica*), which is generally thought to have been introduced into this country after the Crimean War. The large American cockroach (*Periplaneta americana*) is a recent introduction to this country. It is rarely found in houses but frequents such heated places as warehouses, hothouses, and the lion and tiger houses of some zoos. These three species are much alike, though they differ in size and general colour.

External Structure

The body of the fully grown American cockroach is about 0.04 m long and is divided into three dis-

COCKROACH

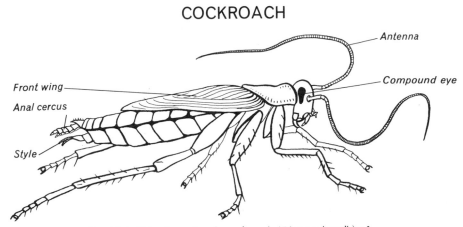

Fig. 18.1 Side view of male cockroach (*Blatta orientalis*) × 3.

tinct regions: head, thorax, and abdomen (see Fig. 18.1). The head is joined by a narrow neck to the thorax, which has three segments and bears the legs and wings; the abdomen has ten segments. The whole body is covered by a reddish-brown cuticle which, besides being impervious to water, is thick and tough so that it protects the body and provides stiffening for the limbs. It facilitates movement in much the same way as the bones of mammals do, but in this case the skeleton is external to the muscles which are attached to it. This exo-skeleton covers the whole body and must, therefore, be provided with joints. It remains thin and flexible in certain places, so that the thickened parts can be moved relative to one another. The body is flattened from above and below, which helps the insects to remain hidden in crevices.

The head, which is pear-shaped, hangs downwards at right angles to the thorax. At each side of it there is a large shining black **compound eye**; between the eyes is a pair of long, many-jointed **antennae** which taper toward the tip and are sensitive. The mouth, which is at the tip of the head, has a hinged flap—the **labrum** or upper lip hanging in front of it. At the sides of the mouth are three pairs of jaws or mouth parts which are shown in Fig. 18.2. Immediately under the labrum are the two **mandibles**, each with a strongly toothed inner edge used for biting, and behind these lie the two **maxillae**. The two maxillae work against each other and break up the food into even finer pieces after it has been chewed by the mandibles. The third pair of jaws, or **second maxillae**, are joined together at the base and each has an outer three-jointed palp.

The thorax consists of three segments, each one of which has a pair of jointed legs attached to its

ventral surface. These legs end in two curved claws with a large pad between them which prevents slipping. The legs are moved three at a time so that the insect always remains balanced on a

HEAD AND MOUTH PARTS

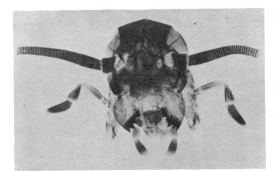

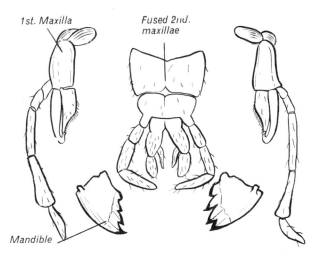

Fig. 18.2 Photograph of the ventral view of the head of the cockroach and drawings of the dissected mouth parts.

NB—K

COCKROACH

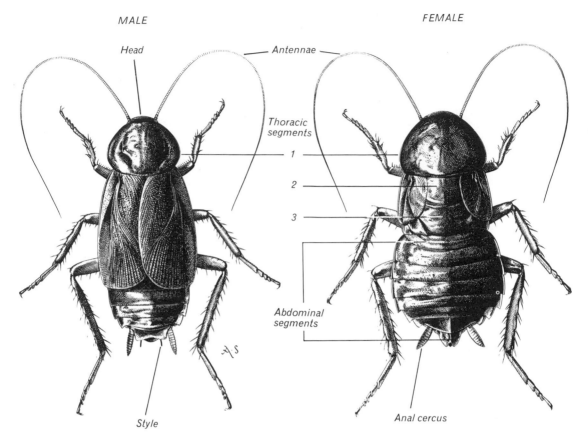

MALE   *FEMALE*

Head — *Antennae* —

Thoracic segments

1

2

3

Abdominal segments

Style   *Anal cercus*

FIG. 18.3 Male and female cockroach (*Blatta orientalis*). (*Courtesy I.C.I. Ltd. Drawings by A. Smith.*)

tripod formed from the first and third legs on one side, and the second leg on the other side. It is difficult to see this movement clearly, for the cockroach runs rather than walks.

The two pairs of wings are joined to the anterior edge of the dorsal surface of the second and third thoracic segments. The front wings or **elytra** are thick and form a protective covering over the thin hinder pair, which, when at rest, are folded lengthwise. The wings of the American cockroach are longer than the body and present in both sexes, but those of the common cockroach are shorter and fully developed only in the male. These differences are shown in Fig. 18.3.

The abdomen consists of ten segments, but the eighth and ninth are partly concealed by the seventh. The anus opens ventrally on the last segment, at each side of which is a short-jointed spindle-shaped **cercus**; the cerci are thought to have a sensory function. In the male the ninth segment has a pair of slender **styles** whose func-

tion is obscure, and in the female the chitinous plate on the ventral surface of the seventh segment is boat-shaped and projects backwards to form the **genital pouch**.

The **spiracles** or respiratory openings occur at the sides of the second and third thoracic and first eight abdominal segments, and can just be seen with a hand lens if the body is stretched and examined from the ventral surface. The method of respiration was described in section 9.11. The blood, which is colourless, and not concerned in the transport of respiratory gases, is made to circulate through the body by the contractions of a long, narrow heart which lies just below the chitin of the dorsal surface and extends through the thorax and abdomen.

Internal Structure

If the dorsal plates of chitin are cut away from the thorax and abdomen, a mass of white tissue is seen. This is the **fat body**, some of whose cells store absorbed food material; others accumulate solid nitrogenous excretory material. The alimentary canal is embedded in it (see Fig. 18.4). The mouth opens into a narrow oesophagus which leads through the thorax into a large pear-shaped crop where some digestion takes place. Lying beside the crop are two large salivary glands which form the saliva that is poured on the food before it enters the mouth. The gizzard, beyond the crop, has thick muscular walls containing six teeth placed lengthwise, with hairs between them which strain the food before it passes into the short, narrow mid-gut from which the food is absorbed. The hind-gut opens at the anus. Attached to the hind-gut at its anterior end are numerous fine Malpighian tubules which may have an excretory function. When the alimentary canal has been removed, the double nerve cord which lies in the mid-ventral line can be seen. There is a large ganglion in the head region which receives nerves from the eyes and antennae, and others in the thorax and abdomen to which paired branching nerves are attached.

COCKROACH

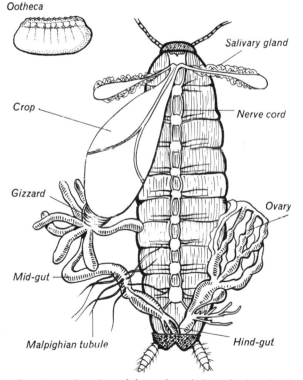

Fig. 18.4 Dissection of the cockroach from the dorsal surface to show the internal organs. The alimentary canal has been displaced to the left side and the left ovary removed. The egg capsule is drawn twice its natural size.

COCKROACH NYMPH

Fig. 18.5 Photograph of the side view of the nymph of the cockroach (*Blatta orientalis*). (*Photograph by Shell.*)

Life-history

The male cockroach has a pair of testes which lie in the abdomen, embedded in the fat body. There are two ovaries in the female, each consisting of eight tubes which have a beaded appearance owing to the rows of eggs which they contain. Pairing takes place as soon as sexual maturity is reached, and the sperms are stored in the female spermotheca which opens into the genital pouch. Fertilisation occurs after the ova have passed from the ovaries into the genital pouch, where the eggs are enclosed in a chitinous **egg capsule** (see Fig. 18.4) formed by special glands in the abdomen of the female. The capsule, which is reddish-brown and about 10 mm in length and 5 mm wide, contains two rows of eight eggs. The female may carry the egg capsule protruding from the end of the abdomen for several days, but eventually she deposits it in a warm, sheltered spot, preferably near a food supply. It is carefully covered with debris such as grit, pieces of wood, and paper. Each female is capable of laying several capsules; the time taken for the eggs to hatch varies according to the temperature and humidity—usually it takes about thirty-two days. The young cockroach, or **nymph** (see Fig. 18.5), eats the same food as the adult which it resembles, though it is smaller, paler in colour, and has no wings. As it grows it is necessary for it to cast off its exo-skeleton and form a new one. This is termed **ecdysis** (moulting), and the new cuticle is formed underneath the old one, which finally splits to allow the insect to climb out. During the first year there are three moults, the insect gradually becoming a darker colour. Three more moults occur before the wings and reproductive organs are fully formed. These changes generally take three years, and after them

LIFE-HISTORY OF THE CABBAGE WHITE BUTTERFLY

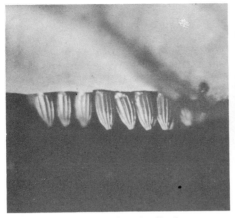
Eggs on underside of leaf

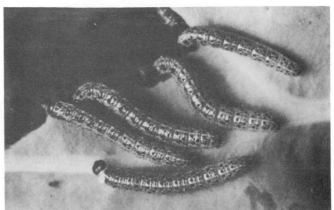
Larvae feeding

Pupa
*(Part of pupal
case removed)*

Adults

Fig. 18.6 *(Photographs taken by Alan Dale and supplied by Flatters and Garnett Ltd.)*

no further increase in size takes place. This gradual assumption of the adult character is termed, in insects, **incomplete metamorphosis**; it differs from that found in other types to be described which have several abrupt changes in form during their life-history, thus showing a **complete metamorphosis**.

18.2 Control

Though cockroaches have not, as yet, been shown to transmit disease to human beings, their presence in houses is most undesirable. They contaminate food, rendering it unpalatable, and their odour is repulsive. Methods of control are not easy, for the insects are quick to hide if exposed to light. Houses which are particularly liable to invasion are generally old, with many cracks and crevices which provide hiding-places for the insects. These should be filled with cement, paste, etc. Special emulsions may be painted on the floorboards, and powders or sprays used to kill off the insects, but care must be taken to see that these do not come into contact with foodstuffs.

CABBAGE WHITE BUTTERFLY

Butterflies and moths belong to the order Lepidoptera (*Gk. lepis, a scale*; *pteron, a wing*) and have two pairs of wings. They may be distinguished from each other by the fact that the antennae of the butterfly are long, smooth, and club-shaped at the tip, whereas those of the moth are shorter, usually taper to a point, and are hairy. The wings of the butterfly, when it is at rest, are often held vertically, while those of the moth are held horizontally.

18.3 Life-history

The cabbage white butterfly (*Pieris brassicae*) is very common in early summer and the eggs are easily reared, so that all stages in the life-cycle can be carefully observed. The eggs are laid in April or May and in August or September, on such plants as cabbages and nasturtiums. They are about 2 mm long, yellow, conical in shape, and have vertical and horizontal ridges upon the surface (see Fig. 18.6). They are laid in batches of twenty to one hundred and are usually fixed to the under side of the leaf. They hatch after seven to ten days; just before this a black dot can be seen at the narrower end of the egg. This is the head of the larva, or **caterpillar**, which bites its way through the egg shell by means of its mandibles. After hatching, the caterpillar eats the egg shell and then starts feeding on the leaf. At this stage the tiny caterpillars have black shiny heads and yellow bodies with a few white hairs. After a few days they stop feeding, remain still for two days, and then moult. This moulting is followed by a rapid increase in size, and four or five moults take place before the caterpillar is fully grown; throughout this time it continues to feed voraciously. After the third moult the caterpillars separate, and each feeds along the edge of a leaf so that the mandibles can work one on each side of the margin of the leaf.

The fully grown caterpillar, which is about 0.04 m long, has a grey-green body with three yellowish stripes running one along its upper surface and one along each side. There is also a number of black, wart-like protuberances covered with short hairs. The body consists of a head and thirteen segments (see Fig. 18.7). The head is spherical and has no large compound eyes like the adult, but has a ring of six small, black, simple eyes on each side toward the ventral surface, so that the caterpillar can see objects only if they are directly beneath its head, e.g. the leaf on which it is feeding. The two short antennae are difficult to see between the hairs. As in the cockroach, the mandibles lie behind the labrum with the small first maxillae beneath them. The second maxillae are joined to form a plate from which a small tubular

structure—the **spinneret**—hangs downward. From this issues a pale-coloured silk which the caterpillar spins to form a pathway on which it can gain a firm foothold.

Each of the first three (or thoracic) segments after the head bears a pair of five-jointed legs ending in a forwardly curved claw. These legs, besides being used for crawling, are also used to grip the leaf while the mandibles are biting. On the third, fourth, fifth, and sixth of its ten abdominal segments there are pairs of **pro-legs**. These are present only at this stage of the butterfly's life and are not jointed. They are soft, fleshy projections ending in a cushion-like pad with a half-circle of hooked bristles, so that they help the caterpillar to cling firmly. On the last segment there is another pair of similar structures termed **claspers**. There is a spiracle on each side of the first thoracic, and first eight abdominal segments. These can be seen with a hand lens as they are bounded by a thin, elliptical, brown line.

The caterpillar is fully grown after about a month, and at this stage it appears restless, stops feeding, leaves the leaf, and crawls to a more suitable place for the next or **pupal stage**. It usually climbs up some strong, dry support such as a fence or wall, or trunk of a tree. There it spins a mat of silk, into which it fixes its claspers and then supports itself in a horizontal position by silken threads passed round its body and fixed on either side to the support. The body gradually becomes shorter and thicker, and after a day or two the final moult takes place. As the cuticle splits it is pushed off backwards through the girdle of silk and reveals the pupa, which is quite unlike the caterpillar. At first the body is soft, but the cuticle quickly hardens and the pupa assumes its typical shape (see Fig. 18.6), and greenish colour with yellow and black spots. The cuticle has several projections; the most prominent of these is on the dorsal surface, and there are two lateral ones just below it. The girdle of silk passes between these so that the pupa cannot fall out of it. Many of the parts of the future butterfly—the antennae, eyes, proboscis, legs, and wings—can be seen pressed close to the thoracic region, and the segments of the abdomen are visible also. The pupa does not feed, and remains still except for occasional twitchings of the abdomen. Inside the pupa many changes are taking place, for most of the larval tissues are being broken down and reorganised to form the adult organs. These changes take from ten to fourteen days, and then the cuticle splits in the region of the thorax, and the adult butterfly or **imago** drags itself out. At first

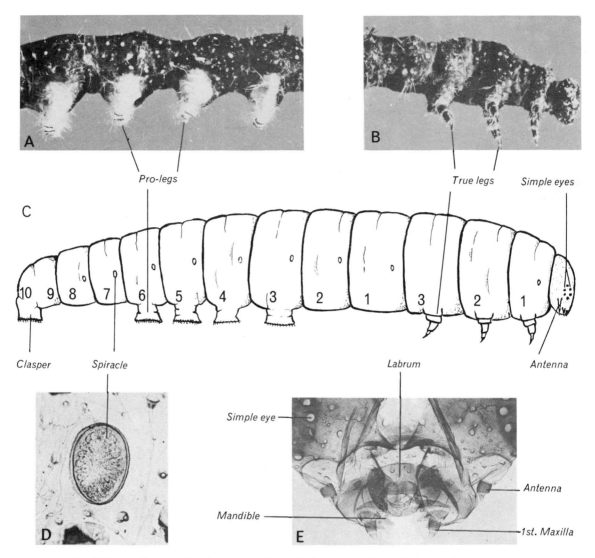

Fig. 18.7 A—Photograph of segments 3 to 6 of a peacock butterfly larva. The pro-legs are extended since the cuticle was dried when inflated with air. B—Photograph of the head and thorax treated as in A. C—Diagrammatic side view of a butterfly larva or caterpillar. D—Photomicrograph of a spiracle much enlarged. E—Photomicrograph of part of the larval head of the cabbage white butterfly.

the wings are small and crumpled, but they soon expand and harden.

The body of the imago is divided into a head, thorax, and abdomen. The head (see Fig. 18.8) has a large compound eye at each side, and just above these are two long, many-jointed antennae with swollen tips. They are very sensitive to touch and probably to smell as well. The mouth parts are very different from those of the larval stage, for there are no mandibles, and the maxillae have be-

come modified to form a long tube—the **proboscis** —up which nectar is sucked when the insect visits a flower. When not in use, the proboscis is coiled spirally under the head. The second maxillae are represented by short labial palps which have a sensory function.

The three segments of the thorax are not easy to distinguish, for they have a thick covering of hairs. There is a pair of legs on the ventral surface of each segment and each one ends in two curved claws with a pad between them. The two pairs of wings are joined to the dorsal surface of the second and third segments. They are covered with a powdery substance which, when examined under a microscope, is seen to consist of scales of varying shapes. The scales cover the whole wing surface and overlap one another like tiles on a roof. Similar scales are found on the body. The wings are a creamy-white colour on the upper surface and pale greenish-yellow on the under surface. The upper surface of the front wing has a small black marking at the tip, and, in the female, there are other black spots. In both sexes these wings have black spots on the under surface in addition. The hind wings are similar in both sexes, and have a dark patch on the front margin which is more conspicuous on the upper than on the under surface. The abdomen is so densely covered with hairs that it is not possible to distinguish the ten segments.

Mating takes place on the wing; the male is attracted to the female by scent. He grips the female with his claspers and passes sperm into her oviducts; the eggs are fertilised internally. There are two broods of the cabbage white butterfly during the year. The winter is passed in the pupal form, and the butterflies which emerge from these pupae form the first brood, which lays its eggs in April or May. These eggs produce the second brood, which lays its eggs in August or September, and form the pupae which survive the winter.

18.4 Importance and Control

The cabbage white butterfly is one of the few butterflies which harms our food crops. The caterpillar is such a voracious feeder that it quickly reduces the leaves of cabbages, cauliflowers, brussels sprouts, and similar plants to skeletons, and fouls any uneaten leaves with its excreta. In small gardens, plants should be examined every week during the egg-laying season and the eggs crushed. It is possible to spray larger areas with insecticides. Soapy or salt water may be used, and there are many powders which can safely be dusted on the leaves.

Many birds assist the gardener in the removal of caterpillars, and one ichneumon fly lays its eggs in the body of the caterpillar. The ichneumon larvae have no mandibles, but feed on the tissues of the caterpillar, which is not killed at once. The life-histories of the parasite and host are so balanced that the time taken for both to reach the pupal stage is the same. The larvae crawl out of the dying caterpillar and form a cluster of yellow cocoons from which the flies emerge at the same time as the undamaged cabbage white butterflies, so that they are ready to infect the next brood.

HEAD AND MOUTH PARTS

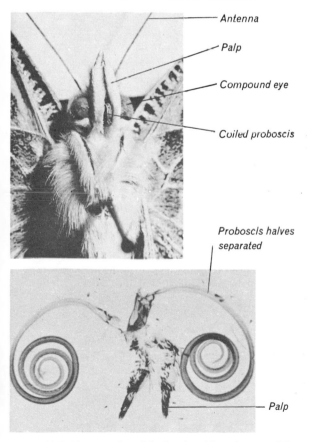

FIG. 18.8 Photographs of the head and mouth parts of the painted lady butterfly.

HOUSE-FLY

18.5 Life-history

The house-fly (*Musca domestica*), like the cabbage white butterfly, undergoes a complete metamorphosis, though in a much shorter period. The eggs are laid in any moist, warm material which will provide food for the larvae—e.g. horse-dung or decaying organic rubbish such as that found in dustbins. The eggs, if examined under a good hand lens, are seen to be glistening white and resemble grains of polished rice (see Fig. 18.9). They are spindle-shaped, about 1 mm long, and are laid in batches of 120 to 150, concealed in crevices in the rubbish. They hatch in from eight hours to three days, depending on temperature, into tiny, white, legless larvae or **maggots**. These burrow into the rubbish, on which they feed voraciously,

HOUSE-FLY LIFE-HISTORY

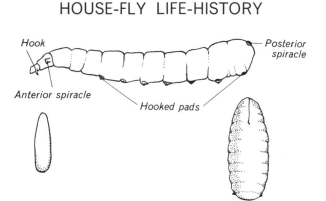

FIG. 18.9 Early stages in the life-history of the house-fly. The egg (bottom left) is ×20, the side view of the larva ×9, and the puparium (bottom right) is ×9.

growing rapidly, so that in hot weather they reach their full size in about forty-two hours, though in cooler weather this growth may take five days or more. The fully grown larva is about 1 cm. long and has an ivory-yellow waxy appearance. It con-

sists of twelve segments and is broader at the hind end (see Fig. 18.9). The mouth is on the narrow anterior end, which is very sensitive, and projecting from the head are two black, hook-shaped spines by which the larva drags itself along. On the lower surface of each of the sixth to twelfth segments there is a small spiny pad which also helps the larva to move. There is a pair of spiracles on the second segment and another pair on the last segment. The larva feeds on the semi-liquid decaying matter, and if its food becomes dry it moves to a moister and darker region, thus avoiding the light which is liable to dry the food. The larva moults twice, and then moves to a drier spot to pupate. The body becomes barrel-shaped and the larval cuticle hardens and becomes dark brown, though it is not shed. The pupa is formed within a new cuticle which is inside this hardened brown one, after which most of the internal tissues break down and are reorganised. These changes take about four days, and when they are complete the imago pushes off the end of the pupal case by means of a dilatable sac on the front of the head,

HOUSE-FLY

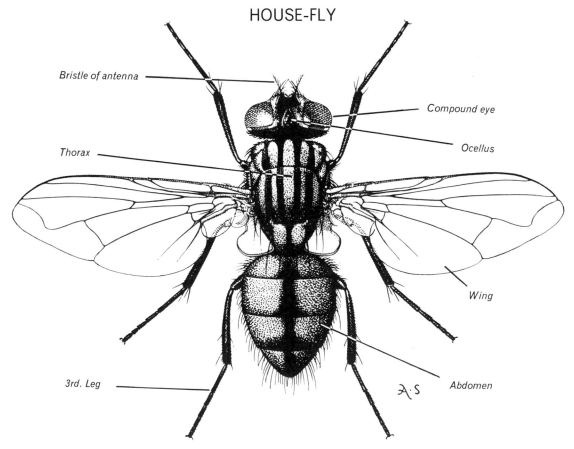

FIG. 18.10 House-fly (*Musca domestica*). (*Courtesy I.C.I. Ltd. Drawing by A. Smith.*)

Fig. 18.11 Photographs of the head and proboscis of house-fly. (*By Shell, and Flatters and Garnett Ltd.*)

which is withdrawn when the imago has emerged and moved to the surface. The newly emerged imago is pale and has shrivelled wings, but it soon darkens and the wings expand.

The body is about 5 mm long, grey in colour, with the thorax marked on the dorsal side by four narrow, black stripes (see Fig. 18.10). There is a large, reddish compound eye at each side of the head, those in the male being larger and the space between them narrower than in the female. Arranged in a triangle toward the top of the head are three simple eyes. The antennae are at the front of the head, and consist of three short joints with a compound bristle on the last joint. There are no mandibles or first maxillae, and the second maxillae are modified to form a proboscis which ends in two pads (see Fig. 18.11). These pads are folded together and the whole proboscis bent under the head when not in use. When the insect is feeding the pads are spread over the food. The house-fly can feed on solid food such as sugar by dissolving it in saliva, which is discharged over it. The pads and proboscis contain channels along which the liquid food is drawn.

The hairiness of the thorax makes it difficult to see the three segments, each of which bears a pair of jointed legs on its ventral surface. Between the claws at the end of each leg there are two pads which secrete a sticky substance, thus enabling the fly to walk upside down on the ceiling and along slippery surfaces (see Fig. 18.12). The house-fly has only one pair of wings, joined to the dorsal surface of the second thoracic segment, and for this reason it is placed in the order Diptera (two-winged flies). The wings are transparent with a network of veins (tracheal branches), and when not in use they are folded flat against the abdomen. There are two **halteres** shaped like drumsticks on the third thoracic segment, which vibrate rapidly during flight.

The abdomen is short and only four segments are visible.

18.6 Importance and Control

House-flies are a very serious danger to health, for they are carriers of disease. Since they feed on and lay their eggs in manure and other decaying material, the bacteria from this may easily be picked up by their hairy bodies or pass into the alimentary canal. They also feed on human food, and this they infect by brushing bacteria off their bodies as they crawl over it, or by spreading the

HOUSE-FLY FOOT

Fig. 18.12 (*Photograph by Flatters and Garnett Ltd.*)

proboscis, pads, and saliva over it as they feed, or by depositing faeces. In this way the flies may spread typhoid, dysentery, and infantile diarrhoea. It is in our own interest, therefore, that we should kill house-flies, though it is even better to prevent them from breeding in the first place. This can be done to a large extent by keeping all rubbish covered; where this is not possible, the rubbish should be packed tightly, so that the eggs, larvae, and pupae are killed off by the heat produced by fermentation. Rubbish heaps may be also treated with chemicals which kill the larvae, pupae, and adults but do not affect the manure. Similar insecticidal sprays may be used in the house, but care must be taken to see that they do not come into contact with food. At all times, food and cooking and eating utensils should be covered so that they cannot be contaminated by flies. The old practice of keeping meat and fish in a safe with perforated zinc panels did not prevent micro-organisms from causing decay; it prevented blow-flies from laying their eggs on it and while so doing brushing bacteria from their bodies on to the meat. When the larvae hatch they feed on the meat, making it unpleasant, although not necessarily unfit to eat.

MOSQUITOES

18.7 Adults

Mosquitoes, which are commonly called gnats, belong to the same order (Diptera) as the house-flies but differ from them in many respects. The females are blood-suckers, the males feed on plant juices.

There are two tribes of mosquitoes—the culicines (e.g. *Culex pipiens*) and the anophelines (e.g. *Anopheles gambiae*, see Fig. 18.13), which transmit malaria. The culicine mosquitoes may be distinguished from the anopheline by their position when at rest: they hold the abdomen horizontally, whereas the anopheline mosquitoes stand with the abdomen pointing upward (see Fig. 18.14). The

MOSQUITO

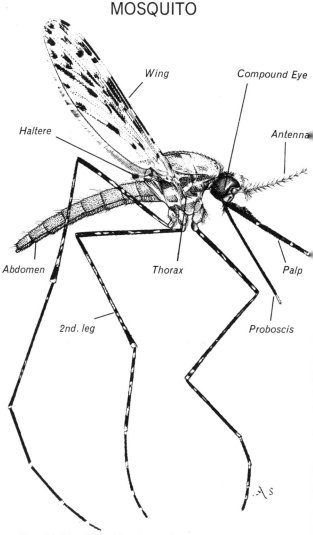

Fig. 18.13 Right side view of an adult female malarial mosquito (*Anopheles gambiae*). Legs of only the right side are shown. (*Courtesy I.C.I. Ltd. Drawing by A. Smith.*)

ANOPHELINE CULICINE

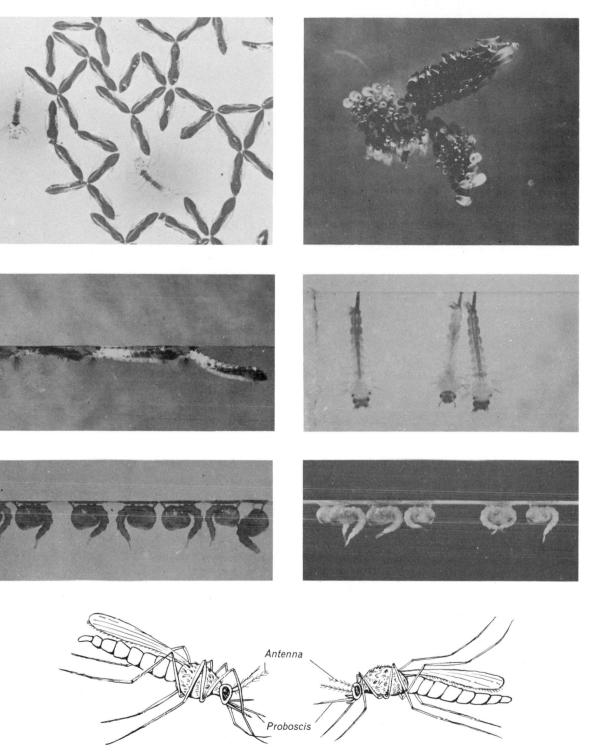

FIG. 18.14 Eggs, larvae, pupae, and imago of the anopheline and culicine mosquito. (*Photographs by Shell.*)

anopheline mosquitoes are all potential carriers of malaria, but they can transmit the disease only after becoming infected by biting a malaria patient.

The mosquito has a small head with two large compound eyes, a pair of antennae which are bushy in the male but slender in the female, and a proboscis formed from the mandibles and maxillae. The proboscis of the female is used to pierce the skin and suck blood. When it is forced into the victim, saliva is passed down it to prevent the blood from clotting and so blocking the narrow tube.

18.8 Life-histories

The mosquito undergoes a complete metamorphosis, the early stages of which take place in water (see Fig. 18.14). The eggs are laid in still pools, and among vegetation in slowly flowing water. The female lays 200 to 300 eggs which are small and cigar-shaped. The eggs of the culicine mosquito are stuck together to form a small raft which floats on the surface of the water, whereas those of the anopheline mosquito are laid separately. The larva hatches in two or three days and emerges by breaking off the cap at the lower end of the egg. It moves by wriggling the abdomen, but if undisturbed it remains floating at the surface. The culicine larva floats head downward, the anopheline with its body parallel to the surface. The head, thorax (which is not segmented), and the nine segments of the abdomen can be seen clearly with a hand lens. The head has a pair of eyes, short antennae and mandibles, and two brush-like structures which draw small organisms toward and into the mouth. The culicine larva obtains oxygen through part of the eighth segment, the siphon, which is open at the end and

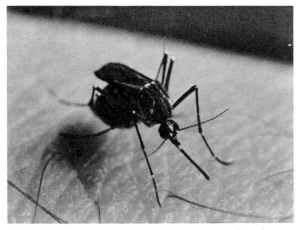

FIG. 18.15 Yellow fever mosquito (*Aedes aegypti*) on human skin preparing to feed. (*Photograph by Shell.*)

terminates in five small valves. These valves are spread out on the surface film of the water and air enters through the spiracle, but if the larva is disturbed the flaps are closed and it sinks rapidly. The ninth abdominal segment bears a tuft of hairs and four small plates which contain tracheae and act as gills, as well as forming a rudder. The anopheline larva obtains its oxygen in a similar way but has no siphon; air enters directly into the two spiracles on the eighth segment when the body lies parallel to the water surface.

In two or three weeks the larva moults four times, and at the last moult it changes its shape completely and becomes a pupa. This has a rounded anterior part formed from the head and thorax, in which the eyes, wings, and legs can be seen through the cuticle, and a narrow, curved abdomen which ends in a pair of tail plates used in swimming. The pupa, though it does not feed, is active, unlike that of most insects, darting below the surface if it is disturbed. There are two respiratory tubes on the head, and these are lined with hairs to prevent water from entering. After a few days the cuticle splits along the back between the two respiratory tubes and the imago emerges.

18.9 The Malarial Parasite

The relationship between the malarial parasite and its host is not a casual one like that between the disease-causing bacteria and the house-fly. The organism which causes malaria is Plasmodium, which belongs to the same group as Amoeba (Protozoa). It undergoes part of its life-cycle in the mosquito and part in the blood of man. If a person suffering from malaria is bitten, the parasite passes with the blood into the stomach of the mosquito, where it undergoes a series of complicated changes, which result in the production of more parasites in the salivary glands of the mosquito. If the mosquito bites another victim, then these parasites are passed with the saliva into the blood. Once in the human blood-stream the parasites undergo a period of development in the liver before invading the red blood corpuscles, where they complete another part of their life-cycle. As nourishment is absorbed from the red corpuscles these break up, and toxins are released into the blood-stream, causing the fever characteristic of malaria.

18.10 Control of Mosquitoes and Malaria

The most efficient method of controlling mosquitoes is to remove their breeding grounds. The drainage of land is an expensive proposition, but it is possible to spray water which cannot be

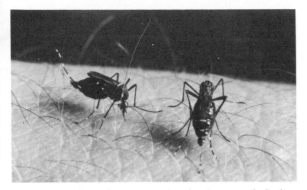

FIG. 18.16 Yellow fever mosquito (*Aedes aegypti*) feeding on human blood. (*Photograph by Shell.*)

drained with crude petroleum. This lowers the surface tension of the water so that the larvae, which can no longer hang on to the surface film, sink and die through lack of air, or as a result of the poisons in the petroleum. The adult insects can be destroyed by spraying walls and other places frequented by them with various preparations of D.D.T. or Gammexane.

It is probable that more people throughout the world die or suffer chronic illness through malaria than from any other cause. The most ancient remedy, and one still in extensive use, is quinine (a product obtained from cinchona bark), but during World War II a synthetic product, mepacrine, was discovered, and later another, paludrine, was found to be very effective. If these drugs are taken regularly by people exposed to infection, the malarial symptoms do not appear, since the drug arrests the development of the

parasites at a stage before they appear. The drugs can also be used to cure the disease itself. In spite of a determined effort to wipe out malaria it is still a disease of major importance.

HIVE-BEE

18.11 Castes

The insects already described have all been harmful to man or to his crops, but the hive-bee is of economic importance, not for the honey which it produces but for the part which it plays in pollinating various crops. It is also of great interest because of the complex social life which it leads, and, apart from the silkworm, it is the only domesticated insect kept by man in this country.

A hive of bees is a highly organised society containing three kinds of castes or individuals (see Fig. 18.17). Each hive contains one **queen**, who is the egg-laying female, several thousand **workers**, who are sterile females, and, during the summer months, two or three hundred **drones**, or males. The queen may be distinguished by her long abdomen and short wings; the drones have a broader abdomen and larger wings and eyes; the workers are the smallest in size of the three castes.

The head of the worker bee has two large compound eyes, with three simple eyes between them. The simple eyes are probably used for examining objects closely, while the compound eyes have wide, long vision. The two antennae project from the front of the head, with a large joint near the head and eleven smaller joints. The mouth parts are adapted for sucking nectar from flowers and

CASTES OF HIVE BEE

	DRONE♂	QUEEN♀	WORKER♀
Eyes	Meet in mid-line	Do not	Do not
Abdomen	Does not extend beyond wing tips	Extends beyond wing tips	Does not

FIG. 18.17.

MOUTHPARTS

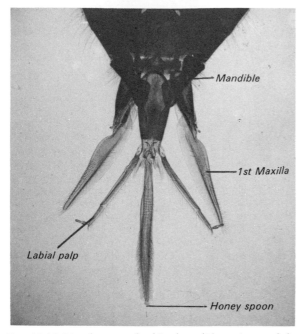

Mandible

1st Maxilla

Labial palp

Honey spoon

FIG. 18.18 Mouth parts of a hive-bee. (*Photomicrograph by Flatters and Garnett Ltd.*)

manipulating the wax of the combs. At each side there is a mandible with a blunt, toothless edge, which is used for kneading the wax and manipulating the pollen to make a paste on which the workers feed. The first and second maxillae are modified to form a proboscis which, when not in use, is bent under the head. The first maxillae are blade-shaped, and form a protective sheath for the fused second maxillae whose two labial palps make a tube round the honey spoon up which nectar is sucked (see Figs. 18.18 and 18.19).

Each segment of the thorax bears a pair of jointed legs on the ventral surface and each pair

HONEY SPOON

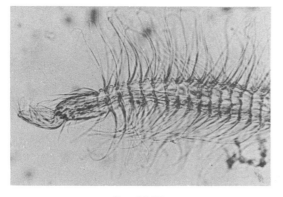

FIG. 18.19.

has a special adaptation for some other function besides locomotion. The first leg (see Fig. 18.20) has a depression lined with hairs, through which the antenna is pulled to clean it; the second has a prong which is used to dig pollen out of the basket on the third leg. The third leg (see Fig. 18.21) also has a very hairy joint used to brush pollen off the body, and a comb by means of which the pollen is transferred to the basket. The second and third segments of the thorax each has a pair of delicate transparent wings attached to their dorsal surface. The two wings on each side are

FIRST LEG

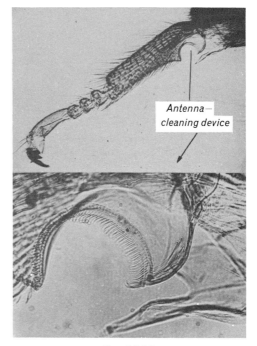

Antenna—cleaning device

FIG. 18.20

held together by minute hooks so that in flight they function as one.

The abdomen is joined to the thorax by a narrow waist, which is partly hidden by short hairs covering the body. It is usually possible to distinguish five segments. At the tip of these segments is the sting, which is a complicated piece of mechanism by which the skin of the victim can be pierced and poison injected into the wound.

18.12 Sequence of Events in the Hive

The life in the hive can most easily be followed by tracing the history of a swarm of a few thousand workers which has left the hive with the old queen. The swarm eventually settles in a dense cluster on the branch of a tree or other object. There the bees remain motionless for some time, and it is now that the swarm is easily 'taken' by the bee-keeper. He uses a flat basket from which the bees are shaken on to a sheet in front of the new hive, into which they soon crawl, all following the queen. If the swarm is not 'taken', then certain bees leave the swarm and act as 'scouts' to find a suitable place for a new hive, and when they return the whole swarm flies off and makes a new home in a hollow tree or other suitable place.

Once inside the new hive some of the workers begin to clean it out and remove any dead animals. When this is finished they join the other workers, who have already started the formation of cells, in which the honey will be stored and new bees reared. The wax from which the cells are made is secreted by special glands on the under side of the abdomen and the workers can form it only if they have been well fed before they swarmed. The wax plates are cut off the abdomen by the large joints on the third legs, and then moulded by the mandibles and built into masses or combs. Older workers now begin to hollow out this mass into cells (see Fig. 18.22). Each comb has two vertical layers of cells placed back to back and arranged so that each cell slopes slightly upwards. As soon as a cell is finished, the queen enters it head foremost,

THIRD LEG

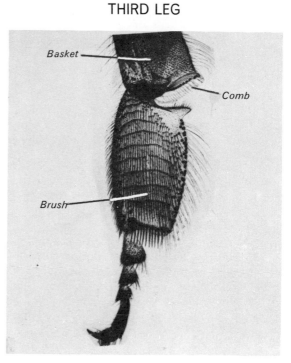

Basket

Comb

Brush

FIG. 18.21 (*Photomicrograph by Flatters and Garnett Ltd.*)

WORKER BEES ON A BROOD COMB

FIG. 18.22 Can you find the cell containing the egg? Note the capped and uncapped brood cells and the honey-containing cells. (*Photograph by Dr. C. G. Butler.*)

examines it, backs out, inserts the tip of her abdomen into it, and lays an egg. She continues this routine of egg laying throughout her life and may, during June, lay as many as 3 500 eggs a day, and during her life of four or five years she will lay hundreds of thousands of eggs. The eggs are kept warm by the bodies of the workers which cluster over the combs, and each egg hatches, in about three days, into a white legless larva which lies curled up at the bottom of the cell (see Fig. 18.23). The larva is fed by 'nurse bees', first on pap, or royal jelly, which they secrete from a special salivary gland, and later on a mixture of pollen and honey made into a soft paste, or bee bread, which the workers place in the cell. The larva is fully grown in about five days, and by that time it has shed its skin several times and nearly fills the cell. It then stops feeding, and the workers close the cell with a cap formed from a porous mixture of pollen and wax. The larva then spins a little silk, which forms an imperfect cocoon, and remains still for two days before it pupates. Within the pupa the reorganisation of the tissue takes place, and this stage lasts about a week and then the pupal skin is cast off and the young worker bee uses its mandibles to bite its way out of the cell. The young workers' first duty is to act as 'nurse bees' to the developing grubs, for only they can secrete the royal jelly. The salivary glands which form this do so only during the second and third weeks after emergence. Thereafter, when the workers are foraging for pollen and nectar, they secrete carbohydrate-splitting enzymes.

The pollen is stored in cells close to the brood combs. The nectar which is sucked up the proboscis is passed through the oesophagus to the stomach, where it loses water and becomes changed chemically into honey, which is regurgitated from the mouth into the storage cells. When the cell is full, the honey is left to thicken for a few days, then, to keep it from fermenting, a drop of acid from the worker's sting is added and the cell closed with a cap of wax (see Fig. 18.22).

After the worker has deposited a load of nectar she performs a peculiar dance on the combs, and other workers crowd round her. This dance announces that nectar has been found, and recent research work indicates that the pattern of the dance tells the watching bees its distance and direction from the hive.

As the number of workers in the hive increases, preparations are made for swarming, the most important of these being the rearing of the drones and a new queen. The cell makers begin to form rather larger cells, and in these the queen lays an

HIVE-BEE

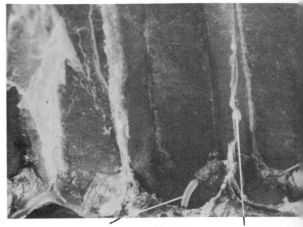

Egg   EGG   Wall of wax c

LARVA   Cell cap

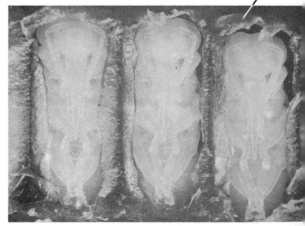

PUPAE

FIG. 18.23 Photographs of egg in worker cell (about ×12), fully grown worker larva spinning cocoon (about ×8), and pupae of workers (about ×8). (*Photographs by Dr C. G. Butler.*)

unfertilised egg which hatches into a drone larva. This is fed for three or four days on royal jelly and then on bee bread. The drones do no work in the hive and are fed by the workers. After enough drones have been reared, the workers alter a few cells in which fertilised eggs have already been laid, and in these royal cells the young queens develop and are fed on royal jelly throughout the whole period of their development. While they are growing, the swarm leaves the hive with the old queen. The first young queen to hatch is helped by the remaining workers to kill off the other developing queens, and then leaves the hive for the nuptial flight during which she is fertilised by one of the drones. She returns to the hive and starts to lay eggs.

The orderliness of the life in the hive used to be attributed to the sterility of the numerous workers, but recent research has shown that the craving of the workers for 'queen substance' is important in maintaining the cohesion of the hive. This substance which is produced by the queen in glands behind her jaws is distributed all over her body by the constant 'grooming' she receives from the workers (see Fig. 18.24). It is eagerly sought by the workers, some of whom obtain it by licking their queen and then transfer it to others. It is possible that the 'queen substance' is trans-ferred when the worker feeds another bee or when she touches its body with her antennae. The 'queen substance' prevents the workers making queen cells and inhibits the development of their ovaries. A reduction in the abundance of 'queen substance' enables the workers to realise the loss or failure of their queen and removes the normal inhibition for producing more queens. The probable breakdown in the distribution of 'queen substance' amongst the very large number of bees present in the colony before swarming results in the workers tolerating the eggs and larvae in the queen cells.

In the autumn the workers refuse either to feed the drones or to allow them to reach the stores of honey. Soon the starving drones die and are dragged out of the hive. A few may be stung to death by the workers who pass the winter clustered round the queen in the hive, feeding on the stores of honey and pollen.

18.13 Importance to Man

The importance of the hive-bee in pollination has already been mentioned. Apiarists often arrange to move their hives to fruit orchards in the early summer so that their bees can benefit from the food supplies provided by the blossom. This ensures a greater 'set' of fruit for the fruit grower.

QUEEN BEE ATTENDED BY WORKERS

FIG. 18.24 (*Photograph by Dr. C. G. Butler.*)

APHIS INFESTATION

FIG. 18.25 Colony of Rubus aphid (*Amphorophora rubi*) on raspberry shoot. (*Photograph by Shell.*)

OTHER IMPORTANT INSECTS
18.14

There are many more insects which are of great importance to man, because they destroy food crops, spread disease, or damage his possessions.

The aphids (see Figs. 18.25 and 18.26), which include the common green-fly and black-fly, are so numerous and so harmful to plants that they are familiar to everyone. Most aphids lay their eggs in autumn on twigs, and from these, in the spring, hatch nymphs, which rapidly complete their metamorphosis to adult females. These are able to reproduce **parthenogenetically** (without fertilisation). Successive generations of these females are formed throughout the summer, and they all feed by sucking juices from plants. The later generations include both males and females, and lay fertilised eggs which survive the winter. Such pests can be controlled by sprays, and recent research work has shown that plants can be watered with substances which, when absorbed, kill off any aphids which suck their juices.

There are many insects which, like the mosquito, transmit specific diseases. The African tsetse flies suck the blood of various backboned animals, e.g. cattle, antelopes, birds, etc., and in so doing may take parasites from the blood of these animals. These may include the trypanosomes (another type of Protozoa) which cause sleeping sickness, and the insect thus becomes capable of infecting a new host as it feeds.

Not all the groups which do damage have been mentioned. You can probably think of many other examples for yourself, e.g. clothes moth, wood-boring insects, etc.

APHIS LIFE-HISTORY

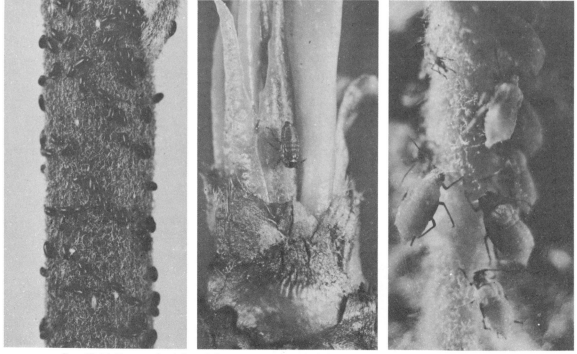

FIG. 18.26 Eggs and adults of the green apple aphid (*Aphis pomi*) and a nymph of the peach aphid (*Appelia tragopogonis*). (*Photographs by Shell.*)

ADAPTATIONS OF INSECTS FOR LIFE ON LAND

18.15

1. A flexible external cuticle reduces loss of water by evaporation as well as providing protection against mechanical damage and bacterial invasion.
2. A tracheal system maintains an efficient supply of oxygen to and removal of carbon dioxide from all parts of the body.
3. A rapid means of locomotion, in the adult stage, increases the insect's chances of survival by helping it to escape from predators.
4. The specialised sense organs—eyes, antennae and palps—enable the adult to find its own food easily and also the particular food on which to lay its eggs.
5. The large number of eggs laid, their protective coloration and concealed position help to ensure that some will escape the notice of predators.
6. Mating results in the introduction of a large number of spermatozoa into the body of the female and in many insects these remain viable for some time (e.g. bee) so that egg-laying can be spread over a longer period. Internal fertilisation provides the watery medium needed for spermatozoa.
7. Metamorphosis reduces the competition for food between larval and adult forms and thus increases the chance of survival.

QUESTIONS 18

1. What are the most important changes which occur during the development, from the egg to the adult, of: (*a*) a named amphibian such as the frog, (*b*) a named insect, and what is the importance of these changes? (Annotated diagrams may replace written work in this question.)　　　　　　　　　　　　　　　　　　　　[*S*]

2. Describe the life-history of a social insect such as the honey bee or the ant or the termite. What are the advantages of this social life compared with the life of a solitary insect?　　　　　　　　　　　　　　　　　　　　　　　　　　　　　[*C*]

3. Give a fully illustrated account of the life-history and habits of a named insect with a complete metamorphosis.　　　　　　　　　　　　　　　　　　　[*L*]

4. Describe the life-history of the honey-bee. Briefly state the use made by the honey-bee of the following: (*a*) pollen, (*b*) nectar, (*c*) water, (*d*) wax, (*e*) resin (propolis)　　　　　　　　　　　　　　　　　　　　　　　　　　　　　[*O*]

5. Make labelled drawings to show the external structure of (*a*) a caterpillar and (*b*) a butterfly. What is the significance of each of these stages in the life history?　　[*O*]

6. Describe briefly with the aid of labelled drawings the life histories of a named butterfly and of the honey-bee. Give an account of the feeding habits of either (*a*) the larval and adult butterfly, or (*b*) adult bees.　　　　　　　　　　　　　[*O & C*]

7. What adult forms occur in the hive of the honey-bee? Briefly describe the role that each of these forms plays in maintaining the social life in the hive. Why are bees important to man?　　　　　　　　　　　　　　　　　　　　　　　　　　[*S*]

8. Name *one* insect which carries a *named* disease to human beings, and make a labelled drawing to show the external appearance of the adult stage of this insect. Outline the life-history of this animal, without the use of diagrams. Indicate two ways by which the spread of the disease can be reduced by (1) personal precautions, (2) organised activity of bodies such as the local authority.　　　　　　[*A*]

19 Bacteria and Viruses

Bacteria have already been mentioned in sections 5.5 and 8.7 as micro-organisms important in the circulation of nitrogen and carbon in nature. There is no doubt that all life is just as much dependent upon the chemical changes brought about by these organisms as it is on the photosynthetic activities of green plants. The decomposition of the complex organic compounds of dead creatures by bacteria present in the soil is as vital to life as is new growth made directly or indirectly with sunlight energy. This fact alone underlines the importance of bacteria to man, but it should be remembered that they frequently live parasitically within the bodies of man and his domestic animals, and occasionally cause disease.

Viruses are disease-producing or **pathogenic** agents which reproduce only within the living cells of organisms. They are smaller than the smallest bacteria and, with very few exceptions, are not visible with the best optical microscopes as are bacteria. The viruses have been detected with the aid of a comparatively new instrument, the

VARIOUS BACTERIA

FIG. 19.1 1. Staphylococci (in groups). 2. Streptococci (in chains). 3. Pneumococci encased in capsules. 4. Escherichia coli. 5. Tuberculosis bacilli. 6. Hay bacilli with flagella. 7. Hay bacilli with endospores. 8. Anthrax bacilli. 9. A spirillum. All are approximately ×2 000.

COCCI

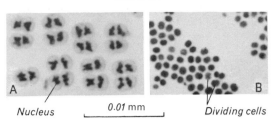

Nucleus   0.01 mm   Dividing cells

FIG. 19.2 A—Bacteria which grow in flat sheets and show nuclei. B—Staphylococci. (*Photomicrographs by Dr. C. Robinow.*)

electron microscope. The importance of viruses lies in their power to cause disease in man, his domesticated animals, and his crop plants.

19.1 What are Bacteria?

Bacteria are minute non-cellular organisms of various shapes (see Fig. 19.1):

1. Spherical ones are called **cocci** (sing. coccus).
2. Rod-shaped ones are called **bacilli** (sing. bacillus)
3. Curved or corkscrew-shaped ones are called **spirilla** (sing. spirillum).

The smallest are cocci with a diameter of approximately $0.75 \, \mu m$; the largest are bacilli which may reach a length of $8 \, \mu m$. Their cytoplasm is surrounded by a wall of complex composition and, in some, a nucleus has been demonstrated (see Fig. 19.2). Long cytoplasmic threads (flagella) which enable them to move about in fluids are found in some, particularly the bacilli (see Fig. 19.5) and spirilla. They are usually included in the plant kingdom though they have neither cellulose walls nor chloroplasts. Certain species of bacteria are able to form resistant spores (endospores) within their bodies, which can remain alive in

LARGE & SMALL BACILLI

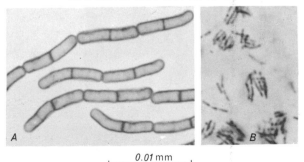

0.01 mm

FIG. 19.3 A—*Bacillus megaterium* from a young culture. B—*Tuberculosis bacilli* from the sputum of a human being. (*Photomicrographs by Dr. C. Robinow.*)

boiling water for periods varying from a few minutes to several hours (see Fig. 19.4). Most active bacteria are, however, readily killed by heating above 60°C to 70°C for a few minutes.

Reproduction takes place by simple division of the bacterium into two, a process which, given ideal conditions, can be repeated every half-hour. Thus, in twenty-four hours a single bacterium might produce 2^{47} individuals.

$$2^{47} = 140 \, 737 \, 488 \, 355 \, 328$$

Such potential reproductive powers are never realised because, after a short time, there is always a lack of food and an accumulation of their own excretory products which slows down the rate of division. If bacteria are not motile and if they are grown on a solid medium, a single individual can develop into a compact colony which is easily visible to the naked eye.

SPORES OF BACILLUS MEGATERIUM

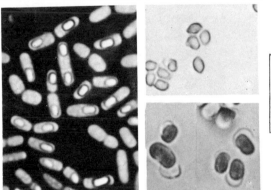

0.01 mm

FIG. 19.4 Left—Spores formed within the bacillus. Right—Free and germinating spores. (*Photomicrographs by Dr. C. Robinow.*)

Bacteria vary a great deal in their food requirements, but the majority, being saprophytic or parasitic, require a supply of complex organic compounds from their environment. To most bacteria it matters little whether there is free oxygen around them or not, but some (obligate anaerobes) cannot develop or are even killed in free oxygen, while others (obligate aerobes) must have free oxygen.

Bacteria are ubiquitous; they are found in the soil, in water, in dust, in air, on much of the food we eat, and in animal bodies, particularly in their alimentary canals. In the large intestines of many herbivores the bacteria present are responsible for the digestion of cellulose and the making of vitamins. Their presence can be detected in experiments such as the following.

ELECTRONMICROGRAPHS - BACILLI

Escherischia coli

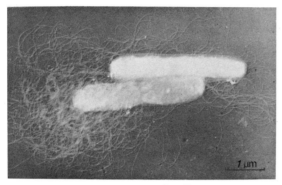

Proteus vulgaris—flagellated

FIG. 19.5 (*Electronmicrographs by Dr. E. Challice.*)

Prepare a bacteria-free food material such as meat-broth agar, which will support the growth of many different bacteria. This is done by boiling meat in water, filtering off the solid matter, and adding 1.5 g agar to every 100 cm³ of fluid in a flask. (Agar is a carbohydrate obtained from seaweeds.) Plug the mouth of the flask lightly with cotton-wool, and dissolve the agar by heating in a water bath, which prevents charring. Continue boiling for two hours, allow to cool, when it will solidify, and then put in a pressure cooker at 1 atm or 103 kN/m² above atmospheric pressure for fifteen minutes. This treatment kills even the bacterial spores. Whilst still hot, and just before it solidifies, pour it into glass petri dishes which have previously been sterilised in a hot-air oven at 160°C for three hours and quickly replace the lids. This operation is best done in a room free from draughts, which might carry bacteria on to the medium as it is being poured into the dishes. Allow the meat-broth agar to solidify, remove condensed moisture from within the dishes by warming in an oven at about 40°C, and treat the dishes as follows, making sure that each one is labelled.

Dish	Treatment
1	Leave untouched as a control
2	Leave the lid off for 15 minutes
3	Inoculate with a minute drop of milk
4	Inoculate with scrapings from the teeth
5	Scatter a few fragments of soil on the agar surface and then shake them off
6	Scatter minute quantity of dust on agar surface
7	Make four finger-prints on the agar surface

GROWING BACTERIA

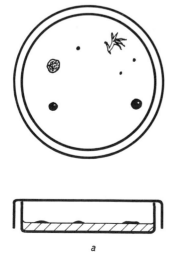

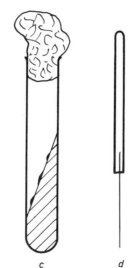

FIG 19.6 (*a*) Petri dish showing several bacterial colonies and one fungal colony growing on agar. (*b*) Wire loop used to spread material on an agar surface. (*c*) Bacterial colonies on an agar slope. Note the cotton-wool plug which prevents further bacteria from entering. (*d*) Wire for making a stab culture. (*e*) A stab culture—bacteria growing in the crack.

a   b   c   d   e

In dishes 3 to 6 the material is applied with a loop of nichrome wire (wire from the element of an electric fire will do), which is sterilised in a bunsen flame and cooled before each operation.

The seven dishes are then incubated at 30°C to 37°C (leaving them over a radiator will do) for one or two days, and examined. If the first dish is free from bacterial colonies, then we know the sterilisation of the medium and the petri dishes was successful, and any bacteria in the other dishes must have come either from the air when the lid was lifted, or, more likely, from the inoculation material. Bacterial colonies appearing as circular slimy patches of various colours usually occur on all dishes except the control. It is not unusual to find saprophytic fungi growing in dishes 5 and 6.

That these colonies are composed of bacteria can be checked by taking a very small piece on a sterile nichrome wire and mixing it with a drop of water on a grease-free microscope slide. Dry it quickly by waving it over a bunsen flame being careful to avoid contamination of the hands, and stain it by immersing in 1 per cent gentian-violet solution for two to three minutes. Wash in running water, dry, and examine, preferably using a microscope with an oil immersion lens. This experiment can also be done using Oxoid dehydrated media, or Oxoid 'agaroids' (available in a pack of 6) together with sterile disposable plastic petri dishes. Whichever method is used, it is important not to subculture colonies which appear.

Bacteria can also be grown on the surface of agar slopes, or within the agar itself in test-tubes plugged with cotton-wool (see Fig. 19.6). In the latter cases the bacteria are introduced by plunging an infected wire into the agar, when anaerobic ones will grow in the crack.

19.2 What are Viruses?

The smallest viruses, such as those of foot and mouth disease and poliomyelitis, have a size of $0.1\,\mu m$. The largest viruses, e.g. vaccinia and tobacco mosaic viruses, reach a maximum size of $0.25\,\mu m$, and are just visible as structureless specks with the best optical microscopes. Virus particles vary in shape; some are spherical, some are spherical with a small tail, and some are rod-shaped. They can reproduce themselves only within the living cells of their hosts, and may be transmitted from one to another by insects which feed on the hosts, by the movement of airborne particles or by leaf contact. Aphids, mosquitoes, lice, and fleas are examples of insects which transmit viruses. Several viruses, tobacco mosaic virus for instance, have been isolated from infected plants in a

VACCINIA VIRUS

|← 1μm →|

FIG. 19.7 *Vaccinia virus* shadowed with gold. (*Electronmicrograph by I. M. Dawson and A. S. Macfarlane.*)

crystalline form which still retains its power to produce the disease when reintroduced into a healthy plant. This has given rise to the view that the viruses are on the borderline between true living things like bacteria and inert chemical substances. Another view is that they are extreme examples of parasitism, having progressively lost the power to make essential enzymes, and come to rely increasingly on the host for a supply of these. If this view is correct, the smallest viruses would be those which have dispensed with most 'metabolic apparatus'.

Several electronmicrographs of bacteria and viruses illustrate this chapter, and it should be clearly understood that these are not by any means the same as photographs taken through an optical microscope. In an electronmicroscope a beam of electrons replaces the light of a normal microscope, and the lenses are electro-magnets which focus the beam of electrons. The great disadvantage of an electronmicroscope is that electrons are readily absorbed by all forms of matter. The specimens to be examined, therefore, have to be very thin and absolutely dry, so as not to give off water vapour into the instrument, which is kept evacuated of air during use by means of a high-speed pump. Thus, what appears on the fluorescent screen or photographic plate of an electronmicroscope needs very careful interpretation. Nevertheless, with the high magnifications obtainable on this instrument, much valuable information has been obtained and it is now a well-established 'tool' in research.

19.3 Decay and Food Preservation

All dead organic material will undergo putrefaction or decay unless special steps are taken to prevent it. The objectionable odours given off by decaying matter such as animal corpses and rotting vegetation are evidence that chemical changes are occurring, and it is now known that these are being carried out by bacteria and fungi. Further evidence of the presence of living organisms in decaying matter is given by the fact that the centre of a heap of rotting grass cuttings is warm (see sec. 9.1).

Before 1860 there was a widespread belief that dead material spontaneously generated the micro-organisms responsible for its decay. It was largely due to experiments carried out by **Louis Pasteur** (1822–1895) that this idea was discredited. Pasteur, a chemist by training, had been studying fermentation, and was well aware of the micro-organisms which caused it. Using the flask shown in Fig. 19.8, he demonstrated that broths, if properly sterilised, could be kept indefinitely, since the airborne putrefying organisms could not get beyond the lower bend in the tube. Shortly after the tube was broken, however, the broth began to putrefy.

Later, working on diseases which attacked silkworms, on anthrax, and on hydrophobia (rabies), Pasteur demonstrated that micro-organisms could cause disease as well as being the agents of putrefaction.

It is instructive to leave organic materials such as hay, horse-dung, and minced meat to rot in water for a few days and then examine the material microscopically. Various shapes of motile and non-motile bacteria will be found.

Nowadays food is often produced a long way from where it is consumed, and much of it is processed in one way or another before it reaches the consumer. Thus, methods of food preservation are important. Food preservation involves the suspension of decay caused by naturally occurring bacteria and fungi, which not only 'eat' the food but make it unpalatable, and sometimes highly poisonous with their excretory products or toxins. There are a great many ways of preserving foods, some better suited to certain foods than others. The more important methods are described below.

Sterilisation

Modern methods of canning and bottling foods are the direct outcome of Pasteur's work. Canning usually involves cooking the food at a high temperature, placing it in a previously sterilised metal can, and reheating. Whilst it is still very hot, an air-tight and therefore bacteria-proof lid is sealed on, and the can of food is allowed to cool. Almost all the air will have been driven out during cooking, and the steam will condense on cooling, so that the ends of the can will be drawn in. Any bacteria or fungi in the food are killed by the heat treatment, even the resistant spores being unable to withstand temperatures in excess of 100°C for very long, and once the lid is sealed on, no more bacteria or fungal spores can enter. Thus, the food remains undecayed.

Certain bacteria are responsible for the occasional outbreaks of food poisoning. The most dangerous, *Clostridium botulinum*, a soil bacterium which causes botulism, thrives only in anaerobic conditions, and gives out into the food a highly poisonous toxin which can cause death by blocking impulses at the nerve muscle junction. Man inadvertently 'invented' botulism when he started

DECAY

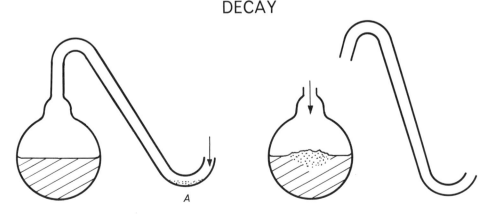

FIG. 19.8 Pasteur's flask. Sterile broth in the flask on the left remained undecayed since airborne micro-organisms collected at A. Breaking the tube off the flask allowed micro-organisms to fall directly on to the broth and cause decay.

to preserve meat and vegetables by canning, but it is now rare. The toxin is the most poisonous substance known—1 mg can kill 20 million mice—but is readily destroyed by a brief period of boiling. Thus foods suspected of being contaminated can easily be made safe. One type of *Staphylococcus* is also responsible for food poisoning, but in this case the toxin is heat stable and not lethal. Pastries, milk products, and meat foods are the chief sources of this type of food poisoning. The best insurance against food poisoning by these organisms is to observe scrupulous cleanliness in the handling and preserving of food.

Bottling is a process usually applied to fruits, and is similar to canning in that heat treatment kills the putrefiers in the food and an air-tight seal employing a rubber washer prevents the entry of further organisms. Most fruits contain acids, and bacteria are much more easily killed by heat when in acids. As a result of this it is usually possible to sterilise fruit efficiently without boiling it.

It should be noted that the exclusion of air or oxygen from the food during canning and bottling is not a vital part of the preservation as a rule, for most bacteria can survive and some thrive without it.

Milk, which sometimes contains pathogenic bacteria, is subjected to a heat treatment known as **pasteurisation**. The taste of milk is spoilt by boiling, so it is heated to a temperature of not less than 72°C for at least 15 seconds, and then suddenly cooled to below 12°C. This process, if correctly carried out, will kill all pathogenic bacteria, but there are always other bacteria left which cause the milk to turn sour and clot.

Modern aseptic (*Gk. a, without; septos, putrefying*) surgery is another application of sterilisation. All instruments, towels, swabs, gowns, masks, and rubber gloves used during a surgical operation are first sterilised, either by boiling in water or by treatment with steam under pressure in an autoclave. Many other precautions are taken to prevent bacteria getting into the tissues during an operation, but these are not examples of sterilisation and will be mentioned later.

Refrigeration

Keeping food cold is a useful way of preserving it, since bacteria and fungi cannot cause any appreciable amount of decay, or reproduce, at low temperatures. In the region of 0°C their metabolism is so slow that they can be described as dormant. Refrigeration is widely used to preserve carcasses of meat during shipment, but it is also useful in the home. Domestic refrigerators usually keep the food at a temperature slightly above 0°C and slow down the metabolism of bacteria and fungi in meat, milk, and many other foods, so preventing decay. They are also used for keeping living bacteria-free foods such as lettuce, tomatoes, and eggs fresh by slowing down their metabolic rate and reducing the rate at which their water content is evaporated. When foodstuffs are removed from refrigerators their putrefying organisms immediately start to cause decay at an appreciable rate.

The biological principle involved here is that all living organisms respond to changes in their body temperature by altering the rate at which their metabolism occurs. This can be detected even in humans, whose general body temperature remains fairly constant; it is most noticeable that the beard grows more quickly in warm than in cold weather since hair follicles are near the surface of the skin, whose temperature tends to follow external temperature changes.

Preserving fresh fruit, vegetables, fish, etc., by freezing deeply and quickly keeps them in a near natural condition for long periods. Such foods are very popular because they save the housewife's time, as they are already prepared, and add variety to the diet. There is nowadays a very wide range of these 'convenience' foods, including complete meals.

Dehydration

No organism can survive in an active form for long without water. Thus, if food can be dried without spoiling it, and then kept dry, it will be preserved. Perhaps the best examples of this method are provided by flour and the various brands of dried milk sold for babies. You can doubtless think of many others, such as breakfast cereals and polished rice. One advantage of dehydrated foods is that they are economical to transport as compared with the corresponding fresh food which often contains a high proportion of water. Many vegetables and even complete meals can now be obtained in this form.

Most dry foods will contain resistant bacterial and fungal spores which will germinate and start to putrefy the food as soon as it is moistened. It is, therefore, important to make sure that dry foods are contained in waterproof packages, or alternatively are in a dry place.

Osmotic Preservation

If the food is converted into, or immersed in, a highly concentrated solution without spoiling it, it will be preserved. This is because bacteria or fungal spores which start to grow in the food will die as a result of water loss by osmosis. This movement of water occurs because the cytoplasm of the organism is a weaker solution than that in which it is growing (see sec. 6.3). Honey, jams, and salted beans and ham are obvious examples of this method of preserving food. Dried fruits such as sultanas, raisins, and currants are best included in this type of food preservation, since they do contain some water with the sugar and other substances in their partially dried cells. Thus, by drying, their cell contents are converted into a concentrated solution which preserves them. The skin of grapes is a bacteria-proof layer, but it is unlikely that this would remain intact during the picking and drying and cleaning which take place before they are sold as dried fruit.

Other Methods

The smoking of fish and ham preserves the food by producing a hard dry outer layer, impregnated with substances which are poisonous to bacteria. Vinegar, an acid fluid in which very few micro-organisms can live, is used for pickling various vegetables such as onions, cauliflower, and gherkins. Concentrated orange juice (as used for babies) and lemon juice have minute amounts of sodium bisulphite added, which prevents the growth of micro-organisms.

In these three examples a substance is added to the food which is poisonous to the putrefiers and yet is harmless to man. The same principle was found in the use of **antiseptic** substances such as carbolic acid, iodine, and methylated spirit. Antiseptics were first used in surgery by **Joseph Lister** in 1865. Lister was stimulated by the work of Pasteur, and saw the connection between the putrefaction of meat broth and that of surgical wounds which would not heal properly. At first his antiseptics were harmful to the tissues of the patient as well as killing the bacteria, but later he introduced milder antiseptics, and before he died in 1912 the practice of washing surgical wounds in antiseptics had been replaced by aseptic methods. Antiseptics (e.g. Dettol and acriflavine) still find many uses today both in the home and in hospital. Formalin is used to preserve biological material.

You are advised at this stage to read sections 5.5 and 8.7, which deal with the circulation of nitrogen and carbon in nature. Several steps in these circulations are carried out by putrefying and other types of bacteria.

DISEASE PRODUCERS—PATHOGENS

Bacteria, viruses, and fungi include the majority of disease-producing organisms (pathogens), but there are a few others amongst the lower invertebrate groups.

Some human diseases caused by bacteria

Boils	Tonsillitis	Tuberculosis
Tetanus	Pneumonia	Whooping cough
Scarlet fever	Bacillary dysentery	Typhoid
Cerebro-spinal fever	Gonorrhoea	Syphilis
Bubonic plague	Cholera	Leprosy
Diphtheria		

Some human diseases caused by viruses

Common cold	Influenza	Measles
German measles	Mumps	Typhus
Poliomyelitis	Smallpox	Warts
Yellow fever	Trench fever	

Other human diseases (P = protozoan, F = fungal, T = threadworm)

Amoebic dysentery (P)	Trichinosis (T)
Tinea or Ringworm (F)	Sleeping sickness (P)
Malaria (P)	Ancylostomiasis (T)

Some diseases of plants (V = virus, F = fungal)

Potato-leaf roll (V)	Potato scab (F)
Sugar-beet curly top (V)	Maize streak (V)
Club root of cabbage, sprouts, etc. (F)	Swollen shoot of cocoa (V)
Tulip mosaic (V)	Rust fungus of wheat (F)
Tobacco necrosis (V)	Mildews of strawberry, gooseberry, etc. (F)
Potato blight (F)	

Some diseases of animals (B = bacterial, V = virus)

Tuberculosis of cats, dogs, and cattle (B)	Anthrax of horses and cattle (B)
Myxomatosis of rabbits (V)	Distemper of dogs (V)
Foot and mouth disease of cattle (V)	Rinderpest or cattle plague (V)
Fowl pest (V)	

19.4 Immunity to Disease

There exist many natural barriers to the entry of foreign organisms or materials into the body. The stratum corneum of the skin, the hairs around the body orifices, the sebum with its bactericidal effect, the tears and the secretions of the digestive glands all play a part in keeping the body free from infection. If these barriers are breached by bacteria, viruses, or substances secreted by them, or even by proteins from a different individual of the same species, then a further line of defence—the antibody-producing system—comes into action. The foreign material or **antigen** stimulates special cells of the body to make specific proteins (globulins) known as **antibodies**. These antibodies react in various ways with the antigens to counteract their poisonous effects. For instance, they may cause certain bacteria to clump together or **agglutinate**, which renders them more readily engulfed by the phagocytic white corpuscles—the last line of defence. Antibodies against a particular disease-producing organism, or the ability to make more of these antibodies very quickly, are retained by individual organisms for long periods, often for many years, and this constitutes immunity.

Immunity is not always acquired naturally; in some cases it is advantageous to induce it. Details of how this is done for diphtheria and smallpox are given below.

Immunisation against diphtheria

Diphtheria used to be a serious disease of young children in Britain, but now as a result of mass immunisation it is a rarity. The bacterium which causes it (*Corynebacterium diphtheriae*) lives in the nose and throat and excretes a highly poisonous antigen known as the diphtheria toxin. It is this toxin which causes the high fever characteristic of the disease, but it also provides the means of immunisation.

Toxins obtained from the bacterium grown on artificial media are filtered to free them from any living bacteria, and are incubated with a small quantity of formalin, which reduces their poisonous properties. The mixture is then treated with alum solution and is injected into the child to be immunised. The child's body responds by producing antitoxins (antibodies) for diphtheria and so builds up an immunity to the disease. It has been found that doses of 0.2 cm³ and 0.5 cm³ given with a month's interval between are most satisfactory and confer a high degree of immunity.

If a person develops the disease he can be helped to recover by an injection of antitoxins obtained from the blood serum of a horse, into which increasing amounts of bacteria-free toxins have been injected. Since no living bacteria are injected, there is no possibility of bacteria rapidly increasing and producing large amounts of toxin which might lead to an attack of the disease in the horse. In response to the gradually increasing doses of toxins, the horse makes more and more antitoxins which circulate in the blood and can be isolated ready for injection. Antitoxins can also be injected into a person as a preventative measure, in which case immunity will be produced passively.

Immunisation against Smallpox—Vaccination

Smallpox is a world-wide disease which occurs in epidemics of variable severity. It is a virus disease characterised by fever, headaches, vomiting, and a rash of vesicles or pustules, thickest on the hands, feet, and face. Immunisation, known as vaccination, is usually carried out soon after the age of three months. It involves the introduction of a small quantity of **glycerinated calf lymph** into a scratch made on the arm or leg. This fluid is prepared by injecting cow-pox (vaccinia) virus into a healthy calf, which then develops cow-pox, a milder disease characterised by vesicles which form on the udder. The lymph fluid from these vesicles is collected and mixed with glycerine, which preserves it and helps to sterilise it.

The result of such vaccination is usually a slight malaise accompanied by the formation of a vesicle at the place where the scratch was made. The vesicle eventually disappears, leaving a scar, and the child will have developed an immunity to smallpox which may last up to seven years. The explanation of this is that the two viruses, vaccinia and smallpox virus, are essentially the same, and that the latter is somehow modified in the body of the calf so that it produces only a mild form of disease, cow-pox.

Edward Jenner, in the late eighteenth century, noticed that people who milked cows often developed vesicles of cow-pox on their hands, and that, when there was a smallpox epidemic, these people never suffered from the disease. As a result of this he commenced deliberate vaccination as a means of preventing smallpox in 1786.

19.5 Chemotherapy and Antibiotic Substances

After the discovery of antiseptic substances which kill bacteria it was natural that a search should be made for similar substances which do not, at the same time, kill the cells of the host amongst which the bacteria are living. Many such substances have now been found and tested. Their use in the cure

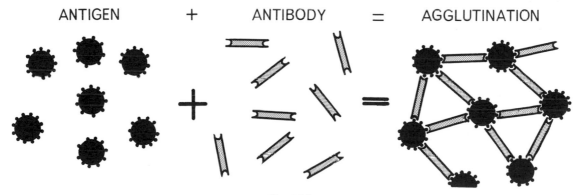

ANTIGEN + ANTIBODY = AGGLUTINATION

FIG. 19.9.

of disease is known as chemotherapy (*Gk. thera-peuo, I cure*). Quinine, salvarsan, sulphanilamide, sulphaguanidine, mepacrin, and paludrin are some examples of these substances.

Antibiotic substances (e.g. penicillin, strepto-mycin, aureomycin) are obtained from naturally occurring moulds, and either kill or inhibit the growth of certain bacteria (see Fig. 19.10). Such substances are less toxic to the tissues of the body than most other drugs and have, therefore, been widely used in recent years.

Finally, several viruses are known which destroy bacteria, and in the future, when we have learnt more about them, they may prove a potent weapon in the treatment of disease. Such viruses are called **bacteriophages**.

19.6 Plant Diseases

Fungi and viruses are the main causes of plant diseases and they are responsible for immense losses in our crops. The Irish famines of the eighteen-forties were due to the ravages of the potato-blight fungus, and in 1947 more than three-quarters of the Indian wheat crop was ruined by a rust fungus.

One virus, the 'tulip break' virus, causes an attractive variegation of the petals. This condition was recorded in the early seventeenth century and must be the oldest record of a plant virus disease.

The potato is host to a large number of viruses as well as fungi. One of them, the potato-leaf-roll virus, causes the leaves to roll up and become brittle, which impairs photosynthesis and reduces the weight of tubers harvested. This is also quite common in tomatoes, which belong to the same family as the potato. The viruses are transmitted from plant to plant by sap-sucking aphids (see sec. 18.14) and spread throughout the plant. Thus, tubers taken from an infected plant will produce infected plants when grown the following year.

It is not possible to cure most virus diseases, so control measures have to be directed against the aphids. In Scotland, many of the potato-growing areas remain almost free from aphids because of the relatively high average wind velocities and

DISCOVERY OF PENICILLIN

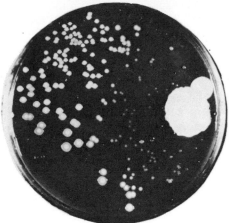

FIG. 19.10 The original petri dish in which Sir A. Fleming first observed the effect of penicillin on colonies of *Staphylococcus aureus* in 1927. Note the large colony of *Penicillium notatum* on the right side of the dish. (*Supplied by the Wright-Fleming Institute of Microbiology.*)

low temperatures. Under these cool, windy conditions aphids do not fly about from plant to plant, and so do not transmit any viruses there may be. In these areas it is usually possible to produce a field of almost virus-free plants, especially if the crop is periodically examined, and any plants which show signs of virus attack are removed and burnt. The tubers which are harvested can then be certified free from virus. This accounts for the popularity of Scottish 'seed' potatoes. In lowland districts where aphids are abundant it is advisable to buy new 'seed' each year so as to prevent viruses severely reducing the crop.

Finally, it should be emphasised that bacteria are essential to the well-being of every organism, since they keep the soil fertile by their chemical activities. Also, even though many of them are parasitic, it is a rarity for the host to be killed. This is hardly surprising, for when the host dies the parasite will also die. Perhaps the best view of these occasional deaths due to parasites is that the intimate host/parasite relationship has become unbalanced.

QUESTIONS 19

1. Which three main types of micro-organism can cause disease in man? Name one disease caused by each type. Give an account of the part the blood plays in defence against disease. How does vaccination help to combat smallpox?　　　[O & C]

2. What are (*a*) viruses, and (*b*) bacteria? State the principles involved in sterilisation, pasteurisation, and vaccination.　　　[O & C]

3. Describe different ways in which bacteria can be of use to man. Explain how you would set up and keep a culture of bacteria in the laboratory.　　　[A]

20 Heredity and Evolution

For centuries man has been impressed by the enormous number of different plants and animals on the earth, but it is only in the last hundred years that he has generally come to believe them to have arisen from primitive ancestral organisms by a process of continual but gradual change over as long a period as perhaps two thousand million years. This belief has much evidence to support it and is known as the theory of organic evolution. Organic evolution obviously involves relationship by descent from parent to offspring, and it will, therefore, be necessary to study the inheritance of characters from parent to offspring before we can hope to understand evolution, which covers thousands and thousands of generations of inheritance.

HEREDITY

Why should a blonde child be born of two dark-haired parents? Why should seeds from a red-flowered plant produce plants with several different flower colours? Why can we not grow Cox's Orange Pippin apples from seed? Such questions about inheritance are always being asked, and emphasise man's fascination with the problems of heredity. A basic knowledge of heredity is easily acquired by anyone who really has the desire to learn, and here we make a start with that process.

20.1 Mendel's Work

The first breeding experiments of any importance were carried out on garden peas (*Pisum sativum*) by Gregor Mendel, an Augustinian monk living at Brünn in Austria (now Brno in Czecho-Slovakia). The work was described in 1866 in an obscure natural history journal and it was not until it became widely known to the scientific world in 1900 that its true worth was realised. It forms the foundation of all the modern work on heredity.

Breeding Experiments involving One Pair of Alternative Characters

Mendel observed seven pairs of clear-cut alternative characters in pea-plants (stems were either tall or dwarf, cotyledons were either yellow or green, and flowers were either axial or terminal, for instance) and he avoided experiments on characters which showed a gradation from one extreme to the other. He selected for his experiments those plants which had bred true (i.e. all offspring resembled their parents) for a particular character during two generations when allowed to self-pollinate, and hence self-fertilise, without the interference of insect pollinators. (The garden pea carries out self-pollination, provided insects are excluded by putting muslin or polythene bags over the flowers.)

Pea plants can be cross-pollinated by removing the anthers before they burst, enclosing the flower in a bag, and dusting its stigma, when ripe, with pollen from another plant (see sec. 16.4). In this way Mendel cross-pollinated pea plants which were opposite in their characters, e.g. he pollinated tall plants with pollen from dwarf plants and vice versa. Whichever way the cross was made, the first filial generation (usually written F_1) all resembled one of the parents. The character which appeared in all these offspring (in this case tallness) he called **dominant** and the character which did not appear he called **recessive**. When he allowed the F_1 plants to self-pollinate and grew plants from the resulting F_2 seed he found that 787 of the F_2 plants were tall (2 m approx. high) and 277 were dwarf (0.25–0.5 m), a ratio of 787 : 277 or 2.84 : 1.

Mendel further found that the tall plants of the F_2 did not all breed true for tallness when allowed to self-pollinate, as the dwarf ones did for dwarfness. Of 100 of the tall plants tested, 28 produced all tall offspring and the other 72 produced a mixture of tall and dwarf offspring in approximately a ratio of 3 tall : 1 dwarf. These results are summarised in Fig. 20.1.

When he compared the F_2 results with those of other breeding experiments he was obviously struck by the approximation of the ratios of the two parental types to 3 : 1 (see table below). Clearly he had discovered a common pattern of inheritance for seven pairs of alternative characters in the garden pea.

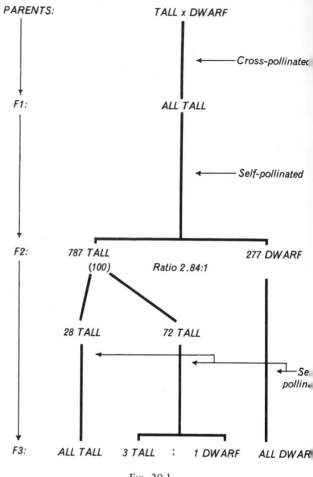

Fig. 20.1.

Experiment No.	Character	Number of F_2 Plants with Dominant or Recessive Character	Ratio
1	Seed shape	5 474 round : 1 850 wrinkled	2.96 : 1
2	Cotyledon colour	6 022 yellow : 2 001 green	3.01 : 1
3	Seed coat colour	705 grey : 224 white	3.15 : 1
4	Shape of ripe pods	882 inflated : 299 constricted	2.95 : 1
5	Colour of unripe pods	428 green : 152 yellow	2.82 : 1
6	Position of flowers	651 axial : 207 terminal	3.14 : 1
7	Length of stem	787 tall : 277 dwarf	2.84 : 1
Totals for all characters		14 949 : 5 010	2.98 : 1

When scientists make such discoveries they rarely stop there, but proceed to make hypotheses which provide an explanation of the data collected and then they try to test the hypotheses by experiment. The explanations given below follow closely those given by Mendel, but use terms such as gamete, zygote, and allelomorphic which were unknown to him.

1. There are **factors** present in pea plants which cause characters such as tallness and dwarfness to develop. Each pair of factors is known as an **allelomorphic** pair. (Let T represent the factor for tallness and t the factor for dwarfness.)

2. An allelomorphic pair of factors is represented twice in a zygote but only once in a gamete (e.g. male nucleus of pollen grain and female nucleus of ovule). This is sometimes referred to as Mendel's first law or 'purity of gametes'. Thus, **gametes must be pure in respect of factors from each allelomorphic pair** and there must be a process in gamete formation in which the number of factors is halved.

3. Where two different factors of a pair exist together in a zygote (Tt), one, the dominant, causes its character to develop to the exclusion of the other, the recessive. Such zygotes (Tt) are called **heterozygotes** or hybrids, as opposed to the **homozygotes** containing only one type of factor (e.g. TT or tt).

4. Heterozygotes (Tt) produce two types of gamete, (T) and (t), in equal numbers, and at fertilisation there is no tendency for any one of the four possible gamete combinations to occur more often than the others, i.e. **fertilisation is at random.**

These explanations of the breeding data from tall and dwarf peas are given below in conventional abbreviated form. When you are writing out breeding data you should follow this type of plan. It is important to note that the difference

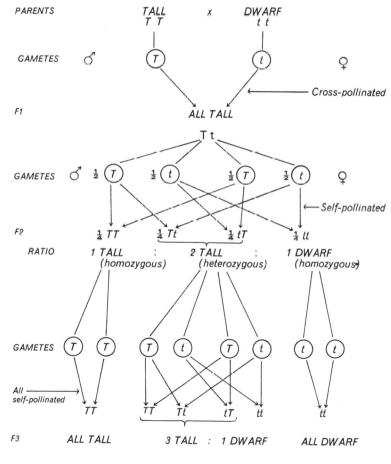

Fig. 20.2.

between the homozygous and heterozygous tall plants cannot be observed directly; only by breeding from them can they be distinguished. In other words, the **genotype** of the heterozygous plants (Tt) is different from that of the homozygous tall plants (TT), but their **phenotypes** (i.e. their obvious characters) are the same since they are both tall.

Can you see how the equal numbers of the two types of male and female gamete and the random fertilisation, which obviously causes a doubling of the number of factors, provide an explanation of the theoretical 3 : 1 ratio in the F_2 generation and the deviations from it which actually occur? The next section on simple probabilities will help you to reason this out.

20.2 Simple Probabilities

A model will help us with this problem. Obtain 400 small, uniform spherical beads, 200 of one colour (say black) and 200 of a different colour (say white). Place 100 of each colour in each of two large bottles which have in the cork a short closed tube large enough to admit one bead to view when the bottle is inverted (see Fig. 20.3). Now shake one bottle well and invert it. What is the chance of the bead appearing being a white one? You will see immediately that you are

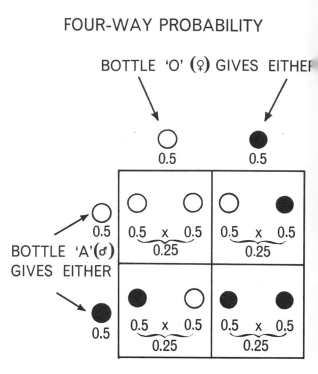

FOUR-WAY PROBABILITY

BOTTLE 'O' (♀) GIVES EITHER

FIG. 20.4.

PROBABILITY BOTTLE

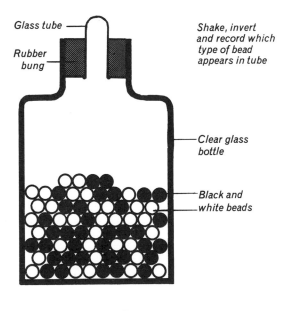

Glass tube

Rubber bung

Shake, invert and record which type of bead appears in tube

Clear glass bottle

Black and white beads

FIG. 20.3.

equally likely to select a white one as a black one, since there are equal numbers of each in the bottle. The chance or probability of selecting a white one must be 1 in every 2 attempts, which is expressed mathematically as 1/2 or 0.5. Had there been 50 white beads and 150 black beads in the bottle the probability of selecting white would have been 1/4 or 0.25.

Now let us use both bottles to make an accurate model of tall heterozygous pea plants carrying out self-fertilisation. Label one 'O' for ovary producing two types of female gamete, and one 'A' for anther producing two types of male gamete. Suppose we now shake both bottles together, invert and record the type of bead appearing from each bottle, and repeat this trial; what results would you expect?

From the diagram above it is clear that there are four possible ways in which these two events can happen, and the probability of any one of these four occurring must be 1/4 or 0.25, since each is as likely as any other. You will see that this probability is the product of the individual probabilities of each of the two independent events.

Now let us test out our model by making 200 'throws' with each bottle and see whether the

RESULTS OF 200 THROWS

EVENT BOTTLE 'O' ♀ BOTTLE 'A' ♂		NUMBER OF OCCURRENCES	ACTUAL PROPORTIONS	EXPECTED PROPORTIONS
○	○	43	$^{43}/_{200} = 0.215$	0.250
○	●	55	$^{55}/_{200} = 0.275$	0.250
●	○	50	$^{50}/_{200} = 0.250$	0.250
●	●	52	$^{52}/_{200} = 0.260$	0.250
TOTALS		200	$^{200}/_{200} = 1.000$	1.000

Fig. 20.5.

probabilities we have deduced for the four different events actually occur. One set of results we obtained is shown in Fig. 20.5. How do your results compare?

This example shows us that the deduced probabilities were the right ones, but that in practice there are almost always bound to be deviations from these. The larger the number of throws the less significant these deviations are likely to be.

A similar, but more tedious model to operate can be made from tossing two coins.

Now let us apply what we have learnt from our model to the problem of the F_2 offspring resulting from self-pollinating the F_1 tall pea plants. Each time pollination results in fertilisation there have been two independent events—the selection of one of two possible types of female gamete and one of two possible types of male gamete. Clearly, there are four possible ways in which fertilisation can happen and so our model fits the problem accurately. The black and white beads represent gametes containing either a T factor or a t factor produced by either the ovary or the anther. Each 'throw' with the two bottles represents one case of fertilisation and results in a seed which can grow into a new plant. From the point of view of the phenotype and genotype of pea plants it does not matter whether the T factor comes from the male or female gamete and hence the heterozygous (Tt) offspring are present in double the

quantity of the other two genotypes. Further, since the TT and Tt genotypes both show the tall character, three of the four sets of offspring will be tall and 1 set (tt) will be dwarf.

What about the deviation from the 3 : 1 ratio? In Mendel's experiment 787 tall : 277 dwarf gave a ratio of 2.84 : 1; does this really fit our theoretical expectation of 3 : 1? If we let the white beads represent the T factor for tallness and the black beads the t factor for dwarfness, then the ratio obtained in 200 throws was

white/white (TT)   43 ⎫
white/black (Tt)   55 ⎬ 148 : black/black (tt) 52
black/white (tT)   50 ⎭

or 2·85 : 1. Thus deviations from the expected ratio occur in our model simply by chance, and this can be taken to indicate that the deviations Mendel observed from the 3 : 1 ratio do not mean that his explanations were incorrect, since they also would be due to chance.

Since 1900, when Mendel's work became widely known, breeding experiments have been carried out on all major groups of organisms which reproduce sexually, and his principles have been found to hold good even though other patterns of inheritance have been discovered. Some examples of characters inherited as simple Mendelian factors, dominant or recessive are given in the table on the following page.

Species	Dominant Character	Recessive Character
Sweet pea	Purple flower	Red flower
Sweet corn	Yellow grain	White grain
Groundsel	Hairy leaves	Hairless leaves
Fruit fly (Drosophila)	Normal (long) wings Dark red eyes Striped body	Vestigial wings White eyes Black body
Man	Pigmented eyes Normal blood Ability to taste phenylthiourea (PTU)	Albino (pink eyes) Haemophilic blood Inability to taste PTU

Many cases are now known where the dominance is incomplete and the heterozygous condition is intermediate between the two homozygous conditions. For example, the four o'clock plant (*Mirabilis jalapa*), which is red or white in the homozygous state, has a heterozygous form which is pink.

Mendel's factors, which were originally envisaged as hypothetical bodies causing characters to develop, are now known as **genes**, and the study of inheritance, since it deals for the most part with genes, is known as **genetics**.

20.3 Environment and Inheritance

Not all characters of organisms are determined entirely by genes; many are influenced considerably by the environment of the individual. For instance, a person who leads an outdoor life will usually have a sun-tanned skin. The brown pigment is laid down in the skin in response to ultra-violet light but, as is well known, this is not a permanent character, and children of sun-tanned parents do not inherit the brown skin. Genes, however, must be involved in the ability to respond to ultra-violet light, for some people never do develop much pigment, their skin merely becomes red and sore and they have to protect themselves from the sun. Furthermore, the pigment in the skin of negroid races is inherited and under genetic control: they do not lose their pigment when they grow up in a less sunny environment.

Many human characters, such as height, weight, and intelligence, as measured indirectly by the familiar tests, are partly determined by the environment and partly by genes passed on from the parents. Either factor can play a decisive role in determining the character and it is only by careful experiment that it is possible to decide what part each has played. It is important to remember that characters acquired during an individual's life by interaction with the environment are not inherited, because the acquiring of the character does not alter the genes which are passed on to the next generation.

Characters such as height and pulse rate in man and seed weight in french beans show a wide range of variation from one extreme to the other (known as continuous variations), but those listed in the table above are said to be discontinuous because they fall into one of two clearly distinct classes.

20.4 Breeding Experiments with the Fruit-fly, Drosophila

If you wish to carry out breeding experiments yourself it would be wise to work with *Drosophila melanogaster*, the fruit-fly, which can produce a new generation in about 11 days if kept at 25°C in an incubator. In a warm place in a laboratory they will take rather longer. Their life-history is almost identical with that of the house-fly. Here are some details which will help to give you a start in Drosophila work.

Food

Water	200 cm³	Quaker oats	30 g
Agar	3 g	Black treacle	1 dessert spoonful
		Antimould solution	2 cm³

The antimould deters the growth of many common fungi which would otherwise interfere. Several materials are satisfactory. For instance, methyl-p-hydroxybenzoate, which should be prepared as a 10 per cent solution in 95 per cent alcohol.

The mixture should be cooked until it reaches a consistency which is easy to pour. One quarter fill small wide-mouthed bottles or 10 cm × 25 cm specimen tubes and then push into the food a folded paper towel or portion of one. The bottles should be covered until cool and dry inside, when two drops of thick yeast suspension should be added and a foamed polystyrene or tight cotton-wool plug applied. The containers are then ready to receive flies.

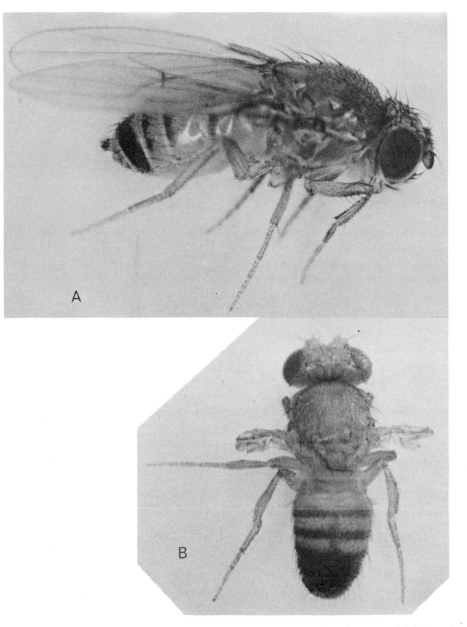

Fig. 20.6 A—Wild-type normal winged female, dorso-lateral view (×24). B—Vestigial winged male, dorsal view (×24). (*Photographs by M. I. Walker*)

Handling Techniques

If you think a little you will soon realise that you need to separate males and females and to select females for breeding experiments before they have mated with their own type in their stock tube. Obtaining such virgin females depends upon the fact that flies do not mate until several hours after hatching from the pupa, and the techniques involved are described in Fig. 20.7. Included in this is the etherising technique and the method of sexing the flies. Carbon dioxide, if readily available, is preferable to ether as there is little danger of overdosage.

HOW TO OBTAIN VIRGIN FEMALE FLIES

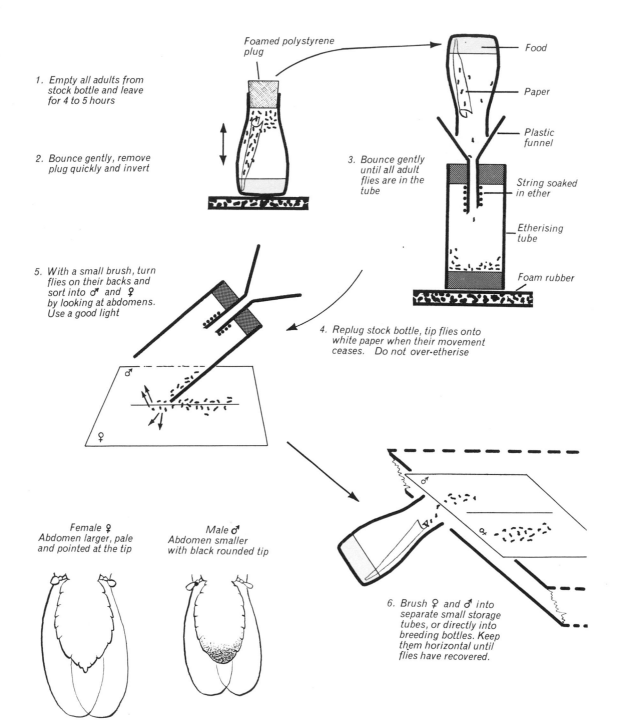

1. Empty all adults from stock bottle and leave for 4 to 5 hours

2. Bounce gently, remove plug quickly and invert

Foamed polystyrene plug

Food

Paper

Plastic funnel

3. Bounce gently until all adult flies are in the tube

String soaked in ether

Etherising tube

Foam rubber

5. With a small brush, turn flies on their backs and sort into ♂ and ♀ by looking at abdomens. Use a good light

4. Replug stock bottle, tip flies onto white paper when their movement ceases. Do not over-etherise

Female ♀
Abdomen larger, pale and pointed at the tip

Male ♂
Abdomen smaller with black rounded tip

6. Brush ♀ and ♂ into separate small storage tubes, or directly into breeding bottles. Keep them horizontal until flies have recovered.

Fig. 20.7.

Suitable Matings

Use about five pairs of flies for each.

1. Normal wing (wild type) × Vestigial wing
   ♂   ♀

   or with sexes reversed.

   Mate up F_1 flies amongst themselves in new tube, collect, and count types in F_2. Remember to remove parents from tubes as soon as plenty of larvae are visible.

2. Dark red eye (wild type) × Brown eye
   ♀   ♂

   or with sexes reversed.

   As for experiment 1. Scarlet eye is another type which can be used to demonstrate simple Mendelian inheritance.

20.5 Chromosomes and Genes

Because the male gamete is mainly nucleus and sometimes only the head enters the egg at fertilisation, it is in the nucleus that one would suspect genes might be carried. Furthermore, the halving of the gene number in gamete formation is paralleled by a halving of the number of chromosomes

CHROMOSOMES IN MEIOSIS

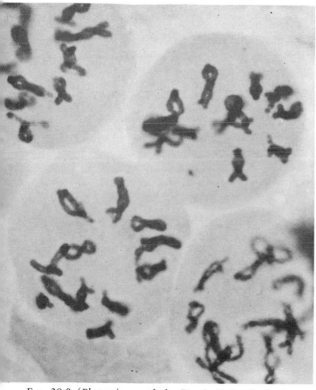

FIG. 20.8 (*Photomicrograph by Dr. L. F. La Cour. From 'Human Biology' by M. F. Martin. E.U.P. Ltd.*)

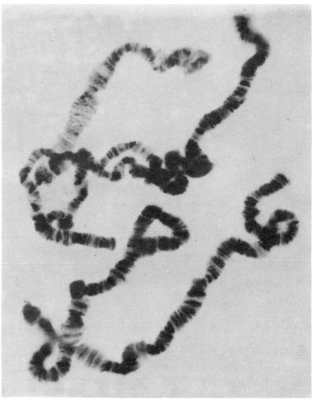

FIG. 20.9 Photomicrograph of the unusually large chromosomes from the salivary glands of a fruit-fly, *Drosophila melanogaster*, highly magnified. (*Photomicrographs by Dr. L. F. La Cour. From 'Human Biology' by M. F. Martin. E.U.P. Ltd.*)

in a special type of nuclear division known as meiosis (see Fig. 20.8) during gamete formation, and at fertilisation both the number of genes and the number of chromosomes are doubled. Such evidence is strongly suggestive that genes could be born on chromosomes within the nuclei of cells. There is a great deal of further evidence for this suggestion and today most biologists accept it as a well-established fact.

Genes are made of D.N.A. (deoxyribonucleic acids), a group of compounds with as infinite a variety as is found in the proteins. It seems likely in view of much recent evidence that each of the thousands of genes in an organism acts as a code or blueprint which can be duplicated and passed to the cytoplasm, where it can be used in building particular types of proteins. Many of these proteins would be enzymes, which could carry out a particular chemical change in the cell. Several cases are known of one gene controlling the production of one enzyme in an organism, and it appears that it could be generally true that genes work through the medium of enzymes to produce characters. The nearest we can get to actually seeing genes is in the very large salivary-gland chromosomes of Drosophila larvae where we can see hundreds of transverse bands in stained preparations (Fig. 20.9).

ORGANIC EVOLUTION

In the air, in the sea, on the land, and in the soil there is such an abundance of life that few of us can be fully aware of it. Only on the vast Antarctic plateau, on the highest mountains, and in desert areas covered by shifting sand dunes have organisms failed to cope with the environment.

The infinite variety of animals and plants never ceases to amaze biologists. In Great Britain, for instance, there are 4 species of oak, 2 of which have been introduced by man, 128 species of grasses, 13 species of bats, 186 species of caddis-flies, and 150 species of terrestrial and freshwater snails. How have all these come to exist today in places which suit them? Were they all created individually, or have they developed by a process of gradual and cumulative change or evolution from primitive ancestors? This issue was bitterly contested from a religious standpoint in the late nineteenth century, but eventually the opposition died down because the evidence in favour of organic evolution became overwhelming. It was also realised that the idea of organic evolution was not incompatible with religious beliefs, as was at first thought.

Organic evolution is an attractive theory. It explains similarities between organisms as being due to their descent from a common ancestor, and differences between them as being the result of cumulative variations from parent to offspring having taken place in different directions over countless generations.

20.6 The Genetic Basis of Evolution

Mendelian inheritance alone clearly cannot provide the cumulative variations of evolution referred to above. In fact, it can produce nothing in the way of an absolutely novel character, though it can produce new combinations of already existing characters in a population of a species. Consider the following:

All organisms which have been thoroughly investigated are thought to have several thousands of gene pairs, so the number of different combinations of characters thrown up by Mendelian recombination would be almost infinite.

How then do genuine novelties occur in populations? A full answer to this question is not yet possible, but you should know something of the rare occurrence of novelty by **mutation**. Very occasionally, when cells divide a gene is 'copied' imperfectly and a new gene or mutant (*L. mutare, to change*) gene is formed. Often this mutant gene will be recessive and disappear at death without ever having caused its new character to appear, since it will have been masked by a dominant gene. If it is a dominant mutant gene then its new character will show from the first. The frequency with which a gene mutates varies, but once in every million cell divisions is an average frequency. Thus a particular gene mutating is a rare event. If it occurs in a body cell it will affect only a limited region of the body, i.e. only those cells derived by cell division from the original mutant cell. If, however, it happens in a reproductive cell it could be present in a large number of gametes and so appear in every cell of individuals in the next generation, and if recessive it might show its character in the following generation when it could appear in the homozygous state. Although the cause of mutations is unknown, it is known that many sorts of radiation can speed up the mutation rates in cells and for this reason we should be careful, especially when young, to what radiations we expose ourselves or our animals and plants. At present, in Britain, X-rays used in medical practice are the major source of artificial radiations.

Let us now consider a mutation in the peppered moth (*Biston betularia*), which was first detected in 1850 when a form with uniformly dark wings was caught in Manchester. This was quite different in wing pigmentation from the normal form with mottled wings (Fig. 20.10). The dark form gradually became more and more common, until now it is rare to find the mottled form in this district. Here, clearly, is an example of a complete change in the genotype of a population—evolution. How could this change in the population have occurred? We still cannot explain the cause of the original mutation, but perhaps we can

1 pair of genes in an organism can produce 2 possible phenotypes.
2 pairs of genes in an organism can produce 2×2 possible phenotypes.
3 pairs of genes in an organism can produce $2 \times 2 \times 2$ possible phenotypes.
7 pairs of genes in an organism can produce $2^7 = 128$ possible phenotypes.
20 pairs of genes in an organism can produce 2^{20} or more than 1 million phenotypes.

PEPPERED MOTHS

Fig. 20.10.

understand how the dark form spread through the whole population. If one suggests that the dark form had a reproductive advantage over the normal form, i.e. for every offspring left by the normal mottled form two were left by the dark form, it is easy to see how a population could change in the course of a number of generations.

Read Fig. 20.11, where such an imaginary population is given as an example.

Obviously the actual advantage experienced by the dark form of peppered moth cannot have been as great as 2 to 1, because the change in the population was much slower than in our example. What might have been the nature of the advantage experienced by the dark moths, if indeed they did experience any? Could it have been that the birds which fed on peppered moths found the dark forms more and more difficult to detect against the progressively darkening background of a smoke-blackened area, and conversely found the lighter mottled forms more readily? This suggestion has been checked by a field experiment using both forms and has been found to be substantially correct. We could, therefore, say that there has been a **natural selection** of the dark form in the Manchester district by the environment.

Natural selection is, in fact, the name given to a mechanism of evolution. In this process the

NATURAL SELECTION

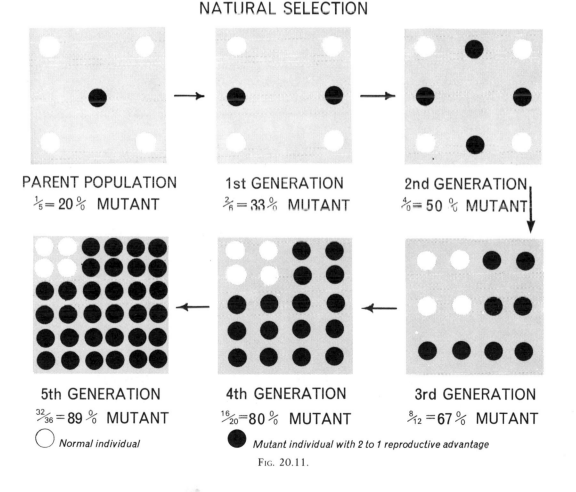

PARENT POPULATION
$\frac{1}{5}=20\%$ MUTANT

1st GENERATION
$\frac{2}{6}=33\%$ MUTANT

2nd GENERATION
$\frac{4}{8}=50\%$ MUTANT

5th GENERATION
$\frac{32}{36}=89\%$ MUTANT

4th GENERATION
$\frac{16}{20}=80\%$ MUTANT

3rd GENERATION
$\frac{8}{12}=67\%$ MUTANT

○ Normal individual

● Mutant individual with 2 to 1 reproductive advantage

Fig. 20.11.

novelties are produced by mutation, the mutant character is recombined with many combinations of other existing characters by Mendelian inheritance, and the environment operates a selective influence on these by causing differential reproductive rates in the different types. Neither mutation, nor Mendelian inheritance, nor the selective influence of the environment alone can produce evolution, but together they can be potent in changing frequencies of genes in a population.

Natural selection results in organisms generally being well adapted to their surroundings. For instance, the various arrangement of hairs forming the pollen basket, pollen brush, and comb on the third leg of a worker bee and the haustoria and reduced leaves on the stem of dodder would be considered as adaptations which have evolved by mutation and selection. It is important to remember, though, that this is a consequence of the environment selecting from random mutations and recombinations, and not the result of any purposeful act on the part of the organisms concerned. In other words, mutations are not directed, but evolution often is. This should be obvious when we remember that genes are often lethal or nearly so in nature, for example, the gene for vestigial wing in fruit flies and the gene for haemophilia in man which prevents the blood from clotting at a wound, though with modern medical help its effect can be reduced to acceptable levels. If one considers a population which has gene frequencies in balance with its surroundings (i.e. a well-adapted population) it is rather unlikely that a new mutant gene will do anything but upset the balance. It is rather as if one were to exchange a part in a well-balanced mechanism such as a clock with a new one chosen quite at random. Is it likely that the new part would fit, let alone improve its time-keeping properties? Evolution is so often a very slow process because for long periods of time the environment is constant and the gene frequencies of populations, even allowing for mutation rates, are kept steady.

A remarkable example of rapid evolution is the new species of cord grass, *Spartina townsendii* (Fig. 20.12), first noticed on the shore of Southampton Water at Hythe in 1870. It was larger and more vigorous in its growth than either the native *Spartina maritima* or the introduced American species, *Spartina alterniflora*, first recorded from the Itchen estuary in 1829. The new species has been shown to be a fertile hybrid between *S. maritima* and *S. alterniflora* and has spread very rapidly along the south coast estuaries where it has colonised acres of shifting tidal mud; it has also been used widely in land reclamation schemes. This hybrid has also formed in the Bay of Biscay where again the American species was introduced.

SPARTINA TOWNSENDII

Fɪɢ. 20.12.

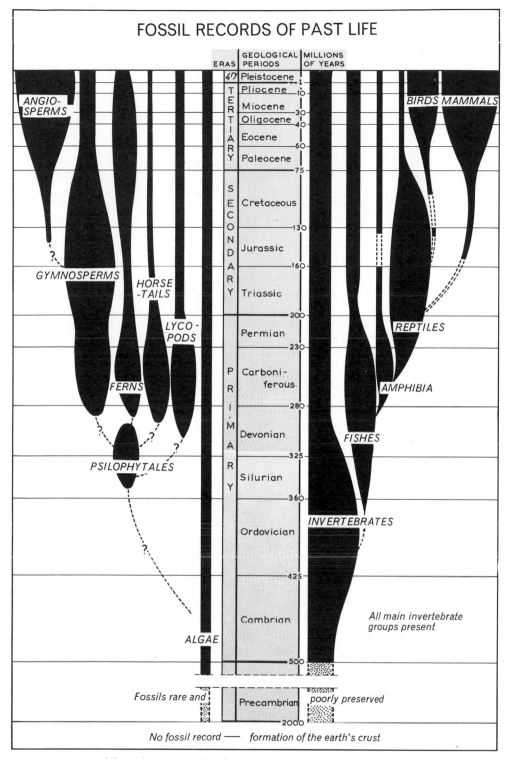

Fig. 20.13 To follow the course of evolution read from below upwards. Dotted lines indicate gaps in the fossil record or the possible origin of the group. The thickness of the branches gives an approximate indication of the abundance of the group. Bryophytes and fungi have been omitted since few well-preserved fossils of these groups are known.

20.7 Indirect Evidence for Organic Evolution

There is an overwhelming mass of indirect evidence for organic evolution, but it requires very careful assessment. It is presented here in abbreviated form under three side-headings.

Evidence from the Fossils of Sedimentary Rocks

As soon as the earth's crust cooled to form solid igneous rocks, and water was present in the liquid state, weathering of the rocks commenced and the fine particles resulting from this process collected under water as a sediment. The sediment became more and more compressed owing to further layers being deposited above, and so formed the strata of **sedimentary rocks**. As a result of earth movements these sedimentary rocks became folded and broken and themselves weathered, so that today the original simple layered arrangement is nowhere perfectly preserved. By careful study, geologists have been able to place the layers in the order of their formation and they can also estimate the approximate age of each layer from the thickness of rock above. The names and ages of the different geological periods during which sedimentary rocks were laid down are given in Fig. 20.13.

Most sedimentary rocks have been shown to contain unmistakable remains of animals and plants, whose bodies were buried as the rock was formed and are now known as **fossils**. Usually it is the hard parts such as bones, shells, scales, and cell walls that are preserved, which is hardly surprising in view of the way soft parts of dead organisms decay.

Vertebrate fossils form the clearest indirect evidence for evolution. Fish, which first appeared in the Ordovician period, could have given rise to amphibians whose earliest representatives are late Devonian. Similarly, some amphibians could have been the ancestors of the reptiles whose earliest forms appeared in the late Carboniferous. The earliest mammalian fossils were fragments (teeth and jaws) from the Jurassic rocks, and Archaeopteryx, a fossil bird showing clear impression of feathers, was found in the same layers. The time of first appearance of these classes of vertebrate suggest that the relationship is:

Fish⟶Amphibia⟶Reptiles⟶Birds
⟶Mammals

This is further supported by a series of fossils, with some gaps it must be admitted, forming a transition between the classes. One fossil, Seymouria, from the Permian rocks of Texas, shows such a combination of reptilian and amphibian features that it oscillates from one class to the other whenever further fossil material comes to light!

In the Cambrian rocks, fossils of all the main invertebrate groups are present and so we get no help from this source in establishing relationships.

Perhaps the most convincing fossil evidence is found in cases where, in successive strata from one locality, a series of fossils is known which exhibit gradual change. Certain fossil horses of the Tertiary era form such a series, and details of their limb bones, their size, and their age are given in Fig. 20.14. How can such facts be explained convincingly if one does not accept organic evolution?

Evidence from Comparative Anatomy

All vertebrates, both living and fossil types, have a vertebral column made up of a number of vertebrae, a skull and a pair of skeletal girdles providing a base for the limb or fin bones to move on. All vertebrates except fishes have pentadactyl limbs (see Fig. 13.6). All living vertebrates have a ventral muscular heart, a hollow dorsal central nervous system, and a kidney made of tubules; in fact they are all constructed on the same general plan. How can such facts be reconciled with each other except by giving an explanation involving evolution from a common ancestor which possessed all these characters?

It is instructive to consider the three evolutionary efforts at solving the problems of flight with the pentadactyl limb (Fig. 20.15).

Evidence from the Geographical Distribution of Organisms

The distribution of animals, particularly on volcanic islands, provides further circumstantial evidence for evolution. The fauna of the Galapagos Islands, 600 miles west of the South American coast, contain many animals similar to, but distinct in certain features from, those on the mainland. Further, there are many cases of slightly different species being present on different islands of the group. The evolutionary explanation of such facts is that the mainland and island species had a common ancestor, but since isolation of the island form or forms no interbreeding has taken place and each stock has evolved independently.

Isolation is a potent factor in producing evolutionary change. When a new variation arises in an island population it may become more frequent, but because the population is isolated the new variation cannot spread to others of the same

THE EVOLUTION OF THE HORSE

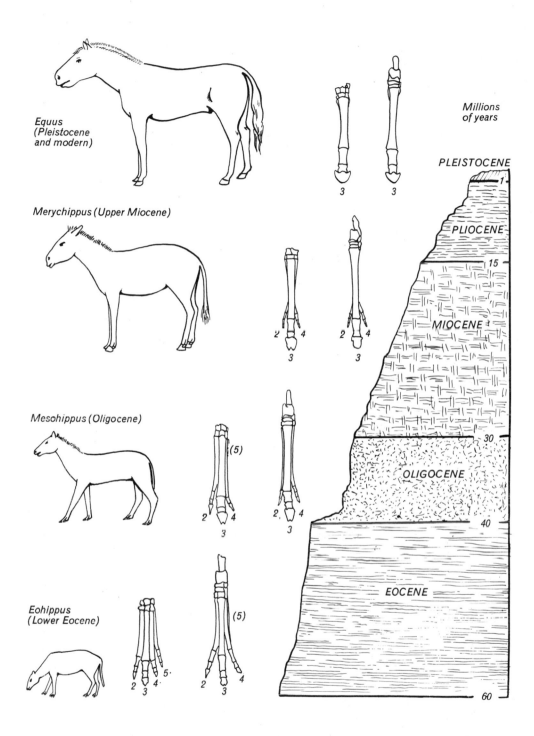

Fig. 20.14 The four fossil horses have been chosen from the large number known, to show progressive increase in size and loss of digits. Skeletons of the left fore-limbs are shown on the left, and of the left hind limbs on the right. There were also progressive changes in the skulls and teeth.

FORE-LIMBS ADAPTED FOR FLIGHT

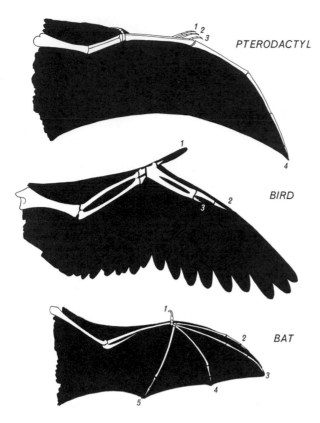

PTERODACTYL

BIRD

BAT

Fig. 20.15 The fore-limb of the pterodactyl had only four digits, of which the fourth was greatly elongated as a support for the wing. The second digit of a bird fore-limb is the only one to be well developed, but there exist obvious vestiges of two others. All five digits are present in the fore-limb of a bat; four of them are long and support the wing membrane, the other is short and ends in a claw.

species. Thus, each isolated population may, given sufficient time, develop along a different evolutionary line. Such is the case in the British wren, whose forms in the mainland, the Outer Hebrides, the Shetlands, St. Kilda, and Iceland are each characterised by different wing lengths and different colouring. At the moment they are each given variety rank, but in time they may evolve sufficient differences to justify each form having specific rank.

20.8 Historical Aspects of Evolution

No account of evolutionary theory would be complete without mention of natural selection as it was jointly suggested by Alfred Russell Wallace and Charles Darwin in 1858. In the following year Darwin published his *Origin of Species by means of*

Natural Selection. Their ideas can be briefly summarised:

1. All organisms overproduce since offspring are always more numerous than parents.
2. Despite this, the numbers of any one species usually remain more or less constant for a very long time.
3. Therefore there must be a struggle for survival.
4. All offspring exhibit slight differences (**variations**) from their parents.
5. Those with variations which are advantageous in a particular environment will be better fitted to live there and will survive and reproduce to the exclusion of the less fit ones.

Darwin, more so than anyone else, collected evidence for evolution, and during his five-year global voyage as naturalist on board H.M.S. *Beagle*, when he saw the infinite variety of plants and animals, he gradually formed his ideas on a possible mechanism of evolution. It says much for his clear insight into the life of whole populations that the main idea of his theory of natural selection has stood up to being tested experimentally on a variety of organisms. He did not, however, understand the interaction of mutation and Mendelian inheritance as a source of variation, and he emphasised the destructive aspects rather than the creative aspects of the process.

20.9 The Course of Evolution

In the Cambrian period which started 500 million years ago all life was aquatic; there were no terrestrial animals or plants and the land was bare. In the sea, all the invertebrate groups were represented. Their forms were often strange; there were the trilobites, some of which looked like giant woodlice, and the eurypterids or seascorpions which 150 million years later became the terror of the Silurian seas. The Ordovician period saw the invertebrate groups expand along several different lines. The first vertebrates, the ostracoderms, made their appearance in the Ordovician. They were small, heavily armoured fish without jaws, and their line is represented today by the lampreys. They were followed toward the end of the period by a new group of fish which possessed the vertebrate type of jaws.

Great changes happened in the Devonian period. The first land plants (Psilophytales) appeared with their spore-bearing bodies rising out of the shallow water in which they must have lived. Some of them developed the first true leaves and a primitive vascular system. These plants had developed from the marine plants (mostly algae) which had been at an evolutionary standstill for nearly 200 million years. Then came a vast array of large cone-bearing trees, which lived in the swamps of the Carboniferous period and eventually died, producing our coal measures. The terrestrial plants provided food and shelter for animals, and the latter soon started to colonise the land. Clumsy amphibians which crawled about inefficiently on their bellies appeared on land, and in so doing had taken two very important steps toward making terrestrial life in vertebrates possible. Their gills had been replaced by lungs and their fins by limbs. Also in the Carboniferous period came the insects already differentiated into modern forms, like dragon-flies and cockroaches.

During the Permian period, some of the amphibians were able to dispense with a watery environment for their eggs and larvae, and started to lay eggs which could hatch on land. These were the first reptiles. For 130 million years throughout the Triassic, Jurassic and Cretaceous periods the reptiles thrived and dominated the land with the many branches of their flourishing family tree. One of these branches, the dinosaurs, includes Brontosaurus, the largest animal known except for the blue whale. Some even returned to the sea (ichthyosaurs) from which their ancestors had emerged, others (pterodactyls) took to the air, supporting their wings with an enormously elongated little finger. During this period, when the reptiles were dominant, the first flowering plants appeared and with them the first butterflies and bees.

There were in the Mesozoic era two insignificant groups whose progeny were later to dominate the land and the air: firstly, the small furtive mammal-like reptiles which must have successfully avoided the large reptilian carnivores, and secondly, a group of feathered types exemplified by the famous Jurassic fossil, Archaeopteryx, which is probably the ancestor of modern birds.

About 75 million years ago all except the four modern reptile groups (crocodiles, snakes and lizards, turtles, and the tuatara of New Zealand) mysteriously died out, leaving the earth to be dominated by the mammals and birds in the Tertiary era. During this time not only did the many and varied mammalian and bird types of today appear, but also most of the modern flowering plants.

The most recent great event in evolution took place only about two million years ago when the first human types developed from an ape-like ancestor. Man certainly cannot have evolved from any one of the four modern apes, but it is likely that he shares with them a common ancestor. Man now dominates the earth largely because of his relatively enormous brain (a feature characteristic of infantile mammals), which has provided the necessary nervous material for activities such as insight learning, higher thought, and communication by speech (see sec. 15.5). His ability to hold things and manipulate them, to feel them with sensitive fingers which are all opposable to the thumb, and to hold them for close visual examination with both eyes which face forward, have been important factors in his domination of the earth, which is proceeding at an ever-increasing pace. This domination is the result of his control of agriculture and food storage, and his use of a bewildering array of sophisticated tools which he has designed and made. Such tools include both sensory ones (e.g. microscopes, radar devices, television), and motor ones (e.g. submarines, hovercraft, aeroplanes and computers). Tool-making is certainly an important factor in man's success and, unlike other characters acquired during a lifetime, it can be passed on from one generation to the next. So man has really developed what amounts to a new, more rapid and more powerful type of evolution!

QUESTIONS 20

1. Robert and Betty received a consignment of ladybird pupae from Ruritania. These were placed separately in glass tubes. When adults matured some were red and others black. Robert took the red ones and mated pairs together in separate tubes. In due time adults were reared, but in some instances all the offspring of a particular pair were red, in others there were three times as many red ones as there were black.

Betty treated the black ones similarly, but the offspring obtained were always black. Explain these results, giving full genetical details. [O]

2. In Lord Carstonmore's family for several generations some children were born with a characteristic shoulder blemish. Lord Carstonmore himself was the first peer without this blemish, though his brother did have it. Give a simple genetical explanation of the inheritance of this blemish (not involving sex-linkage). Are Lord Carstonmore's children likely to have this blemish? [O]

3. In a certain mammal, brown hair is dominant to white. Mating between various pairs of brown individuals produced results of three kinds: (i) All the offspring were brown and further breeding revealed no whites. (ii) Three-quarters of the offspring were brown and one-quarter were white. All the white but only some of the brown bred true in further generations. (iii) All the offspring were brown, but in further breeding only half bred true, the other half behaved like those in group (ii). Using the symbols B for brown and b for white, explain these results in Mendelian terms.

[O & C]

4. (*a*) Using tall (dominant) and short peas, how could you demonstrate the law of the segregation of germinal units (Mendel's first law)?

(*b*) Briefly state what you mean by the terms below and give an example of each of these: (i) dominant character, (ii) recessive character, (iii) hybrid. [S]

5. In field buttercups prickly fruits (dominant) and smooth fruits (recessive) are found. How would you carry out a breeding experiment to demonstrate Mendel's law of the segregation of germinal units using field buttercups and what results would you expect to obtain? Give full precautions to be taken at each stage of the experiment. [S]

6. In man, brown eye colour appears to be dominant over blue eye colour. If a brown-eyed man marries a blue-eyed woman, and they have three children, two blue-eyed, and one brown-eyed, how can this be explained in terms of cell division, genes, chromosomes, sperms, and eggs? (A complete diagram, with notes, will be acceptable as an answer.) [S]

7. A school has for many years had a Rabbit Club, the members of which keep only pure-bred white-coated rabbits and pure-bred grey-coated rabbits. Unfortunately the doors of the cages were left open for a time and the rabbits mated with each other. All the offspring born were found to have grey coats. (*a*) Which coat colour is dominant? (*b*) If two of the grey-coated offspring are mated together, what results would the Club expect to find, as far as coat colour is concerned, if there are eight offspring in the litter? Explain carefully how you have reached your conclusions. [*M*]

21 Food Production: Conclusion

'There is far too little food for the world's hungry mouths, and the yearly increase in production is not enough to keep pace with the increase in the number of human beings. A third of the world's people enjoy health, wealth, and well-being to a comparatively high degree; the other two-thirds live on intimate terms with poverty, hunger, disease, illiteracy, and premature death.' (Extract from the Report of the Director-General of the Food and Agricultural Organisation of the United Nations, Oct. 1950.)

'We, the peoples of the United Nations, determined . . . , to reaffirm faith in fundamental human rights, in the dignity and worth of the human person, . . . and for these ends . . . , to employ international machinery for the promotion of the economic and social advancement of all peoples, have resolved to combine our efforts to accomplish these aims.' (Extract from the Charter of the United Nations, signed by fifty-one member states on the 26th June 1945.)

The first quotation states the world food problem, a problem which it is all too easy to ignore when one has an adequate diet. In the second quotation one sees man setting a high value on the life of the individual, and resolving to co-operate to ensure that everyone shall have the basic needs. In this co-operation man is undoubtedly unique amongst animals, for animals behave in such a way as to preserve the race rather than the individual. Unfortunately, man's progress in this direction is slow, and hindered by national rivalry, distrust, old habits, and economic problems.

In parts of the world where diets are inadequate there is usually a shortage of animal protein, the most expensive type of food. It is only when there is a famine that there is shortage of energy-producing foods, but famines are not infrequent in parts of China and India. The countries of Western Europe, because of their comparative wealth, have been able to monopolise almost the whole of the exportable beef, mutton, and pig meat of the world. The big meat producers such as N. America, Argentina, Australia, and New Zealand have, of course, no shortage, and the same is true of many smaller states which are self-supporting in this way. While such inequalities exist there is always liable to be unrest and even war, and the answer to such problems must be to produce more food.

In the United Kingdom, one of the most densely populated countries in the world, there is at present no food shortage, but there is a food production problem. Less than six per cent of our population are engaged in agriculture, and we are the least self-supporting country as a food producer. We therefore have to rely on making enough attractive goods to sell abroad to enable us to buy our food and other raw materials. The problem is likely to become more acute in the future, since people are starting productive work later in life, retiring earlier, and living for longer in retirement. In less than fifty years we could be relying on the productivity of only one-quarter of the population if the present trend continues.

Since 1939 the population of the United Kingdom has increased by about 14 per cent: we produce about half of our food requirements or

on the value of fertilisers is that which was started in 1843 in the Broadbalk Field at Rothamsted (see Fig. 21.1). Here wheat has been grown in eighteen plots every year since 1843, and the yields recorded. The treatment of certain plots, together with the average yields up to 1925 when

PERCENTAGES OF TOTAL U.K. FOOD SUPPLIES PROVIDED BY HOME AGRICULTURE

Food Product	Pre-war average	1955	1964	1965
Wheat and flour (as wheat equivalent)	23	34	47	47
Oils and fats (crude oil equivalent)	16	16	10	11
Sugar (refined value)	17	23	27	30
Carcase meat and offal	51	61	69	70
Bacon and ham (excluding canned)	32	47	38	38
Butter	9	7	6	8
Cheese	24	33	42	43
Shell eggs	71	90	98	98
Milk for human consumption (as liquid)	100	100	100	100
Potatoes	96	94	95	96

Source: Ministry of Agriculture, Fisheries and Food.

about two-thirds of what could be grown in this temperate climate. The table above shows the percentages by weight of the total supplies of certain foods which we produce ourselves in relation to pre-war averages.

We obtain 98 per cent of our energy and 94 per cent of our protein from terrestrial vegetation either directly as plant food or indirectly in animal food. Thus we are a soil-based community, and we feed largely upon what we can obtain from the soil by our labours and ingenuity. This applies also to most other peoples of the world.

Since 1939, initiated by the stimulus of war, food production in the United Kingdom has more than doubled. Some methods by which food production has been and can be increased are described in the rest of this chapter.

21.1 Care of the Soil and its Fertility

The principle of adding fertilisers to the soil to replace mineral salts which have been absorbed by a crop or washed away in drainage water was explained in section 5.4. Such fertilisers, containing nitrogen, phosphorus, and potassium, have been used extensively for many years.

Perhaps the most convincing experimental work

FIG. 21.1 The Broadbalk Field at Rothamsted Experimental Research Establishment. Notice the fallow strip running across the plots. (*Photograph by Rothamsted Experimental Research Establishment*)

BROADBALK FIELD—WHEAT SINCE 1843

Plot No.	Treatment	Relative Yield
2	Dung 70 Mg/ha	100
3	No fertilisers	35
5	Superphosphate 439 kg/ha Potassium sulphate 502 kg/ha Sodium sulphate 251 kg/ha Magnesium sulphate 251 kg/ha	41
6	As for plot 5, but in addition ammonium sulphate 251 kg/ha	65
8	As for plot 5, but in addition ammonium sulphate 735 kg/ha	97

[1 000 kilogrammes (kg) = 1 Megagram (Mg) or metric ton or tonne.
1 hectare (ha) = 10^4 m^2 = 10 000 square metres = 100 metres square.]

fallowing every fifth year was first started, are given in the table above.

There is no doubt that the proper use of artificial fertilisers could increase the yield of crops, and hence the amount of animal protein available for human consumption in many parts of the world. It must be remembered that fertilisers are expensive, but the need for them can be kept to a minimum by efficient farming practice, particularly in the use of leguminous crops.

A more recent advance in our knowledge of soil fertility is the recognition of 'trace elements' (see sec. 6.3) which need to be present in minute quantities for the normal growth of plants, or of animals which feed upon them. 'Heart rot' of sugar beet, for instance, is caused by the absence of the usual minute amounts of boron, and 'pine' in sheep and cattle has been traced to cobalt deficiency in the pastures on which they feed.

21.2 Breeding of Plants and Animals

It has been possible by careful interbreeding and selection to produce new varieties of animals and plants. In this way we have produced, for example, cattle which give more milk or fatten more quickly, sugar beet which yields more sugar, wheat which ripens earlier, wheat which is resistant to the attacks of rust fungus, and clovers and ryegrasses which are more nutritious as cattle food. Dr. G. H. Bell has recently bred the new Proctor barley, as a result of which the United Kingdom now produces more than eight million tonnes per year and exports one million tonnes, while only six years ago it produced half that quantity and imported about one million tonnes from abroad. For this work he was awarded the first Mullard gold medal and a £5 000 prize by the Royal Society.

The production of new varieties of clovers and ryegrasses has led to the practice of **ley farming**,

which is now widely recognised as a valuable method. It involves the abolition of permanent pasture (there are very few parts of the United Kingdom where permanent pasture remains as productive as a good ley), and the substitution of a temporary pasture or ley. The ley is obtained by sowing a mixture of seeds of red and white clover, Italian and perennial ryegrasses, cocksfoot, and timothy grass, which have been selected for their ability to produce high yields of proteins and energy-producing foods. The plants can be fed directly to cattle by grazing, or cut, and fed as hay or silage. After a variable length of time (1–6 years) the ley becomes invaded by grasses and weeds which are not so highly nutritious; it is then profitable to plough it up and perhaps take a corn crop and a root crop off it before seeding it again as a ley. A comparison between the yields of a good permanent pasture and a good ley is given below.

	Yield	Starch (or its Equivalent)	Protein (or its Equivalent)
	Mg/ha	Mg/ha	Mg/ha
Permanent pasture	3.8	1.2	0.2
Ley	7.5	4.5	0.9

21.3 Pest Control

Vast quantities of human food are lost every year through the activities of insects, fungi, viruses, and weeds, which attack our crops when they are growing and even when they are stored before being consumed.

The cheapest ways of controlling these pests are by cultural control, an example of which is the control of potato-virus diseases as described in section 19.6. Another example is the frit-fly of oats, whose larvae feed on young oat plants before the four-leaf stage, and on ryegrasses. By planting oats early in the year it is usually possible to get the plants through the delicate four-leaf stage before the grubs hatch out, and the crop suffers no damage.

One advantage of crop rotation is that the pests of one crop do not find the same crop there the following year, so that the population does not have the chance to build up.

Chemical control of pests is now well known. Substances such as DDT, derris, and malathion are widely used against aphids, weevils, frit fly of oats, and many other pests of crop plants. Malathion is a systemic insecticide which permeates the tissues of a plant and remains effective for several weeks. 2.4-D and MCPA are herbicides much favoured in agriculture to control broad-leaved weeds in cereal crops (see sec. 12.4) and Dalapon is effective against the persistant couch grass. Zineb, a zinc-containing fungicide, has been successfully used to combat downy mildew of lettuce, hops, and vines. Copper fungicides, discovered quite accidentally by Millardet in 1885, have been used in increasing quantities ever since. In fact the chemical aspect of crop protection is now a considerable industry in its own right.

Perhaps the most interesting examples of pest control are those in which an organism (which feeds on the pest) is introduced. One example of this, an ichneumon fly, was mentioned in section 19.4. A remarkably successful example of this biological control is exhibited by the moth, *Cactoblastis cactorum*, which was introduced into Australia from the Argentine and controlled the prickly pear (*Opuntia species*). In 1925, thirty million acres of good grazing land in Queensland and New South Wales were rendered useless by a dense growth of this weed, which was invading new land at the rate of one acre per minute, but owing to the voracious feeding of the larvae of this moth the prickly pear was no longer a serious pest in 1933. The spread of the moth was largely achieved by attaching eggs to the weeds in new areas.

Another example is the control of greenhouse white fly (*Trialeurodes vaporariorum*), which feeds on tomatoes and cucumbers grown in the United Kingdom and covers them with a sticky secretion which smothers the leaves. The control is carried out by introducing into affected greenhouses a minute chalcid wasp (1 mm long) known as *Encarsia formosa*, which lays eggs inside the bodies of the young white flies. The young flies are known as scales because they remain flattened against the leaf with their mouthparts piercing its

cells and drawing up fluid food. The wasp's eggs hatch and its larvae feed on the body of the white fly, turning the white scale-like object black. This wasp was first discovered in greenhouses at Wisley in 1914. White fly can also be controlled by fumigation with hydrocyanic acid gas.

21.4 Mechanisation

The replacement of horses by tractors has meant that more land can be cultivated in a given period of time and thus the farmer can make better use of the spells of fine weather. The use of a combine harvester for grain, vegetable harvesters, etc., means that crops can be taken straight from the field for storage. Many crops are harvested, and transported to 'food factories' for processing, all in the space of a few hours. This quick 'turn-over' enables the farmer to grow more than one crop per year on his land. Such 'intensive farming' both increases the amount of food produced and reduces its cost. The machinery is, however, expensive, but there are in the United Kingdom some 1 300 farmers' machinery syndicates through which farmers have the use of such equipment without tying up their own capital.

21.5 World Food Production

The limit of food production on present lines is set by the fact that it is practicable only on about a third of the earth's surface, the remaining area being either too dry, too cold or too mountainous for cultivation. However, the rapid advances in science and technology have, on the global scale, led to an expansion in both the yield from the land at present under cultivation and an increase in the area cultivated. It has been estimated that this expansion could be extended to satisfy the needs of the world population for the next forty years, but the increase in population inevitably leads to the reduction in the area of cultivated land per head.

It is true that in the past advances in food production have come from the results of scientific research, and there is no reason to think that these advances will cease. Many other factors, however, are involved. For instance, those engaged in agriculture must be educated in new methods and those engaged in feeding the families must be given the knowledge that will enable them to select a nutritional and economical diet. Such education is not easy, for it takes time and demands understanding of the 'culture patterns' of the peoples involved. Most important of all is the international co-operation necessary for the development of the resources of the world, which are not at present fully used.

Some progress toward this international co-operation has been made by the International Council of Scientific Unions (I.C.S.U.), which has launched an International Biological Programme (I.B.P.), entitled 'The Biological Basis of Productivity and Human Welfare', to promote the world-wide study of:

(*a*) Organic production on the land, in fresh waters and in the seas, and the potentialities and use of new as well as existing resources;
(*b*) Human adaptability to changing conditions.

It is hoped that such an organisation will foster international co-operation in the basic biological studies upon which the practical agriculture and public health work of the United Nations Food and Agriculture Organisation (F.A.O.) and the World Health Organisation (W.H.O.) depend.

CONCLUSION

This book has included a study of the life processes in organisms, from the simplest to the most complex. In all of them, the cell with its nucleus and cytoplasm contains the chemical system in which these processes take place. Only one main difference exists between the cytoplasm of higher plants and animals: the possession of the green chloroplast in plants. This piece of metabolic equipment in their cytoplasm raises their synthetic ability and makes them the prime producers of organic compounds, on which all other organisms, whether parasite, saprophyte, carnivore, or herbivore, have to rely.

Life is enormously complicated: it involves the replication and maintenance of complex molecular patterns in a hostile environment. The molecules of proteins are a vital part of living organisms. They form enzymes which catalyse metabolic changes, and together with fats they build membranes in the cytoplasm which help to separate one type of chemical factory from another. The complex molecules of D.N.A. in the chromosomes are responsible for transmitting characteristics from parent to offspring. In fact, life is impossible to envisage without these compounds, but how protein and D.N.A. molecules first appeared on this planet is a matter for speculation, though there is no doubt that this was a necessary preliminary to the formation of the first life as we know it today.

The reader may have come to realise, with some irritation perhaps, that there are exceptions to everything biological; nothing is absolute. Man has, for convenience, classified the objects around him as living or lifeless, but he cannot be certain where to place viruses, for although they possess nucleic acids and reproduce, some can be prepared in a crystalline form. Plants possess chlorophyll and are holophytic, but dodder and potato blight are devoid of chlorophyll and feed parasitically. Animals differ from plants in having a compact body, but many corals branch to form a diffuse structure. The duck-billed platypus (*Ornithorhyncus paradoxus*) and the spiny ant-eater (*Tachyglossus aculeatus*) are two Australian mammals which do not give birth to young in the typical manner, but persist in the egg-laying habits of their reptilian ancestors. You can doubtless add many more to this list of exceptional examples.

Man's spirit of enquiry into the living world around him has increased greatly in historic times. Is he any nearer understanding the natural and physical forces at work in the world? Much of the earlier work in biology was concerned with describing the structure of living things, and as better microscopes have been produced this study has been pursued down to finer and finer detail until now, with an electronmicroscope, it is possible to detect a few large molecules. In the twentieth century the biologist has been mainly concerned with finding out how organisms and their parts function, and in the last two decades this study has reached down through organs, tissues, and cells to molecular level. Every new fact established reveals new problems for investigation. One thing is clear, however: the metabolic changes of any physiological process such as respiration, photosynthesis, or excretion are intricately interconnected with those of other processes. It seems highly unlikely that knowledge of metabolism in any one organism will ever be complete, and yet such knowledge is necessary before life can be fully understood.

Man is unique among animals in having learnt to control his environment to a certain extent. There is no doubt that he is more efficient at controlling the physical environment than he is at controlling other life. Nevertheless, he has produced new organisms through speeding up the rate of mutations by exposure of organisms to X-rays. It was by use of this method that the new type of *Penicillium notatum* was produced, which made one thousand times more penicillin than the original type. He can protect individuals from the ruthlessness of natural selection by modern surgery and medical science, and even, by feeding on a special diet from birth, prevent the effects of genes which would normally make a person mentally deficient. He has taught himself to produce more food by selecting more productive crop plants and by waging war on parasites and pests so that the ever-increasing human population can be fed.

There are, however, many dangers inherent in these activities. When man uses modern insecticides, fungicides, etc., he kills millions of organisms as well as those that are pests; many of these chemicals leave indestructible residues which continue to build up in the soil and these too may reduce the soil's productivity. While he is poisoning his own environment, many types of organisms are breeding new varieties which are immune to such chemical attack. In fact as man progresses toward controlling his environment he leaves a depressing record of destruction not only in the soil but amongst the living organisms that share it with him. The destruction of wayside weeds has robbed many butterflies of their food plants; the spraying of food crops has poisoned many birds. What of the effect on man himself? X-rays and chemicals are used to produce mutations in other organisms, but what of their effect and that of other radiations on our own genes? Are we giving sufficient thought to the genetical make-up of future generations?

Ever since man started to till the soil he has been trying to control his environment to his own advantage. Where is all this effort to control nature leading? Will man suffer the same fate as all the other groups which have dominated the earth in the past, or is he different because he has learnt to suspend the ruthless operation of natural selection? Is man's mind a superb computer and is this suspension of natural selection a good or a bad thing? Does the harnessing of nuclear energy involve another hazard?

Man is now venturing beyond his immediate environment into space. Will he find life on other planets, solar systems, or galaxies? Will such life, if it exists, be very different from our own?

Classification of Living Organisms

Classifications of organisms are man-made: modern systems are the product of our knowledge of evolution.

A comparative study of both living and fossil organisms is the foundation of our present schemes of classification, but these are continually being modified as new evidence for the relationship between groups becomes available. The evidence is, of course, subject to different interpretations by individual scientists and this leads to there being more than one scheme of classification. This is especially true of the plant kingdom in which the position of the non-chlorophyll-containing organisms (e.g. bacteria and fungi) is widely debated.

The next two pages give a simple scheme for the classification of plants and animals.

Classification of the Plant Kingdom

1 Schizophyta

Simple plants without well-developed nuclei; they reproduce only by splitting asexually.

1.1 Bacteria

Minute non-cellular organisms. Saprophytic, parasitic, and in rare cases photosynthetic.
e.g. Mycobacterium tuberculosis, Staphylococcus aureus.

1.2 Blue-green Algae

The simplest plants which contain chlorophyll. Minute non-cellular or filamentous forms. Found together in gelatinous masses.
e.g. Anabaena, Nostoc.

2 Thallophyta

Plants without roots, stems, or leaves. They reproduce by spores, some of which are formed sexually.

2.1 Algae

Contain chlorophyll and sometimes other pigments which mask the green colour. Structure varies from non-cellular ones (*e.g. Euglena and diatoms*) to those with complex tissues (*e.g. brown seaweeds*). Other examples are *Spirogyra and red seaweeds.*

2.2 Fungi

Without chlorophyll; saprophytic or parasitic. Body usually composed of branching hyphae.
e.g. Pin moulds, potato blight, yeasts, Penicillium, mushrooms, toadstools, mildews.

2.3 Lichens

Composite organisms containing an alga and a fungus living together symbiotically. Body varies from an encrusting growth on bare rock to a dense branching structure.
e.g. Iceland moss, Rhizocarpon.

3 Bryophyta

Small, green plants without true roots but usually possessing stems and leaves. Two generations in the life-history—an asexual spore-producing one which develops on the gamete-producing one.

3.1 Liverworts (Hepaticae)

The simplest forms have a flat, ribbon-like body which branches and lies flat on the ground, anchored by many thread-like unicellular hairs. The leafy forms have two rows of leaves (without midribs) and two rows of scales on the stem.
e.g. Pellia, Lophocolea.

3.2 Mosses (Musci)

All leafy. Leaves with midribs and arranged spirally on the stem. Commonly grow in compact cushions or carpets.
e.g. Funaria, hair moss (Polytrichum), bog moss (Sphagnum).

4 Pteridophyta

Green plants with true roots and a well-developed vascular system. There are two separate generations in the life-history, the gamete-producing one being small and short-lived, while the spore-producing one is often a large plant which may live for several years.

4.1 Psilophytales

Oldest known land plants, found as fossils at Rhynie, near Aberdeen.
e.g. Rhynia, Hornea.

4.2 Lycopods and Horsetails

Those living at present, such as *clubmosses, horsetails, and quillwort*, are small, unimportant plants, but in the Carboniferous period this group produced large trees such as *Lepidodendron and Calamites* which are fossilised in our coal measures.

4.3 Ferns

Spore-producing plants, are of moderate size and bear their spores on certain areas of the lower leaf surface.
e.g. bracken, male fern, maidenhair fern.

5 Spermaphyta

The most successful group. They produce pollen grains and reproduce by seeds.

5.1 Gymnosperms

Ovules and seeds are naked, no carpels.
e.g. Pine, yew, cedar, larch.

5.2 Angiosperms

Ovules enclosed in an ovary of one or more carpels, seeds in a fruit. This group contains all the flowering plants and is subdivided into the monocotyledons and dicotyledons, some families of which are named below.

Monocotyledons
e.g. Grass family, rush family, lily family, orchid family, sedge family.

Dicotyledons
e.g. Buttercup family, rose family, cabbage family, pea family, parsley family, potato family, daisy family, dead-nettle family.

262

Classification of the Animal Kingdom

1–8 are often referred to as **invertebrates**.

1 Protozoa

Body not divided into cells (non-cellular) and often has only one nucleus. Structurally the simplest animals.
e.g. Amoeba, Paramecium, Euglena, the malarial parasite.

2 Two-layered Animals (Coelenterata)

Body made up of two layers of cells surrounding a central cavity whose only opening is the mouth. Feed by means of thread-capsules.
e.g. Hydra, jelly-fish, sea-anemones, corals.

3 Flatworms (Platyhelminthes)

Small, flattened, unsegmented worms without a body cavity. Only one opening to the alimentary canal, the mouth. Many are parasitic.
e.g. Planarians, tapeworms, liver-flukes.

4 Threadworms (Nematoda)

Small cylindrical unsegmented worms without a body cavity. Two openings to the alimentary canal, mouth and anus. Mostly parasitic.
e.g. Hookworms and Trichina worms in man, eelworms in potato plants.

5 Segmented Worms (Annelida)

Segmented worms with a body cavity and two openings to the alimentary canal. Bristles are usually present in each segment.
e.g. Lugworms, earthworms, leeches.

6 Animals with Jointed Limbs (Arthropoda)

Animals with one pair of jointed limbs on some or all of the segments. A hard external skeleton or cuticle.

6.1 Crustacea

Aquatic animals with two pairs of antennae and many other paired limbs used for a variety of purposes such as feeding, walking, running, swimming, and respiring.
e.g. Water fleas, barnacles, wood-lice, shrimps, crayfish, crabs.

6.2 Myriapoda

Terrestrial. Large number of segments each bearing similar limbs.
e.g. Centipedes (carnivorous, flattened body), millepedes (herbivorous, cylindrical body).

6.3 Insecta

Segmented body divided into head, thorax, and abdomen. Thorax bears three pairs of legs. A highly successful class of animals.
e.g. Silverfish, locusts, grasshoppers, cockroaches, earwigs, termites, dragon-flies, bed-bugs, green-fly, may-flies, lice, caddis-flies, beetles, butterflies, moths, bees, wasps, ants, crane-flies, house-flies, mosquitoes.

6.4 Arachnida

Body divided into two parts. Four pairs of legs on anterior part.
e.g. Spiders, mites, ticks, scorpions.

7 Mollusca

Unsegmented, soft-bodied animals, usually with one or more calcareous shells.
e.g. Snails, limpets, slugs, oysters, mussels, squids, octopus.

8 Spiny-skinned Animals (Echinodermata)

Unsegmented marine animals. Parts of the body arranged symmetrically on radii (usually five radii).
e.g. Starfishes, sea-urchins, brittle-stars, sea-cucumbers.

9 Vertebrata

Segmented animals with a skeleton of bone or cartilage. They have a well-developed head in which the brain is protected by a skull, two pairs of limbs or fins, gill-slits at some time in their life, and a series of vertebrae forming the central axis of their skeleton. Possess a tail.

9.1 Fishes (Pisces)

Completely adapted to aquatic life. Locomotion by a powerful muscular tail and paired and median fins. Gaseous exchanges of respiration are carried out by gills. The heart is two-chambered.
e.g. Skates, sharks, salmon, roach, herring, plaice.

9.2 Amphibians

Only partially adapted to terrestrial life. Eggs must be laid in water. Gills always present in larvae and may be retained in adults or be replaced by lungs. They have a moist, slimy skin, pentadactyl limbs, and a three-chambered heart.
e.g. Frogs, toads, newts, salamanders.

9.3 Reptiles (Reptilia)

Completely adapted to a terrestrial life. Eggs protected by a shell and can be laid on land. Gills never develop, vestiges of gill-slits found only in embryos; skin covered with scales, pentadactyl limbs, and heart usually four-chambered.
e.g. Lizards, snakes, crocodiles, turtles.

9.4 Birds (Aves)

In many ways they closely resemble reptiles; but differ in having feathers, in having the fore-limbs and sternum specialised for flight, and in being without teeth. Fairly constant body temperature ('warm blooded').
e.g. Ducks, owls, gulls, finches, robins, penguins.

9.5 Mammals (Mammalia)

They surpass all other vertebrates in their adaptations to terrestrial life, although some, such as whales and porpoises, have returned to an aquatic life. They possess hair and keep their body temperature fairly constant ('warm blooded'). The young are fed on milk from mammary glands after birth.
e.g. Hedgehogs, bats, cattle, rats, camels, seals, elephants, apes, man.

Miscellaneous Questions

1. List the features which distinguish living organisms from non-living things.

Explain why a simple alga and any free living Protozoan, e.g. Amoeba, are placed in separate kingdoms. [A]

2. How would you determine the amount of air and humus in a sample of garden soil?

From what sources do (*a*) animals, (*b*) leguminous plants, and (*c*) green plants other than legumes, obtain their nitrogen compounds? [A]

3. Write an account of the part played by (*a*) pseudopodia, (*b*) cilia, and (*c*) muscle in the locomotion of named animals. [A]

4. With the aid of a diagram describe a reflex arc. Briefly state how a reflex action differs from a voluntary action.

In what ways do a reflex action in a mammal and a tropic response in a plant resemble and differ from each other? [A]

5. By what means do (*a*) a flowering plant and (*b*) a mammal make provision for the safe development of their embryos? [A]

6. Write an account of the way in which a mammal breathes in oxygen from the atmosphere. Describe how the oxygen of the atmosphere enters a leaf and then passes into a mesophyll cell.

What happens to oxygen in living cells? [C]

7. What is asexual reproduction? Illustrate your answer by reference to a named mould fungus, Hydra, and a named perennial plant. What are the advantages to the herbaceous perennial of reproduction by vegetative propagation instead of by seeds? [C]

8. Define transpiration. Describe an experiment to find out whether a green plant transpires.

List the ways in which a mammal loses water from the body. Why is it necessary for mammals to give off water? [C]

9. What do you understand by irritability in living organisms?

Describe one experiment to find out whether a plant shoot responds to the stimulus of light.

Give an account of the response to bright light of the iris of the mammalian eye. [C]

10. Choose three of the following and show how the organ is suited to the function given in brackets. (*a*) A bird's wing (flying). (*b*) A mammal's ear (hearing). (*c*) A green leaf (photosynthesis). (*d*) A corm or a bulb or a rhizome (vegetative reproduction). [C]

11. Make labelled drawings of (*a*) a palisade mesophyll cell, and (*b*) a mammalian nerve cell. Compare and contrast the structure and function of these two cells.   [C]

12. The lungs and small intestine in mammals and the leaves and roots of plants provide a large surface area. Give two other examples of the occurrence in living organisms of a relatively large surface area. For each one of these six examples suggest reasons why a large surface area is important.   [C]

13. Answer any four of the following:
(*a*) Explain the functions of the contractile vacuole of Amoeba.
(*b*) State the external conditions necessary for the germination of seeds.
(*c*) Distinguish between arteries and veins.
(*d*) State what is meant by a saprophyte and name two examples.
(*e*) What features of a cell of Spirogyra are characteristic of plant cells?   [L]

14. State concisely what is meant by the term photosynthesis. How are the following organisms dependent on this process: (*a*) a mould, (*b*) a caterpillar, (*c*) an adult frog?   [L]

15. Answer four of the following:
(*a*) State the characteristics of pea or bean seedlings that have been grown in the dark for some time.
(*b*) State four differences between the external features of a frog and a rabbit.
(*c*) What are the characteristic features of an insect?
(*d*) How does Amoeba survive adverse conditions?
(*e*) What are the functions of root hairs?   [L]

16. Briefly describe the structure, and state the functions, of the skin of a mammal (e.g. rabbit). Does the epidermis (skin) of a leaf have any similar functions? If so, what are they, and how are they carried out?   [L]

17. (*a*) Some soaked broad-bean seeds were placed in warm, moist sawdust and allowed to germinate until they had radicles at least 20 mm long. Two dozen of the seedlings with straight radicles were carefully removed and separated into two sets. In set A (1 dozen seedlings) each of the radicles was marked with waterproof ink at intervals of 1 millimetre. In set B (1 dozen seedlings) each of the radicles was similarly marked, and the tip of each radicle was cut off. All the seedlings in sets A and B were then replanted with their radicles placed in a horizontal position. Describe and explain what occurs in each set after an interval of three days.
(*b*) (i) In what way does the ear of a mammal make the animal aware of changes in its position with reference to gravity and the maintenance of balance? (ii) What other organs in the body assist in this function?   [N]

18. (*a*) Name two plants in which the seeds are dispersed by birds. Describe, for each example that you name, the way in which seed dispersal occurs and draw diagrams to show the structures which develop to aid in dispersal.
(*b*) (i) Draw a fully labelled diagram of a quill feather from the wing of a bird. (ii) Describe how such a feather is arranged on the wing in relation to adjacent quill feathers, and explain how this arrangement is useful in flight. (iii) How does a tail feather differ from a wing feather?   [N]

19. (*a*) What distinguishes a living organism from a non-living object?
(*b*) Briefly describe how a mammal eliminates waste nitrogenous material.
(*c*) Explain what happens to the waste nitrogenous material of the mammal before it is made available to the green plant.
(*d*) What use is made by the plant of the compounds of nitrogen it absorbs?
     [N]

20. (*a*) (i) Describe carefully how you would measure the rate of water loss from a leafy shoot. (ii) Name the structures through which this loss occurs. (iii) What effect would a damp atmosphere have on the rate of water loss from the shoot?
(*b*) Explain how water loss occurs and is affected by external conditions in (i) Amoeba, and (ii) the skin of a mammal such as man.   [N]

21. (*a*) Describe a possible route followed by an atom of carbon in a molecule of carbon dioxide present in the atmosphere until it appears in a molecule of starch in a potato tuber. For each part of the route, name the compound in which the carbon is found and name the process involved when there is a change from one kind of molecule to another.

(*b*) Describe a simple test by which you would demonstrate the presence of starch in the tuber.

(*c*) When a mammal eats the potato, some of the carbon from the potato starch may eventually be found as animal starch (glycogen) in the liver. Trace the path followed by this carbon and name the compound in which it occurs at each stage. [N]

22. (*a*) (i) With the help of labelled diagrams explain how you would show by a controlled experiment that heat is produced by germinating seeds. (ii) What part is played in germination by oxygen, water, and enzymes?

(*b*) (i) What is the source of heat in a warm-blooded animal such as a mammal? (ii) Explain how the loss of heat from a mammal is reduced when the external temperature is lowered. [N]

23. Describe the adaptations which enable (*a*) a fish to live in water, (*b*) a bird to fly in the air, (*c*) a named animal parasite to live in or on its host. [N]

24. (*a*) (i) Give three differences in structure between red blood cells (corpuscles) and white blood cells (corpuscles). (ii) State one function of each of these blood cells (corpuscles).

(*b*) Compare the composition of the blood flowing through (i) the pulmonary artery and the vein (portal vein) between the intestine and the liver, (ii) the aorta and the vein (renal vein) between the kidney and the inferior vena cava.

(*c*) Two groups of young, newly weaned rats were kept in isolated conditions and given specially prepared foods for a prolonged period. The rats in batch A were fed on foods forming a well-balanced diet; the rats in batch B were given the same foods from which all iron compounds had been extracted. After a period of time the rats in batch B were found to be less energetic than the rats in batch A. Give an explanation for this. [N]

25. Why do all organisms require nitrogen? How, and in what form, is nitrogen obtained by (*a*) Amoeba, (*b*) Mucor, (*c*) a rabbit, (*d*) a flowering plant?

Why do animals produce nitrogenous waste compounds and how do they eliminate them? Give three ways in which a farmer may increase the nitrogen content of his soil. [O]

26. Explain the meaning of the terms breathing and respiration with reference to: (*a*) a named herbaceous plant; (*b*) a named mammal; (*c*) Mucor; (*d*) a tadpole. [O]

27. Distinguish clearly between the terms in five of the following pairs: (*a*) excretion, secretion; (*b*) tendon, ligament; (*c*) ovary, ovule; (*d*) plastid, chlorophyll; (*e*) hypha, root; (*f*) animal, fish; (*g*) Amoeba, organism. [O]

28. What do sugars, starch, and cellulose have in common? (Details of photosynthesis are *not* required.) What part does each play in the life of a flowering plant? How would you recognise their presence in a storage organ? [O & C]

29. What conditions are necessary for photosynthesis in a leaf? State what happens in a growing potato plant to starch produced in the leaves. [O & C]

30. Describe how water is taken into, and lost from, (*a*) Amoeba, and (*b*) the mammalian body. [O & C]

31. List the features most useful in distinguishing between (*a*) living and non-living matter, and (*b*) plants and animals. Mention any cases in which it may be difficult to make these distinctions. [O & C]

32. Write short notes on four of the following: amnion, pasteurisation, neurones, adrenalin, insulin. [O & C]

33. (*a*) Describe the seed of a named plant which shows hypogeal germination. Give a brief illustrated account of the germination of this seed.

(*b*) What are the advantages of vegetative reproduction to a plant?　　　[S]

34. State the composition of the atmosphere. What gases are removed from the air by (*a*) a green plant, (*b*) a mammal? Describe the use made by these organisms of the gases removed.　　　[W]

35. State the characteristics of living things. Tabulate the differences between plants and animals, and show how these differences are illustrated by (*a*) a fish, (*b*) a wallflower.　　　[W]

36. Describe controlled experiments which demonstrate the following: (*a*) The liberation of heat in the germination of seeds. (*b*) The effect of light on the direction of growth of shoots.　　　[W]

37. How does each of the following get the oxygen it needs throughout its life: (*a*) a named fish, (*b*) a named amphibian, (*c*) a named herbaceous plant? For what purpose will the adult fish use the energy liberated from the foodstuffs it oxidises during a period of one year?　　　[A]

38. Describe the stages by which: (*a*) a mammal excretes nitrogenous waste, (*b*) the nitrogen of dead plants and animals is converted into a form which plants can use.　　　[S]

39. Explain how: (*a*) the carbon dioxide of the air is converted into starch, (*b*) starch is digested in a mammal (diagram *not* required).　　　[S]

40. (*a*) Oxygen, carbon dioxide, soluble food substances, and hormones are all carried by the blood. State how each is obtained by the blood, where in the blood each is to be found, and what happens to it.

(*b*) Explain how water and salts are transported in a plant.　　　[S]

41. (*a*) Draw a labelled diagram to show the path of a typical named reflex action.

(*b*) Name four ways in which reflex actions benefit man.

(*c*) How do the following react to a stimulus of light? (i) A plant shoot. (ii) Earthworm. (iii) Amoeba. (iv) Slug (or snail).　　　[M]

42. Describe the position in the body of a mammal, of the (*a*) kidneys, (*b*) liver, and (*c*) pancreas. (Diagram can be used.)

Describe the work of (*a*) kidneys, (*b*) liver, (*c*) pancreas.　　　[M]

Glossary of Biological Terms

Adaptation. Any characteristic of an organism which helps it to survive in its own environment.

Androecium. The collection of stamens in a single flower.

Anther. The terminal pollen-containing part of a stamen.

Assimilation. The incorporation into the protoplasm of simple soluble and diffusible substances.

Atom. The smallest part of an element that can take part in a chemical reaction.

Auxins. Chemical substances which are produced in growing cells and help to control growth in plants.

Calyx. The outer whorl of floral leaves or sepals of a single flower.

Carnivore. There are two distinct uses of the word. 1—Any flesh-eating animal. 2—A member of the group of mammals known as the Carnivora.

Carpal. One of the small wrist-bones found in all vertebrates except fish.

Carpel. One of the floral leaves of the gynaecium of a flower. It consists of three parts—the ovary, style, and stigma.

Cell. A unit of cytoplasm containing a single nucleus, bounded by a membrane, and specialised to perform one main function. The structural unit of higher organisms.

Cellulose. A carbohydrate which is found in the cell walls of plants.

Chlorophyll. The green pigment of plants, which absorbs light and enables them to photosynthesise.

Cilia. Hair-like vibrating threads of cytoplasm which act as locomotory organs.

Ciliate. An organism belonging to a group of protozoa which are more or less covered with cilia.

Coleoptile. Sheath round the plumule of grains such as oats and maize.

Conjugation. A method of sexual reproduction involving the fusion of similar gametes which are not freed from the parents.

Constriction. Compression or narrowing.

Contraction. The shortening and fattening of muscle fibres so that the parts of the body to which they are attached are pulled closer together.

Cytoplasm. The transparent, slightly viscous living material of organisms. A highly complex mixture of water, proteins, nucleic acids, fats, carbohydrates, mineral salts, and other substances, which contains many membranes and other structural features which are not visible under an optical microscope.

Diffusion. The free movement of molecules of gases or dissolved substances from regions of high concentration to regions of low concentration. This movement results in their uniform distribution.

Digestion. The breaking up of food by chemical means into simple soluble and diffusible substances which can be distributed to all parts of the body.

Dilation. Widening.

Element. A simple substance which cannot be split up by chemical means.

Enzyme. A complex catalyst of a protein nature which can be produced only by living cells.

Excretion. The discharging from the body of the waste products of metabolism or their deposition within the body as insoluble substances.

Female. An individual which produces only the passive type of gamete which after fertilisation may develop into a new organism.

Fertilisation. The fusion during sexual reproduction of special nuclei from two gametes.

Fruit. The structure which develops from the ovary after fertilisation.

Gamete. A cell with a nucleus containing half the normal number of chromosomes, which takes part in sexual reproduction. It must fuse with the nucleus of another gamete before it can give rise to a new organism.

Ganglion. Any part of a nervous system containing the cell bodies of nerve cells.

Gonads. The sexual organs of an animal in which the gametes are formed.

268

Gynaecium. The collection of carpels in a single flower.

Herbivore. An animal which feeds on plant material.

Hermaphrodite. Hermaphrodite organisms possess both male and female organs in the same individual.

Holophytic. The typical plant method of nutrition, in which gaseous and liquid inorganic substances are absorbed and built up into complex organic compounds.

Holozoic. The typical animal method of nutrition, in which solid organic compounds are used as food.

Hormone. A ductless gland's secretion which circulates in the blood and assists in co-ordinating the activities of the body.

Humus. A mixture of decomposition products, dead soil organisms, fungal mycelia, and finely divided plant residues that have lost their structure.

Inorganic. Of mineral origin; substances which are not organic.

Ions. Electrically charged particles which exist independently.

Joule. The work done when the point of application of a force of one Newton is displaced through a distance of one metre in the direction of the force.

Kilocalorie. The amount of heat energy required to raise the temperature of 1 kg of water through one degree C. This unit, although it may still be used is an arbitrary one and should be replaced by the Kilojoule (kJ). 1 kcal = 4.2 kJ.

Kilojoule. One thousand joules.

Lignin. Complex material found in large quantities in the cell walls of the xylem of plants. It does not stretch and is impermeable to water.

Lymph. A colourless body fluid, derived from the blood by filtration. It is contained in very fine vessels which join together and empty into the venous system near the heart.

Male. An individual which produces only the motile type of gamete which fertilises the female gamete and stimulates it to grow into a new organism.

Metabolism. The chemical changes that are taking place continually in living cells.

Metamorphosis. An abrupt change in the life-cycle of an animal. A period of rapid transformation from the larval to the adult stage, which involves a complete change of form and often a change in the mode of life and method of feeding.

Molecule. The smallest part of a substance which can normally exist in a free state.

Newton. The force which, when applied to a body having a mass of one kilogram, gives it an acceleration of one metre per second squared.

Omnivore. An animal which feeds upon both plant and animal material.

Organ. A localised part of an organism which contains several tissues and is specialised both in structure and function (e.g. leaf, liver). Where such a specialised part is not localised it is known as an **organ system** (e.g. the blood system).

Organic. 1—Of living origin. 2—Complex carbon compounds many of which are formed only in metabolism.

Osmosis. A movement of water from a dilute solution to a more concentrated one through a semi-permeable membrane.

Ovary. (Animals) The reproductive organ in which female gametes are formed. (Plants) The hollow basal part of a carpel in which one or more seeds develop. Several carpels may be fused together to form a single ovary.

Ovule. A special reproductive structure which contains an ovum and develops inside the ovary of flowering plants. After its ovum is fertilised it forms a seed.

Ovum. The non-motile female gamete or egg cell.

Oxidation. The addition of oxygen to a compound or element, or the removal of hydrogen from a compound.

Parasitic. A special method of nutrition in which a plant or animal obtains its food supply by living in or on another organism, the host, which suffers some loss or harm as a result of the association.

Pericarp. The wall of a fruit. It is formed from the ovary wall after fertilisation.

Peristalsis. Waves of muscular contraction which pass along the alimentary canal and drive the food along it.

Photosynthesis. The building up by green plants of complex organic substances from simple inorganic ones, using the energy of sunlight.

Pollination. The transfer of pollen from an anther to the stigma of the same species.

Protoplasm. An obsolete term for the colourless living material of organisms. Differentiated into nucleus and cytoplasm.

Receptacle. The short stem to which all the floral leaves are joined.

Reduction. The removal of oxygen from a compound, or the addition of hydrogen to a compound or element.

Refraction. The bending of rays of light as they pass obliquely from one medium to another.

Reproduction. The production of new organisms like the parent or parents. *Sexual*—Achieved by the production and fusion of gametes. *Asexual*—Achieved without the production and fusion of gametes, i.e. by spores, buds, or the separation of a vegetative part of a flowering plant.

Respiration. The complex processes by which energy is produced by living organisms. *Aerobic*—Free oxygen is used to oxidise sugar and so liberate energy. *Anaerobic*—Sugar is partially oxidised to liberate energy without the presence of free oxygen.

Saprophytic. A special method of nutrition used by some plants which contain no chlorophyll. They absorb organic food in solution from dead organisms or the products of their decay.

Saprozoic. A special method of nutrition used by some animals in which the food is obtained from dead organisms or the products of their decay.

Secretion. The formation of a substance useful to an organism and its discharge from the cells where it is produced.

Seed. The structure formed from a fertilised ovule. It contains an embryo which may grow into a new plant.

Spermatozoon (sperm). The motile male gamete of animals and of the lower plants.

Sporangium. A simple asexual reproductive organ containing spores.

Spore. A small reproductive body which is produced in large numbers and dispersed. It may, if conditions are suitable, develop directly into a new organism.

Stamen. One of the floral leaves of the androecium of a flower. It consists of a filament supporting an anther which usually contains four pollen sacs.

Stigma. The receptive surface of a carpel.

Style. The region between the stigma and the ovary of a carpel.

Symbiosis. An intimate association between two individuals of different species in which there is mutual benefit.

Testis. The reproductive organ in animals in which the male gametes are formed.

Tissue. A collection of cells and the material around them. It has a uniform texture, and is specialised for a particular function (e.g. bone, skeletal muscle, xylem, cambium).

Translocation. The transport of sugars and other products of photosynthesis about a flowering plant. It takes place mainly in the sieve-tubes of the phloem.

Transpiration. The evaporation of water from a terrestrial plant.

Tropism. The response of the growing region of a fixed plant to an external stimulus, in which the direction of growth is controlled by the direction of the stimulus.

Turgidity. This is the state of plant cells whose walls are fully stretched owing to the absorption of water by the cytoplasm inside. The cells are rigid.

Vegetative propagation or reproduction. A form of asexual reproduction used by many flowering plants, in which the new plant is formed by the separation of part of the vegetative structure of the plant.

Wilting. This is the gradual drooping of the leaves and young stems as they lose water and hence their turgidity.

Zygospore. A special type of zygote which has a thick protective wall and is able to resist adverse conditions. It is found in Algae and Fungi.

Zygote. A fertilised ovum, or the body produced by the fusion of two gametes, which is capable of division and growth to form a new organism.

Index

Page numbers in bold type indicate that the subject is illustrated.